Comprehensive Analytical Chemistry
Volume 74

Biosensors for Sustainable Food - New Opportunities and Technical Challenges

Advisory Board

Comprehensive Analytical Chemistry

Volume 74

Biosensors for Sustainable Food - New Opportunities and Technical Challenges

Edited by

Viviana Scognamiglio
National Research Council (CNR) Rome, Italy

Giuseppina Rea
National Research Council (CNR) Rome, Italy

Fabiana Arduini
University Tor Vergata Rome, Italy

Giuseppe Palleschi
University Tor Vergata Rome, Italy

AMSTERDAM • BOSTON • HEIDELBERG • LONDON
NEW YORK • OXFORD • PARIS • SAN DIEGO
SAN FRANCISCO • SINGAPORE • SYDNEY • TOKYO

ELSEVIER

Elsevier
Radarweg 29, PO Box 211, 1000 AE Amsterdam, Netherlands
The Boulevard, Langford Lane, Kidlington, Oxford OX5 1GB, UK
50 Hampshire Street, 5th Floor, Cambridge, MA 02139, USA

ISBN: 978-0-444-63579-2
ISSN: 0166-526X

For information on all Elsevier publications
visit our website at https://www.elsevier.com

Working together
to grow libraries in
developing countries

www.elsevier.com • www.bookaid.org

Publisher: Zoe Kruze
Acquisition Editor: Poppy Garraway
Editorial Project Manager: Shellie Bryant
Production Project Manager: Radhakrishnan Lakshmanan
Designer: Maria Inês Cruz

Typeset by TNQ Books and Journals

Contents

Section II
New Generations of Synthetic Receptors and Functional Materials for Food Biosensors

13. Intelligent Food Packaging 377

S. Otles, B.Y. Sahyar

14. (Bio)Sensor Integration With ICT Tools for Supplying Chain Management and Traceability in Agriculture 389

M. Durresi

Contributors to Volume 74

G. Alarcon-Angeles, Universidad Autónoma Metropolitana-Xochimilco, Mexico City, México

G.A. Álvarez-Romero, Universidad Autónoma del Estado de Hidalgo, Hidalgo, México

A. Antonacci, IC-CNR Istituto di Cristallografia, Rome, Italy

F. Arduini, Università di Roma Tor Vergata, Rome, Italy

N. de-los-Santos-Álvarez, Universidad de Oviedo, Oviedo, Spain

L.M. Dumitru, Johannes Kepler University, Linz, Austria

M. Durresi, European University of Tirana, Tirana, Albania

A. Escarpa, Universidad de Alcalá, Alcalá de Henares, Madrid, Spain

G.A. Evtugyn, Chemistry Institute of Kazan Federal University, Kazan, Russian Federation

J. Fitzgerald, Dublin City University, Dublin, Ireland

D. Hoffmeister, University of Cologne, Cologne, Germany

M. Irimia-Vladu, Institute for Surface Technologies and Photonics, Weiz, Austria

B. Jurado, Universidad de Alcalá, Alcalá de Henares, Madrid, Spain

G. Kaur, Punjabi University, Patiala, Punjab, India

M.J. Lobo-Castañón, Universidad de Oviedo, Oviedo, Spain

J.H. Loftus, Dublin City University, Dublin, Ireland

M.A. López, Universidad de Alcalá, Alcalá de Henares, Madrid, Spain

G. Marrazza, University of Florence, Sesto Fiorentino (FI), Italy

D.A. McPartlin, Dublin City University, Dublin, Ireland

A. Merkoçi, Catalan Institute of Nanoscience and Nanotechnology (ICN2), CSIC and The Barcelona Institute of Science and Technology, Barcelona, Spain; ICREA — Catalan Institution for Research and Advanced Studies, Barcelona, Spain

R. Miranda-Castro, Universidad de Oviedo, Oviedo, Spain

K.L.M. Moran, Dublin City University, Dublin, Ireland

M. Moreno-Guzman, Universidad de Alcalá, Alcalá de Henares, Madrid, Spain

D. Moscone, Università di Roma Tor Vergata, Rome, Italy

A. Mulchandani, University of California, Riverside, CA, United States

R. O'Kennedy, Dublin City University, Dublin, Ireland

S. Otles, Ege University, Izmir, Turkey

G. Palleschi, Università di Roma Tor Vergata, Rome, Italy

R. Rapini, University of Florence, Sesto Fiorentino (FI), Italy

B.Y. Sahyar, Ege University, Izmir, Turkey; Indesit Company, Manisa, Turkey

N.S. Sariciftci, Johannes Kepler University, Linz, Austria

N.M. Saucedo, University of California, Riverside, CA, United States

V. Scognamiglio, IC-CNR Istituto di Cristallografia, Rome, Italy

K.C. Siva balan, IIAT, Trichy, Tamil Nadu, India

N. Verma, Punjabi University, Patiala, Punjab, India

Series Editor's Preface

I am delighted to introduce volume 74 in the CAC series edited by Viviana Scognamiglio, Giuseppina Rea, Fabiana Arduini, and Giuseppe Palleschi on *Biosensors for Sustainable Food: New Opportunities and Technical Challenges*. This book is an excellent addition to the CAC series and it reinforces our interest in the field of biosensors for trace analysis. We had already three CAC books on biosensors: volume 39 on *Integrated Analytical Systems* edited by Salvador Alegret in 2003, volume 44 entitled *Biosensors and Modern Biospecific Analytical Techniques* edited by Lo Gorton in 2005, and a third one on *Electrochemical Biosensor Analysis* edited by Salvador Alegret and Arben Mercoçi in 2007. Almost after 10 years, volume 74 will be an excellent addition to the series, and our readers will be able to judge the extraordinary progress achieved in this field since the release of our first titles.

The volume that you have now in your hands covers the main pillars of food safety in 14 chapters. Five chapters address trends on biosensing-based technologies for the detection of a variety of food contaminants, such as pesticides, pathogens, and toxins. The second part of the book contains four chapters on new generation of synthetic receptors as well as on new technologies for improving biosensor efficiency, such as the use of emerging nanomaterials or biocompatible integration of electronics into food sensors. The last part of the book covers already commercially available biosensors in the agrifood sector and the new revolutionary achievements in biosensor technology. Robotic-based agriculture using drones and sensors, intelligent food packaging, and biosensor integration with ICT tools for an improved management and traceability in agriculture are excellent examples of the new trends in this field.

Indeed, this book offers a comprehensive and state-of-the-art view on the development and use of biosensors in food and agriculture sectors. Biosensors can offer today an extraordinary potential of use since they are becoming cheap, fast, and easy-to-use technological devices as compared to conventional analytical methods. The book is certainly timely. It is well-known the enormous demand on food products at global scale, especially from China with a high growth rate of per capita livestock consumption. Fast control devices such as biosensors can assure a high degree of quality control as well as a guarantee for food supply.

Finally I would like to thank specially Viviana, Giuseppina, Fabiana, and Giuseppe for the amount of work, time, and expertise devoted as editors of the book. I would like to acknowledge as well to the various well-known authors for their contributions in compiling such a world-class and timely book for the CAC series.

D. Barceló,
IDAEA-CSIC, Barcelona and ICRA, Girona, August 22, 2016,
Editor in Chief of the Comprehensive Analytical Chemistry Series.

Preface

New Frontiers in the Agrifood Field: Sustainability and Precision Farming

Nowadays it is a common knowledge that to sustain the global increasing demand forecast for the next centuries in the food production system, sustainable practices should be adopted to protect the environment and human health. Food production and consumption have a strong impact on the environment in terms of greenhouse gas emission, water and soil contamination, reduction of arable lands, water consumption and many others, which in turn negatively affect also human health. Also these issues produce heavy consequences on the economic development. To address these challenges, it became mandatory for a close cooperation among scientists with different expertise, policymakers and economists to develop new smart technologies, introduce them on the market and formulate adequate regulations.

In this regard, a sustainable food production system could be thought as a chain of procedures with low impact on the environment and able to guarantee a secured supply of healthier and fortified food, supporting at the same time the economic growth. The development and exploitation of biosensors as supporting tools along the entire food production chain meet these criteria allowing the detection of food quality and safety with a reduced level of CO_2 emissions, and organic solvent released in the environment compared with the conventional analytical tools. Indeed biosensor technology has the high potential to guarantee a comprehensive control among all the single steps of the food supply chain, from the crop cultivation and harvesting, to food process, transportation, packaging and distribution.

Sensor technology is at the leading edge of the development in almost every sector. The market for sensors grew from EUR 81.6 billion in 2006 to EUR 119.4 billion in 2011 and can be expected to grow to EUR 184.1 billion until 2016, according to the new World Report entitled 'Sensors Markets 2016' published by Intechno Consulting, Basel (Switzerland). A huge amount of this market is represented by the biosensor revolution promoting sustainable foods in the near future.

In this context, this book will discuss about the last trends on biosensor technologies, including (1) advanced technologies for fast, sensitive and cost-effective determination of different targets to ensure food quality, safety, authenticity and traceability as well as (2) emerging technologies for crop monitoring, accurate analysis of soil nutrients and pesticides or for maximizing the efficiency of water use for a smart agriculture. Thus the emerging innovations in biosensor technology devoted to food quality and safety control will be discussed in the first section 'Biosensors for food safety, quality and security'. The concepts of food quality and safety are crucial issues to get a sustainable and healthy food. Appearance, taste, smell, nutritional value content, functional ingredients, freshness, flavour and texture are all crucial parameters to be considered to this regards, together with its comprehensive estimation of freshness and the evaluation of the correct composition of the natural components (eg, sugars, amino acids and alcohols) and additives (eg, vitamins and minerals). On the other hand, safe foods should respect the legal limits regarding pesticides, heavy metals, pathogens and toxins, to avoid serious damages of human health and well-being, and consequent burdens on health-care system and economic productivity. In this overall scenario, the last trends on biosensors will be described considering the main advantages and limitations of this technology, to minimize threats that cause unsafe or off-quality food.

With the aim to optimize biosensing performances for a smart food chain monitoring, the last trends on synthetic biology and biomimetic chemistry will be addressed in the second section 'New generations of synthetic receptors and functional materials for food biosensors', to foster the development of tailor-made bioreceptors with desired features in terms of stability and sensitivity. These extraordinary disciplines have been largely exploited in the last years to obtain ad-hoc bioreceptors with desired features of robustness, sensitivity and useful detection range. For this reason, the last tendencies on artificial molecules (ie, aptamers) and functional materials (ie, biomimetic surfaces) will be discussed to redesign the configuration of the sensing elements or by producing novel synthetic entities mimicking key properties of natural molecules.

The third section 'New technologies improving biosensor efficacy' deals with the nanotechnology approach, the biocompatible integration of electronics and the (bio)microfluidics to design ad-hoc sensing systems, meeting the requirements for effective monitoring of the food supply chain. The use of emerging nanomaterial will be debated to either develop novel smart analytical methods or improve the existing ones. In addition, the recent progresses in microelectronics and microfluidics will be highlighted to foster the development of accurate, low-cost and ready-to-use biosensors.

The fourth section 'Biosensor trade in agrifood sector' provides significant and up-to-date information about the commercially available biosensors in the agrifood sector. Considering the international concern with food quality and safety, biosensor development occupies an enormous space in the market, also due to the growing interest of companies in realizing analytical systems with an effective commercial success. However, despite the huge literature on biosensors applied in food analysis, only few systems are able to reach the market. This section aims to highlight the main limitations that should be faced, in terms of biological components, mass production, handling convenience, specificity and timesaving.

The fifth section 'New revolutionary frontiers in biosensor technology' debates on the emerging technologies, including laser scanning approaches, robotic-based analytical tools, intelligent food packaging and ICT tools, developed with the aim of supplying the entire food supply chain, managing excess and waste of food and boosting up precision farming.

By profiling the advances accomplished in recent years in synthetic biology, new material design (biohybrids), nanotechnology, micro/nanofluidics, ICT and so on, the main ambition of this book is to strengthen the relationship between high-tech researchers, industrial professionals, and national governments to promote a uniform synergistic research programme to develop innovative, tailor-made biosensors for a sustainable agrifood management.

The precious contribution of scientists, working on multidisciplinary researches on biosensing technology, will shed light on the emerging, vanguard and forthcoming knowledge to overcome the main limitations hindering the transition of biosensors from the bench to the market, promote a sustainable agriculture and lay the foundations for a bio-based economy.

Viviana Scognamiglio
Giuseppina Rea
Fabiana Arduini
Giuseppe Palleschi

Biosensors for Food Safety, Quality and Security

Chapter 1

Biosensor Potential in Pesticide Monitoring

R. Rapini and G. Marrazza*
University of Florence, Sesto Fiorentino (FI), Italy
*Corresponding author: E-mail: giovanna.marrazza@unifi.it

Chapter Outline

1. INTRODUCTION

The pesticide is any substance or mixture of substances used to control the growth of infesting species that can compromise the agricultural production. Infesting species include insects, weeds, little mammals, fungi and others. The official definition of the Food and Agriculture Organization (FAO) says that they are defined as substances intended for preventing, destroying, attracting, repelling or controlling any pest, included unwanted species of plants or animals, during the production, storage, transport, distribution and processing of food, agricultural commodities or animal feeds, for use as a plant growth regulator, defoliant, desiccant, fruit thinning agent or sprouting inhibitor and substances applied to crops either before or after transport. The active molecules, defined as 'the components of a formulation responsible for the direct or indirect biological activity against pests and diseases, or in regulating metabolism/growth, etc.', of a commercial pesticide, are the main

Comprehensive Analytical Chemistry, Vol. 74. http://dx.doi.org/10.1016/bs.coac.2016.03.016

ingredient of a formulation that can be available as emulsion concentrate or solid mixture (ie, soluble or wet powders, granules) and usually need to be diluted prior to use. Moreover, a single active ingredient may be comprised of one or more chemical or biological entities, which may differ in relative activity. A formulation may contain one or more active ingredients [1]. The recent history of agricultural production has been characterized by the use of different kinds of chemical substances to control pests, starting with inorganic compounds, eg, sulphur, arsenic, mercury and lead, and observing a great change after the discovery of dichlorodiphenyltrichloroethane (DDT) as insecticide in 1939 by Paul Müller, that became soon widely used all over the world [2].

Nowadays, worldwide consumption of pesticides can be estimated as 2 million tonnes per year, about 45% of which being used in Europe, 25% in the USA and 25% in the remaining countries, and it has been estimated that around one-third of the agricultural production would be lost without their use [3,4]. Despite the intense usage, pesticides are some of the most toxic, stable and mobile substances released into the environment, and we know that they can have toxic effects also against nontarget organisms, including humans: for this reason, their presence in food is particularly dangerous [5]. Mainly, diffused health problems, derived from the exposure to those substances and eventual bioaccumulation, can range from asthma, skin rashes, chronic disorders and neurological diseases [6–9]. Nevertheless, due to the well-studied, real and perceived impact of those substances on environment and human health, a strict regulation of their application and residue levels in drinkable water and especially in food supplies has been compiled. Maximum residue limits admitted in food products and drinkable water are regulated in the European Union (EU) by EFSA (European Food Safety Authority) or by international bodies such as the Codex Alimentarius [10]. Despite the strict regulations, due to prevent environmental contaminations and risks for human health that can be caused also from food frauds, intense researches have been developed over the years to obtain sensitive and selective approaches for the detection of pesticides.

Considering pollution caused by pesticides, it is known that their intense use leads to a diffused soil contamination of soil, groundwater and even air. Moreover, some toxic compounds resulting from the pesticide formulation or degradation products can accumulate in crops, reaching then directly human alimentation products or passing by farm animals and their derivative products. The environmental fate of pesticides is shown in Fig. 1.

At present, the detection and the quantification of the pesticides are generally based on chromatographic methods coupled with mass spectroscopy [11,12]. Although highly sensitive and selective, conventional analytical methods are time-consuming and laborious when a large number of samples must be screened. Besides, they require expensive equipment, skilled operators and complicated pretreatment. Thus, researchers have been investigating

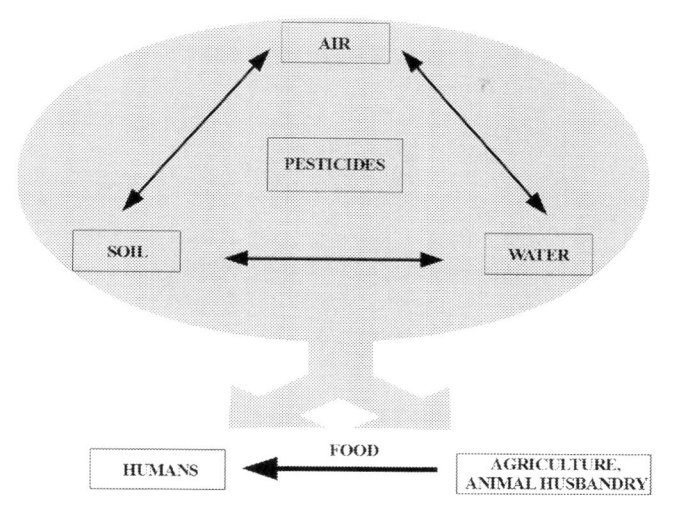

FIGURE 1   The environmental diffusion ways of pesticides.

alternative methods of detection and screening that are cheaper and more user-friendly. Recently, there has been an increasing interest in biosensor technology for fast pesticide detection using easy and rapid procedures. Biosensors have the potential to complement or replace the classical analytical methods by simplifying or eliminating sample preparation and making field-testing easier and faster with significant decrease in cost per analysis.

Focusing on the recent activity of worldwide researchers, without pretending to being exhaustive, the aim of this chapter is to give the readers an overview of recent innovation in the field of pesticide biosensors, after an introduction of the main chemical classes of pesticides, their sampling and extraction from food samples.

1.1 Pesticides: Properties, Persistence and Regulation

Pesticides are characterized by a great diversity of chemical structures, action mechanisms and usages, things that make it very difficult to create a classification. Currently, over 800 active ingredients, belonging to more than 100 substances classes, are present in a wide range of commercially available products [13]. They can be listed according to many criteria, eg, their toxicity, their purpose of application, their environmental stability, the ways by which they penetrate target and nontarget organisms, and, of course, their chemical structures. Considering the aim of this work, we are going to focus on a short description of the main classes of pesticides, dividing them into three major groups: (1) herbicides, substances used to control unwanted plant growth or to kill weeds; (2) insecticides, used to kill infesting insects; (3) fungicides, used to control the proliferation of fungi. It should be considered that some chemical classes might fall into more than one major group. In Table 1, a classification of the most important pesticides and their toxic effects are shown.

Referring to the chemical classes division, the most diffused groups are organochlorine, organophosphorus (OP) and organonitrogen (including carbamates, triazines and other derivatives).

Organochlorine pesticides are reported as probably the most toxic class [31,32], to which belong some of the oldest types of compounds used in this way, characterized by active molecules that can persist in environment for more than 30 years. For this reason, over the last years, compounds such as organochlorines are begun to be replaced by compounds and formulations characterized by a faster biodegradation, as in the case of the above-mentioned OP pesticides [13].

1.1.1 Herbicides

Implementation in agriculture mechanization and increase in crops production require a control of weed growth, and herbicides have played a very important role in this.

Phenoxy acids are many compounds formed by a phenoxy radical linked to a low carbon number alkanoic acid. They can be both soil and foliar applied, and they generally have a short persistence in soil [33].

Urea derivates are one of the most important agricultural herbicide groups. They are usually applied as aqueous emulsions to the surface of soil. Almost all of the urea compounds with good herbicidal action are trisubstituted urea, containing a free imino-hydrogen. The urea bridge is substituted by triazine, benzothiazole, sulfonyl, phenyl, alkyl or other moieties [34].

Triazine compounds are characterized by heterocyclic chemical structures, containing one or more than one nitrogen heteroatoms, with a wide range of uses. They act as inhibitors of electron transport in photosynthesis. Triazines are some of the oldest herbicides, with research initiated on their weed control properties during the early 1950s. Atrazine, one of the most representative compounds of this group and probably the most diffused, has been banned in the European Union in 2003 because of its ubiquity in drinking water, its demonstrated harmful effects on wildlife and its potential health hazards for humans [35].

1.1.2 Insecticides

The importance of eliminating insects and other plant parasites is a crucial aspect, to maintain a high productivity in agriculture.

Carbamate insecticides is a broad group composed by esters and thioesters of the carbamic acid (R_1-O-CO-NR_2R_3, R_1-S-CO-NR_2R_3), having a great variety of chemical, physical and biological properties. They are generally soluble

TABLE 1 Classification of Pesticides Including Main Commercially Available Molecules and Their Toxic Effects

Groups	Chemical Class	Representative Examples	Examples of Toxic Effects
Herbicides	Phenoxy acids	2,4-D, 2,4-DB, dichloroprop, MCPA, MCPB, mecoprop, 2,4,5-T	Carcinogenicity [14,15]
	Ureas (phenylurea, sulfonylurea)	Chlorotoluron, diuron, fenuron, isoproturon, linuron, metoxuron, monolinuron, neburon, chlorimuron-ethyl, chlorsulfuron, metsulfuron-methyl, sulfometuron-methyl, triasulfuron	Embryonic development diseases [16]
	Triazines	Atrazine, ametryn, cyanazine, prometryn, propazine, simazine, terbutryn	Carcinogenicity [17]
Insecticides	Carbamates	Aldicarb, carbaryl, carbofuran, fenoxycarb, methiocarb, methomyl, oxamyl, primicarb	Reversible inhibition of red blood cell AChE and plasma BChE [18], nervous system diseases [19]
	Organochlorines	DDT, chlordane, dicofol, endosulfan, endrin, heptachlor, lindane, methoxychlor	Oxidative stress [18], carcinogenicity, endocrine disturbance [20]
	Organophosphorus	Malathion, parathion, acephate, azinphos-methyl, chlorpyrifos, diazinon, dimethoate, phosmet	Irreversible inhibition of red blood cell AChE and plasma BChE [18], oxidative stress [21], neurotoxicity [22]
	Neonicotinoids	Acetamiprid, imidacloprid, thiamethoxam, clothianidin, dinotefuran, nitenpyram	Nicotinic ACh receptors activation (low effects on mammalians) [23], genotoxicity [24,25], carcinogenicity [26]
	Pyrethroids	Allethrin, bifenthrin, cyfluthrin, cyhalothrin, cypermethrin, deltamethrin, fenvalerate	Neurotoxic effects [27]
Fungicides	Dithiocarbamates	Mancozeb, ferbam, maneb, metiram, propineb, thiram, zineb, ziram	Neurotoxic chronic effects [28]
	Dicarboximides	Chlozolinate, iprodione, procymidone, vinclozolin	Antiandrogenic effects [29]
	Organomercurials	Methyl-mercury, phenyl-mercuric-acetate	Neurotoxic effects [30]

in water and in polar organic solvents and commercially available as wet powders, dust granules and emulsion concentrates. Used also as herbicides, they are usually soil applied and absorbed by root and shoot. These compounds are usually decomposed by soil microorganisms in 3–5 weeks [33].

Three chemically different subgroups compose organochlorine group: DDT analogues, benzene hexachloride isomers and cyclodiene compounds. DDT has a wide spectrum activity on different families of insects and related organisms, and it is one of the most persistent contact insecticides because of its low water solubility and low vapour pressure. Cyclodiene compounds are also characterized by a high environmental persistence. Due to these characteristics and their high toxicity, the majority of these compounds have been banned, or their use has been strongly restricted [33].

Hydrocarbon compounds compose OP group, containing at least one phosphorus atom in their structure. The great structural variety of these compounds makes them the most versatile insecticides, being also characterized by a generally low persistence in biological systems. The group includes both molecules with nonresidual action and prolonged residual action, broad-spectrum pesticides and ones with very specific action, all soluble in water and easily hydrolysed, making them dissipated from soil in a few weeks after application. For these reasons, they are widely used as systemic insecticides. Compounds that have a broad spectrum and nonselective herbicide activity also form OP group. They are extensively used for foliar application, both in agricultural and nonagricultural areas. Their main degradation product is aminomethylphosphonic acid (AMPA), which can be found in plants, soil and water [33].

Neonicotinoids are neuroactive substances derived from nicotine. Acting as nicotinic acetylcholine (ACh) receptor agonist, these compounds cause the paralysis of contaminated organisms followed by their death, and they are active against many sucking and biting insects. Depending on the application form (seeds or foliar treatment) of neonicotinoid insecticides, there are different routes of exposure of nontarget organisms to these pesticides [36].

Pyrethroid insecticides are composed by synthetic derivatives of pyrethrum, a substance extracted from some species of chrysanthemum flowers. They generally show low toxicity for nontarget organisms (birds, mammalians), and they are used as contact poisons, affecting insects' nervous system and depolarizing neuronal membranes. These molecules are degraded in soil and can be used in domestic ambiances to control flies and mosquitoes [33].

1.1.3 Fungicides

Fungicides are used in agriculture to control fungal disease on plant crops and can be applied pre- or postharvest on many kinds of cultivations.

Dithiocarbamates are applied worldwide on a great number of crops, due to their high efficiency in controlling plant fungal diseases, and are characterized by a relatively low mammalian acute toxicity. Their toxicity is given by the degradation (metabolic or not) product ethylene thiourea, known for its carcinogenic and teratogenic action in laboratory animals [37].

Dicarboximide fungicides are rapidly converted to 3,5-dichloroaniline in soil. Their repeated use over several years reduces their effectiveness. It is known that some kind of resistance against their action has developed in many plant species and protected crops. For this reason, dicarboximides are recommended to be used in conjunction with other fungicides [38].

Organomercurials are composed by some of the oldest kind of fungicides, and the most diffused of its compounds are the methyl-, ethyl-, methoxyethyl- and phenyl-substituted molecules. Applied on seeds, aerobic microorganisms can convert organomercurial compounds to aliphatic or aromatic hydrocarbons and mercury (oxide or sulphide), that is accumulated in soil and can be absorbed by plants. In this way, it can be assimilated and accumulated in mammalian tissues. For this reason, the use of organomercurial compounds is strongly decreasing [39].

2. FOODS: SAMPLING AND PRETREATMENT

Continuous development of analytical methods has brought in many different approaches for the detection of pesticides, either using classical instrumental analytical tools or using biosensors. Despite of this, the necessity to measure very low concentrations of active molecules or degradation products in complex and variable matrices, such as fruit, vegetables and biological samples, still makes the necessity of developing effective sampling and pretreatment procedures one of the most important targets for analytical chemists. We are going to focus on this short description on the several stages commonly necessary to perform food analysis.

It is possible to make a list of subsequent actions that are required to prepare a sample for the analysis, using as more as possible validated procedures [5]:

- Sampling
- Fixing, transport, homogenization and storage
- Extraction from the sample and eventual enrichment
- Extract cleanup and preparation for the analysis
- Analysis

The realization of the whole procedure is strictly connected to analyte and sample characteristics: considering food matrices, a primary division should be made evaluating the amount of fat content. The threshold value is usually set at 2%, where the majority of matrices for pesticide analysis are composed by nonfatty foods such as fruits and vegetables, usually consumed raw or with very light postharvest treatment, giving the highest risks for consumers in case of contamination. Detection of lipophilic pesticides, such as organochlorines, is then the main reason for fatty matrices sampling, and in this case health risks come from the possible accumulation of toxic substances in lipid tissues [40]. The sampling and subsequent pretreatment of matrices should conserve their representativeness, considering their origin and amount [41]. In this way, sampling is crucial for food analysis because of peculiarities of samples: a rapid preparation helps to prevent composition changes given by degradation, and it is useful to carry it out at low temperature, as well as prepare a frozen storage if the samples cannot be analysed immediately. It is important to consider that the analytical sample could be just one part of the whole material collected, so a size reduction has to be carried out using appropriate procedures, with different approaches for dry solid (granules, powders), wet solid (fruits, vegetables) and liquid samples [41].

After homogenizing the sample, all the moist materials such as meats, tissues, fruits, vegetables (leaves, roots or tubers) are fine-sliced using mechanical devices such as mortars or mixers. If necessary (especially for moisture-rich samples or fatty samples) a predrying or defatting procedure can be performed.

Target analytes have then to be extracted from the homogenized samples and eventually enriched, because their concentration (this is true especially considering pesticides) can be relatively low in all various environmental matrices. The extraction and the eventual preconcentration involve the transfer of the target pesticide from the primary matrix to a secondary one, eliminating the majority of interfering compounds [42−44]. For analysis of solid samples such as fruit and vegetables, the pesticide would be extracted by using different solvent extraction techniques (eg, accelerated solvent extraction, microwave solvent extraction, ultrasound-assisted solvent extraction or supercritical fluid solvent extraction) [45,46].

Unfortunately, food sample extract contains a large amount of components that can create difficulties in the determinative step. A cleanup procedure is often necessary to remove coextracted interfering species without decreasing target pesticide recovery. The degree of cleanup procedure depends on the technique that is going to be used for the analysis, and many different approaches are available [5], but the most used remains the solid phase extraction [47,48].

The last step of the analytical procedure involves the detection of the analyte; recent and various approaches of pesticides detection using biosensor-based devices will be dealt in the following paragraphs.

3. SENSING OF PESTICIDES FOR FOOD SAFETY

Food safety can be defined as the inverse of food risk, ie, the probability of not suffering some hazard from consuming a specific food [49]. Potential hazardous residues in foods include a great amount of substances, natural and environmental contaminants (eg, mycotoxins, dioxins, etc.), human and veterinary drugs, growth promoters, packaging components and, especially, agro-chemical pollutants such as pesticides [50]. For this reason, the development of fast and sensitive analytical techniques for their detection remains one of the most interesting targets for chemists. The use of (bio)sensors offers the possibility to develop a large amount of simple-to-use devices with low costs, fast and sensitive response and high selectivity. For these reasons, they are particularly suitable for on-field screening analysis, being easily coupled with portable devices.

A biosensor is a device composed of two intimately associated elements: a molecular receptor, that is an immobilized sensitive biological or biomimetic element recognizing the analyte, and a transducer, that is used to convert the (bio)chemical signal resulting from the interaction of the analyte with the molecular receptor into an electronic one. The intensity of the generated signal is directly or inversely proportional to the analyte concentration. In literature, a great variety of biosensors for pesticide detection based on different transducers (eg, electrochemical, optical, thermal, calorimetric or piezoelectric) are reported. However, the recognition elements are the most important part of biosensors since they are responsible for the recognition of target analyte. Thus, we specifically provide an overview on some recent pesticide sensing classified as for the molecular receptors used in assay format. Application of novel nano and hybrid advanced functional material (eg, quantum dots, carbon nanotubes (CNTs), graphene and metal organic frameworks) is another key to develop next-generation sensor systems. Consequently, in this chapter, particular emphasis is given to new (bio)sensing based on nanomaterials.

3.1 Enzyme Biosensors

The enzyme biosensors are based on enzymes in intimate contact with the transducers. The enzyme reacts selectively with its substrate (target analyte). Many works have been already published, and detection of pesticides is included in the group of analyte that can be detected using enzymes [51−55]. Enzyme biosensors can measure the catalysis or the inhibition of enzymes by the target analyte. In this way, the biosensor detects produced or consumed species, respectively. Commonly, only one enzyme is used in enzymatic biosensors, but progress in this field involves combining enzymes to obtain multienzyme systems where several enzymes are incorporated into the same assay.

Biosensors based on a catalytic reaction are superior to the inhibition mode since they can be potentially reused and are suitable for continuous monitoring. Here, electrochemical enzyme biosensors have been reviewed.

A growing class of bacterial enzymes known as phosphotriesterases has recently been characterized. These enzymes catalyse the hydrolysis of OP pesticides with a high turnover rate. The typical one is organophosphorus hydrolase (OPH), an enzyme that can hydrolyse large number of OP pesticides, and this approach can be useful to perform a continuous monitoring of the analyte concentration [56]. OPH acts on the p-nitrophenyl group on the target molecule (many OP pesticides have a p-nitrophenyl substituent, eg, paraoxon, parathion) hydrolysing it to p-nitrophenol, that can be detected with electrochemical or optical devices [57,58]. Recently, several types of OPH-based electrochemical biosensors based on nanomaterials have been introduced.

Choi et al. developed an amperometric biosensor coupling a conductive reduced graphene oxide/nafion hybrid films with OPH enzyme. The biosensor exhibited a sensitivity of 10.7 nA/μM and a detection limit of 1.37×10^{-7} M [59]. A novel biosensor based on OPH and mesoporous carbon and carbon black as an anodic layer has been developed from Lee et al. [60]. The well-ordered nanopores and high surface area of the mesoporous carbon resulted in increased sensitivity and allowed for nanomolar-range detection of the analyte paraoxon. The biosensor had a detection limit of 0.12 μM for paraoxon.

Methyl parathion−degrading enzyme (MPDE) is one kind of OPH. It can detect methyl parathion (MP), which is an extensively used OP pesticide, with high selectivity since MPDE hydrolyses the P−S bond-containing agent. MPDE catalyses the hydrolysis of MP to generate an equimolar amount of p-nitrophenol, which, in turn, is electrochemically oxidized at the working electrode poised at a fixed potential to generate a current that is proportional to the pesticide concentration. An amperometric biosensor for highly selective and sensitive determination of MP has been developed from Du et al. [61]. The biosensor is based on dual-signal amplification: a large amount of introduced enzyme on the electrode surface and synergistic effects of nanoparticles towards enzymatic catalysis. The detection limit of the biosensor is found to be 1.0 ng/mL.

The same research group has then developed an OPH biosensor for detection of MP, based on self-assembly of MPH on Fe_3O_4/Au nanocomposite. The magnetic nanocomposite provides an easy way to construct the enzyme biosensor by simply exerting an external magnetic field, and it provides a simple way to renew the electrode surface by removing the magnet. Under optimal conditions, the biosensor shows rapid response and high selectivity for the detection of MP, with a detection limit of 0.1 ng/mL [62].

The enzyme inhibition−based biosensors have been applied for OP and organochlorine pesticides, derivatives of insecticides, heavy metals and glycoalkaloids. The choice of enzyme/analyte system is because these toxic analytes inhibit normal enzyme function.

Inhibition-based biosensors have been the subject of several recent reviews [63−66]. The use of this approach requires a quantitative measurement of the enzyme activity before and after exposure to a target analyte, and the experimental procedure can be summarized in three main steps: initial measurement of the enzymatic activity, incubation of a sample containing a certain concentration of the target pesticide and measurement of the inhibited enzymatic activity.

The mainly used enzymes as biological receptors in the enzyme inhibition−based biosensors are acetylcholinesterase (AChE), butyrylcholinesterase and urease [67]. The AChE enzyme is the most used because it is characterized by high turnover, and its substrate is soluble in aqueous solution and is not so expensive. Further, several compounds such as OP pesticides and nerve agents inhibit AChE, and a fast and in situ detection of these analytes is very useful. AChE active site is composed by three amino acids, histidine, serine and aspartic acid: the hydroxyl group of the serine can hydrolyse ACh by deprotonation when its quaternary ammonium group (positively charged) electrostatically interacts with the aspartic acid residue. In presence of an OP pesticide, its phosphorus group covalently binds the nucleophilic serine, inactivating the enzyme [68,69].

While many studies have looked at the effect of single pesticides on AChE, the effect of mixtures of pesticides still requires extensive investigation. This is important to evaluate the cumulative risk in the case of simultaneous exposure to multiple pesticides. Mwila et al. studied the effect of five different pesticides (carbaryl, carbofuran (CBF), parathion, demeton-S-methyl and aldicarb) on AChE activity to determine whether combinations had an additive, synergistic or antagonistic inhibitory effect. The data from the assays of the mixtures have been used to develop and train an artificial neural network. The obtained results indicated that the mixtures had an additive inhibitory effect on AChE activity [19].

In recent years, AChE has been immobilized onto various nanomaterial-modified surfaces, such as CNTs, graphene, gold nanoparticles (AuNPs) and quantum dots (QDs), and nanohybrid materials to improve the response and the stability in trace-pesticide detection.

CNTs are of special interest due to their ability both to form three-dimensional (3D)-nets with high adsorptive activity toward enzymes and to establish electric contact with the electrode [70,71].

The CNTs have been often combined with other materials to modify the sensor with novel nanocomposites such as 7,7,8,8-tetracyanoquinodimethane [72]. The biosensor developed by immobilization of AChE in sol−gel allowed the detection of two reference AChE inhibitors, paraoxon-methyl and chlorpyrifos with detection limits of 30 pM and 0.4 nM, respectively.

Wang's group reported an AChE biosensor based on chitosan/prussian blue/multiwalled CNTs/hollow gold nanospheres nanocomposite film formed by one-step electrodeposition. Based on the inhibition by pesticides of the AChE activity, using malathion, chlorpyrifos, monocrotophos and CBF as model compounds, this biosensor showed a wide range, low detection limit, good reproducibility and high stability. A schematic representation is reported in Fig. 2 [73]. A simple and reliable technique has been developed for the construction of an amperometric AChE biosensor based on screen-printed carbon electrodes−modified single-walled CNTs and Co phthalocyanine. This procedure has been proposed to decrease the working potential and to increase

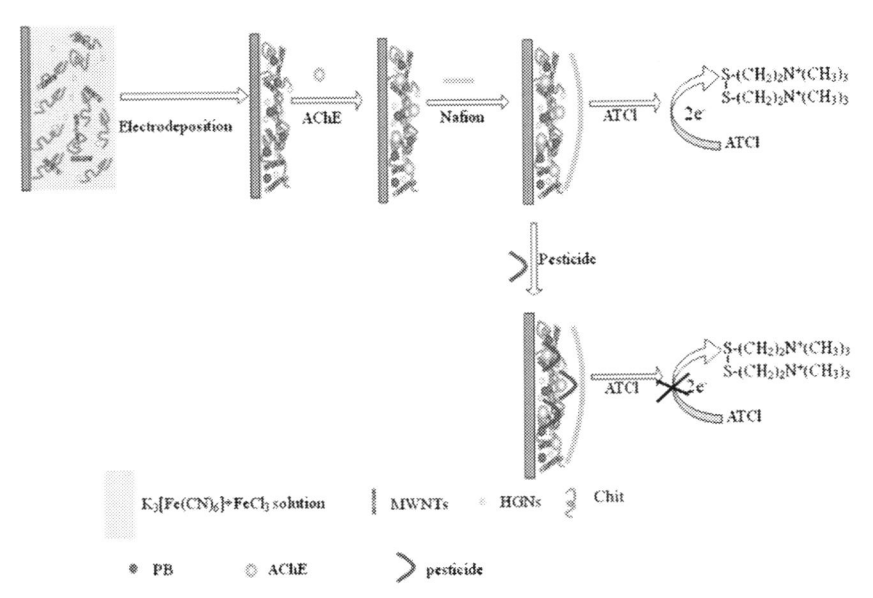

FIGURE 2 Schematic illustration of the stepwise acetylcholinesterase (AChE) biosensor fabrication process and immobilized AChE inhibition in pesticide solution. *Reprinted from C. Zhai, X. Sun, W. Zhao, Z. Gong, X. Wang, Acetylcholinesterase biosensor based on chitosan/prussian blue/ multiwall carbon nanotubes/hollow gold nanospheres nanocomposite film by one-step electrodeposition, Biosens. Bioelectron. 42 (2013) 124–130, with permission from Elsevier.*

the signal of thiocholine oxidation. The reliability of the inhibition measurements has been confirmed by testing spiked samples of sparkling and tape waters.

The immobilization of AChE on multiwall CNTs-β-cyclodextrin (β-CD) composite-modified glassy carbon electrode has been proposed from Du et al. [74]. Due to the good dispersion of porous structures of MWCNTs-β-CD composite, the resulting surface provided a favourable microenvironment for AChE biosensor fabrication and maintained the bioactivity of AChE for screening of OPs exposure. The same research group later proposed AChE biosensor based on nanocomposites using multiwalled carbon nanotube coating gold nanoparticles (MWCNTs-Au). The formed MWCNTs-Au nanocomposites offered an extremely hydrophilic surface for biomolecule adhesion, leading to a stable AChE biosensor [75].

Since conducting polymers show some numerous features for sensing and biosensing, they have recently attracted a lot of attention in this field. Incorporation of CNT, metals and metal oxides in the conductive polymers can enhance electron transfer through a direct or mediated mechanism with improved conductivity and enhanced stability [76–78]. The polymer is used to provide a high surface, protection against the fouling of the metal catalyst, and a scaffold for high dispersion and anchoring of the metal particles.

A simple method to immobilize AChE on polypyrrole (PPy) and poly-aniline (Pan) copolymer doped with MWCNTs has been proposed. The synthesized PAn-PPy-MWCNTs copolymer presented a porous and homogeneous morphology, which provided an ideal size to entrap enzyme molecules [79].

A macromolecular polymer and carboxyl multiwall carbon nanotubes (MWCNTs-COOH) coated with AChE s on the Ag-coated crystal surfaces has been presented from Shang at al. The authors determined pesticide residue in freshly picked radishes 4 and 8 days post application of phoxim and chlorpyrifos. The sensitivity of the method has been compared with that of gas chromatography. The results showed that there was no significant difference between the two methods [80]. The use of nanocomposites allows for detecting the inhibitors at very low detection limit, but very often requires high potential, even higher than the sensor modified with only CNTs.

Some electrochemical biosensors based on enzyme inhibition have been assembled using graphene. Graphene, as a new two-dimensional carbon nanomaterial, has attracted increasing attention during recent years by virtue of its outstanding physical and chemical properties and excellent electro-catalytic ability. The special properties of graphene may provide insight into the fabrication of novel biosensors for virtual applications: the high surface area is helpful in increasing the surface loading of the target enzyme molecules, and the excellent conductivity and small band gap are favourable for conducting electrons from the biomolecules.

Li et al. [81] reported the sensitive amperometric biosensing of OPs using AChE modifying glassy carbon/reduced graphene oxide electrode. This electrochemical biosensor displayed a detection limit of 0.5 ng/mL with good reproducibility and stability.

Xue et al. also reported a graphene–nafion matrix-modified glassy carbon electrode for the determination of OPs. MP is detected over a concentration range of 0.02–20 µg/mL in vegetable samples [82].

Graphene-based composite materials possessing 3D porous architectures are preferred owing to their very large surface areas and low mass transport resistance. Graphene can enhance direct electron transfer between enzymes and electrodes. It has been also reported that the use of graphene associated with metal nanoparticles can form exceptionally stable and cost-effective biosensors.

A sensitive amperometric AChE biosensor based on NiO nanoparticles, carboxylic graphene and nafion-modified glassy carbon electrode has been developed from Yang et al. [83]. The nanocomposite material showed excellent conductivity, catalysis and biocompatibility and offered an extremely hydrophilic surface for AChE adhesion. Under optimum conditions, the biosensor detected MP and chlorpyrifos in the linear range from 1.0×10^{-13} to 1.0×10^{-10} M and from 1.0×10^{-10} to 1.0×10^{-8} M, with the detection limit of 5.0×10^{-14} M. The biosensor detected CBF in the linear range from

1.0×10^{-12} to 1.0×10^{-10} M and from 1.0×10^{-10} to 1.0×10^{-8} M, with the detection limit of 5.0×10^{-13} M.

A green synthesis procedure for preparing zinc oxide nanoparticle—decorated MWCNT—graphene hybrid composite by using solar energy, and its application as a transducer candidate for OP biosensor has been proposed from Nayak [84]. The fabricated biosensor using this composite as a transducer candidate shows a high affinity to AChE. It exhibits a linear response for paraoxon detection from 1 to 26×10^{-6} M with a detection limit of 1×10^{-9} M.

AuNPs are characterized by relevant electrochemical properties such as high superficial area, conductivity and electron transfer rate [85]. Graphene and the AuNPs have been often coupled for designing biosensors.

Wang et al. synthesized a nanohybrid of AuNPs and chemically reduced graphene-oxide nanosheets (cr-Gs). Then, an electrochemical sensor based on AChE/AuNPs/cr-Gs, which has been prepared by self-assembling AChE on an AuNPs/cr-Gs nanohybrid, has been developed for ultrasensitive detection of paraoxon. The nanosize effect of AuNPs scattered on cr-Gs sheets leads to greatly improved electrochemical detection of the enzymatically generated thiocholine product, including a potential negative shift in operation, great electrochemical-signal amplification, and higher sensitivity and stability [86].

A stable AChE biosensor based on self-assembling AChE to graphene nanosheet (GN)-AuNPs nanocomposite electrode for investigation of inhibition, reactivation and ageing processes of different pesticides has been proposed from Zhang et al. [87].

A nanohybrid of AuNPs, polypyrrole and reduced graphene oxide sheets has been achieved by electrochemical deposition of reduced graphene oxide with pyrrole and the introduction of AuNPs. AChE has been further encapsulated in a silica matrix and immobilized on the Au-PPy-rGO nanocomposite by co-deposition. The presence of PPy helped to avoid the aggregation of rGO caused by van der Waals interactions between individual sheets and significantly increased the surface area of the modified electrode. Since AChE molecules have been protected by the circumambient silica matrix, which provided a biocompatible environment and facilitated mass transport, the fabricated AChE biosensor displayed high stability and excellent activity together with a fast response to OP pesticides [88].

An amperometric biosensor based on immobilizing AChE on 3-carboxyphenylboronic/reduced graphene oxide—gold nanocomposites-modified electrode has been developed for the detection of OP and carbamate pesticides from Liu et al. [89]. The biosensor showed good sensitivity owing to the excellent properties of AuNPs and reduced graphene oxide, which promoted electron transfer reaction and enhanced the electrochemical response.

A new method based on specific binding between glycoprotein AChE and boronic acid functionalized Fe/Au magnetic nanoparticles has been presented for the development of AChE biosensor [90]. Based on enzyme inhibition, carbamate pesticide has been detected using Furadan as a model compound. Two linear ranges of 0.05−15 and 15−400 ppb have been obtained with a detection limit of 0.01 ppb. The biosensor also showed acceptable reproducibility and relatively good storage stability. Moreover, satisfactory results have been obtained in real sample analysis.

Pesticides can be detected not only by single enzyme systems, but also by bienzyme systems. Recently, a biosensor composed of bienzymes AChE and choline oxidase and CdTe QDs has been developed to determine dichlorvos, with a linear range of $4.49−6780 \times 10^{-6}$ M and an LOD of 4.49×10^{-6} M [91]. This biosensor has good performance in detecting residues of dichlorvos in real samples, such as apples.

A. Hatefi-Mehrjardi described another bienzymatic sensor. AChE and COx were covalently immobilized on the mercaptopropionic acid self-assembled monolayer on gold electrode to detect carbaryl at nanomolar level [92].

A bienzymatic biosensor (laccase and tyrosinase) for carbamates has been prepared in a single step by electrodeposition of a hybrid film onto a graphene-doped carbon paste electrode. The biosensor proposed was able to determine carbaryl, formetanate hydrochloride, propoxur and ziram in citrus fruits based on their inhibitory capacity on the polyphenol oxidases activity [93].

Recently, the realization and characterization of a tyrosinase-based biosensor for the determination of atrazine has been reported. The developed biosensor is based on the inhibition effect of atrazine towards the catalytic activity of the blue copper protein tyrosinase. Tyrosinase has been immobilized on the electrode surface by either polyvinyl alcohol with styrylpyridinium groups. The developed inhibition biosensor displays a linearity range towards atrazine within 0.5−20 ppm, a limit of detection of 0.3 ppm and acceptable repeatability and stability. This analysis method has been applied to spiked drinking water samples [94].

Inhibition-based methods have, unfortunately, some disadvantages and can give false positives as handling and storage could cause loss of enzyme activity. As a result, baseline testing prior to sample application must be carried out which lengthens testing time. Further, various compounds can interfere with the Op pesticides' interaction with AChE, such as heavy metals, fluoride, nicotine etc., and for this reason in some cases a different approach is useful.

3.2 Immunosensors

As mentioned above, especially for complex matrices, the use of enzyme inhibition−based biosensors for pesticide detection can be affected by the presence of interfering substances, and in this case, the response of the biosensor is given by a sum of effects that contribute to the inhibition of the enzymatic activity, losing the quantitative determination of the analyte. The use of immuno-based biosensors can be useful to solve some of these problems. The interaction between an antibody and its target antigen is characterized by a high selectivity, so antibodies appear as an efficient alternative for the detection of single pesticides or small groups of similar molecules. Furthermore, the use of immunosensors is a very well-known procedure, because of their specificity, low costs and versatility, and there are a lot of examples that can be found in literature about the determination of many kinds of different molecules, including pesticides [95,96]. Above all it is known that the antibody—antigen complex can be dissociated using an appropriate dissociating reagent, and this, concerning the detection of pesticide, gives the possibility to create reusable devices, very useful for continuous monitoring of real samples or on-field analysis. Using an immunosensor, the experimental procedure can be summarized in two steps: molecular recognition of the target and transduction process. Determination of pesticide is usually carried out making use of a competitive interaction or a displacement between the analyte and a tracer molecule (that can also be a secondary antibody, labelled with a tracer) or using a second recognition step in which a secondary labelled antibody binds the primary antibody—antigen complex. The transduction technique is related to the chosen label or tracer molecule, and various examples are reported using, eg, electrochemical, optical, fluorescent or piezoelectric transduction systems [97]. By the way, it is also known that a competitive label-dependent detection is more complex to obtain with respect to a direct immunoassay, and in the last five years, many examples of label-free immunosensors for pesticide detection can be found, due to the advantage given by their rapidity, low cost and real-time response [98,99].

Although many immunosensors have been reported, the development of new antibodies with the desired selectivity and high affinity for any different target and the necessity to be developed using a guest in which to stimulate an immune reaction often find difficulties due to the toxicity of some classes of pesticides [100]. In the last years, many reports have proved that nanomaterials can improve electrochemical performance of immunosensors. A disposable amperometric immunosensor for sensitive detection of chlorpyrifos-methyl (CM) has been developed by combining dual-signal amplification of platinum colloid with an enzymatic catalytic reaction [101]. This method integrated the advantages of nanotechnology, bioconjugation techniques, enzyme amplification and electrochemical detection for monitoring the CM residues.

Many immunosensors for atrazine detection in different format are present in literature.

Tran et al. described a multifunctional conducting polyquinone film coupled with a monoclonal antiatrazine antibody for label-free electrochemical detection of atrazine using square wave voltammetry [98]. Atrazine has been covalently immobilized on interdigitated μ-electrodes. Detection of free atrazine has been achieved through a competitive reaction with immobilized atrazine for the antibody added in solution. The detection method has been based on the use of antibodies labelled with AuNPs. Their presence amplified the conductive signal. This biosensor has been suitable for the detection of atrazine in red wines since no matrix effect related to red wine samples has been observed. A schematic representation is reported in Fig. 3 [102,103]. Impedimetric immunosensor has been developed by immobilizing antiatrazine antibody modified with histidine-tag onto a polypyrrole (PPy) film N-substituted by nitrilotriacetic acid (NTA) electrogenerated on a gold electrode. In the presence of atrazine, the interaction of the analyte with the immobilized antibody triggered an increase of the charge transfer resistance proportional to the pesticide concentration. The detection limit has been 4.6×10^{-11} M [104].

Piezoelectric immunosensors are devices based on materials such as quartz crystals with bioreceptors, such as antigens or antibodies, immobilized onto their surface, which resonate on application of an external alternating electric field. The biospecific reaction between the two interactive molecules, one

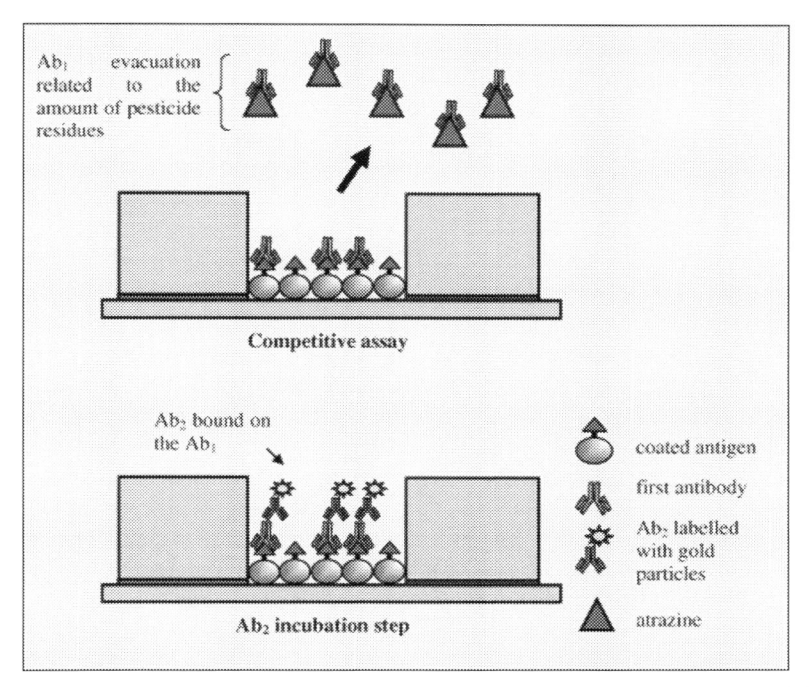

FIGURE 3 Schematic illustration of the conductimetric immunosensor. An amount of the secondary antibody (Ab$_2$) is bound to the specific antibody (Ab$_1$). Previously, the amount of Ab$_1$ bound to the pesticide was evacuated. The amount of gold nanoparticles is indirectly related to the pesticide residue concentration. *Reprinted from E. Valera, J. Ramón-Azcón, A. Barranco, B. Alfaro, F. Sánchez-Baeza, M.-P. Marco, A. Rodríguez, Determination of atrazine residues in red wine samples. A conductimetric solution, Food Chem. 122 (2010) 888–894, with permission from Elsevier.*

immobilized on the surface and the other free in solution, can be followed in real time. Several QCM immunosensors for environmental monitoring have been described, but only few of them were developed for the detection of pesticides. Jia et al. reported a cost-effective protocol to fabricate acoustic micro-immunosensors to measure simultaneously CBF and atrazine. It was found that the proposed methodology allows sequential specific detection of 4.5 and 4.6 μM atrazine, respectively [105]. A multiwall CNT/poly(amido-amine) dendrimer hybrid material and its application in the fabrication of piezoelectric immunosensing platform have been also described. The immunosensor has been used for the determination of amounts of a carbamate pesticide, metolcarb, in spiked apple and orange juice samples analysis [106].

Biosensors based on the field effect transistors and CNTs have become a suitable system for immunosensor applications because they combine the principles of molecular recognition through the recognition layer with the transduction capabilities of the CNTs. They have been used to detect various compounds such as the label-free detection of atrazine with potential application in seawater and riverine water analysis [99].

Among the immunosensors, test strips based on ELISA in combination with lateral flow device for pesticide detection have attracted extensive interest in recent years. Thus, we focused the attention on this kind of immunosensors. In contrast to the above methods, lateral-flow sensors (LFSs) have been applied to detect analytes, due to their unique advantages, such as easy operation, low cost and visualized results, that can be judged directly by the naked eye. Colloidal gold or AuNPs are commonly used in LFSs. The strip tests have the advantages of being efficient and easy to use for on-site testing of food served or in the food industry, allowing the monitoring of the quality of raw food materials at the early stages of food production. Novel technologies have been summarized in a recent review [107]. The test principle involves a flow of fluid containing the analyte through a porous membrane and into an absorbent pad. It is a rapid (within 5 min) and simple immunoassay for qualitative screening, with the advantage of large volumes of samples, but now its application is often restricted by relatively low sensitivity, especially for the analysis of food contaminants such as chemical pesticides. Blažková et al. have reported a strip-based immunoassay for the rapid detection of thiabendazole in a 1% fruit juice matrix with a limit of detection of about 0.08 ng/mL [108]. Since matrix interference of food sample is always a problem for immunochemical method, another similar test strip that can detect carbaryl in a 10% fruit juice matrix has been developed [109]. The LOD of this strip is about 5 ng/mL [110].

With the increasing demand for more efficient and more economical tools for screening of pesticides, multianalyte rapid testing that detects two or more pesticides simultaneously is one of the rational solutions. The recent development of a multianalyte immunoassay has been reviewed by Yan et al. [111]. A semiquantitative strip immunoassay has been developed by Xu et al. for the rapid detection of imidacloprid and thiamethoxam in agricultural products using specific nanocolloidal gold-labelled monoclonal antibodies. The conjugates of imidacloprid—bovine serum albumin (BSA), thiamethoxam—bovine serum albumin and goat antimouse IgG have been coated on the nitrocellulose membrane of the strip, serving as test lines and control line, respectively. In the presence of free imidacloprid and thiamethoxam, both pesticides compete with the immobilized haptens for the binding site of antibodies labelled nanocollodial gold. The visual LOD for imidacloprid and thiamethoxam of this strip are 0.5 and 2.0 ng/mL, respectively, in 2% orange extract.

The multianalyte testing approach makes it possible to detect a class of pesticides by one antibody or a specific reaction mechanism that reacts with that class of pesticides. This approach aims to screen particular sample with the maximum limit of a class of pesticide, which is the total quantity of pesticides. The major classes of pesticides include organophosphate and carbamate pesticides. In general, the multianalyte rapid test of OP and CM are based on the irreversible inhibition of AChE. A gold-labelled antibody test strip that can recognize eight OP pesticides by immunoassay in cabbage, apple and greengrocery has been reported [112]. Although the detection principle of this strip is the same as the single-analyte immunoassay in which free pesticide competes with hapten for the binding to the colloidal gold antibody, this strip is able to detect eight OP pesticides (parathion-methyl, parathion, fenitrothion, EPN, cyanophos, paraoxon-methyl, paraoxon-ethyl and fenitrooxon) with LODs ranging from 2.7 to 100.2 ng/mL.

These multianalyte testing approaches show their potential in the rapid detection of the total pesticide content of OP and CM pesticide. They are suitable for the initial screening of pesticide residues and followed by the confirmation by conventional analytical methods.

3.3 Aptasensors

The main difficulties about the use of bioreceptors in pesticide analysis concern the possibility to isolate biomolecules for targets with high toxicity. In this way, the use of synthetic biomimetic receptors has recently become an interesting alternative in food safety analysis, giving the chemists the possibility to directly synthesize the probe molecule, obtaining high selectivity and specificity [113].

Aptamers are short and single-stranded DNA or RNA sequences, selected in vitro using a technique called SELEX (selection evolution of ligands by exponential enrichment) [114,115] from synthetic oligonucleotide libraries, that are able to bind different kinds of target molecules (eg, other oligonucleotides, biomarkers or small synthetic molecules such as pesticides) with high selectivity and specificity [116]. Compared to antibodies, aptamers show higher detection ranges, a higher stability under different chemical and physical conditions, a prolonged shelf life and acceptable cross reactivity and can be obtained using efficient and cost-effective processes. They can be also easily modified, giving the possibility to obtain various labelled probe elements [117]. As a bio-recognition element, aptamers can be used directly linked to the transducer, both as single-target [118] or multitarget [119] probe (labelled or label-free). Another possible approach is given by the possibility to use aptamer–target interaction to indirectly activate on–off devices in which the interaction with the analyte or the aptamer itself inhibits certain reactivity [120]. Moreover, some studies have shown that aptamers are subjected to significant conformational change caused by the interaction with the analyte that can be used as recognition parameter combining them with appropriate transducers [121,122]. Various aptamer-based bioassays including fluorescent, colorimetric and electrochemical methods have been developed and extensively adopted for an impressively wide variety of applications. A recent review discusses several transduction systems and their principles used in aptamer-based nanosensors, which have been developed in the last years [123]. According to our knowledge, at present time, only a few aptamer-based pesticide-detection sensors have been developed.

Kwon et al. have recently described a sensitive detection of iprobenfos (IBF) and edifenphos (EDI) by using an aptamer-based colorimetric multiaptasensor. Both pesticides IBF and EDI can be eventually detected in a range from 10 nM to 5 nM, respectively. This multiaptasensor has been then implemented in spiked rice samples, obtaining accuracies around 80% and 90% in spiked paddy and polished rice samples, respectively [124]. A schematic representation is reported in Fig. 4. Recently, a DNA aptamer specific for acetamiprid has been described [125]. A simple aptasensor for sensitive and selective detection of acetamiprid has been developed based on electrochemical impedance spectroscopy. To improve sensitivity of the aptasensor, AuNPs were electrodeposited on the bare gold electrode surface by cyclic voltammetry, which has been employed as a platform for aptamer immobilization. With the addition of acetamiprid, the formation of acetamiprid–aptamer complex on the AuNPs-deposited electrode surface resulted in an increase of electron transfer resistance (R_{et}). A wide linear range was obtained from 5 to 600 nM with a low detection limit of 1 nM. Besides, the applicability of the developed aptasensor has been successfully evaluated by determining acetamiprid in the real samples, wastewater and tomatoes [118]. The same aptamers have been then successfully used with an innovative approach, developing an electrochemical assay based on a sandwich interaction with the pesticide acetamiprid. The aptasensor has been able to obtain a linear response in the concentration range of 0–200 nM in water solution [126].

A simple and selective aptamer-based colorimetric method has also been developed for highly sensitive detection of acetamiprid, taking advantage of the sensitive target-induced colour changes that result from the interparticle

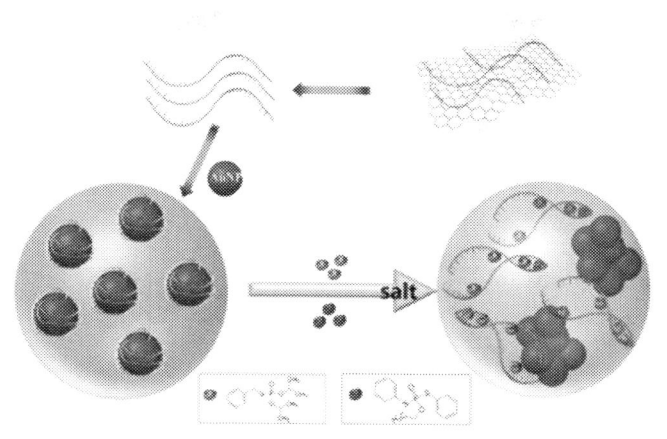

FIGURE 4 Schematic illustration of the aptasensor: sensitive detection of iprobenfos (IBF) and edifenphos (EDI) has been successfully conducted by using a colorimetric method. *Reprinted from Y.S. Kwon, V.-T. Nguyen, J. Gun Park, M.B Gu, Detection of Iprobenfos and Edifenphos using a new multi-aptasensor, Anal. Chim. Acta 868 (2015) 60–66, with permission from Elsevier.*

plasmon coupling during the aggregation of Au nanoparticles. The results showed that the established method could be applied to detect acetamiprid in the linear range between 75×10^{-6} M and 7.5×10^{-3} M, with a low detection limit of 5×10^{-6} M. The practical application of the colorimetric method has been realized for detecting acetamiprid in real soil samples and monitoring its natural degradation process [127]. New silver dendrites pesticide SERS sensor conjugated with aptamers through silver–thiol bonds that targets isocarbophos, omethoate, phorate and profenofos pesticides has been developed [119]. The silver dendrites are filled with a small blocker molecule to allow the conjugated aptamer to change its 3D conformation during the capture of pesticides and to eliminate nonspecific binding on Ag particles. This technique is specific to the four targeted pesticides, and the changes of Raman peak of Ag dendrites can be statistically quantified. The limit of detection of isocarbophos, omethoate, phorate and profenofos pesticides are 1, 5, 0.1 and 5 ng/mL, respectively, and the total analytical time for six samples is 40 min.

3.4 Molecularly Imprinted Polymers

Whereas aptamers are synthetic receptors realized using biological 'building blocks', molecularly imprinted polymers (MIPs) are realized generating specific recognition sites on polymeric nanoparticles to mimic a biological receptor [128–130]. MIPs are synthesized using a template-assisted approach: functional monomers form a complex with the template (that will be the analytical target) and then the polymerization is started, using an appropriate solvent. Removing then the template by extensive washing steps (to break the template–monomers interactions) allows the polymer to maintain specific recognition sites, complementary to the template in size, shape and position of interacting functional groups [131]. The choice of the chemical reagents to synthesize the MIPs must be done to create highly specific cavities designed for the template molecule, and for this reason, a computational modelling study is usually carried out to develop the synthesis [132,133]. At present time, many MIP-based biosensors have been developed [134]. Even if MIPs are characterized by a high affinity to the target, they show some disadvantages involving the low binding capacity (and possible low site accessibility) and a slow binding kinetic, given by the rigidity of the polymeric matrix that forms the particles [135]. Recent advances in MIPs in food analysis are presented in literature [136,137].

Electrochemical sensors based on MIPs have been developed recently. These sensors show high selectivity towards the target molecules: for this reason, MIPs have been described as artificial locks for target molecules. A variety of MIP-based sensors have been reported such as the amperometric sensor based on metolcarb-imprinted film [138] and MIPs/sol—gel sensor for methidathion organophosphorous insecticide recognition [139]. To improve the characteristics of MIPs used as biomimetic-recognition agent, nano-materials are intensively used, as support or signal mediator [140], through electrochemical transducers or in combination with them [139]. Examples of combination between MIPs and enzymes have also been developed [141].

A novel composite of vinyl group—functionalized MWCNTs MIP has been synthesized and applied as recognition element to build an electrochemical sensor for parathion-methyl. The response of the MIP-based sensor has been linearly proportional to the concentration of parathion-methyl over the range of 2.0×10^{-7} to 1.0×10^{-5} M with a lower detection limit of 6.7×10^{-8} M. This sensor has been also successfully applied in the detection of parathion-methyl in pear and cucumber samples [142].

A novel potentiometric sensor with high selectivity in addition to sensitivity has been developed for the determination of lindane, γ-hexachlorocyclohexane (γ-HCCH), based on the modification of MWCNT-imprinted polymer film onto the surface of a Cu electrode. The sensor responds to γ-HCCH in the range 1×10^{-10} to 1×10^{-3} M and the detection limit has been found to be 1.0×10^{-10} M [143].

Tan et al. realized an MIP sensor based on glassy carbon electrode decorated by reduced graphene oxide and AuNPs for the detection of CBF. The sensor exhibited high adsorption capacity and good selectivity for CBF, and it has been successfully applied to the detection of CBF in real vegetable samples [144].

An MIP-based sensor to detect methidathion in wastewater samples has been proposed. The sensing platform has been architected by the combination of a molecularly imprinted technique and sol—gel method on an inexpensive, portable and disposable screen-printed carbon electrode surface. Electrochemical impedimetric detection technique has been employed to perform the label-free detection of the target analyte on the designed MIP/sol–gel integrated platform [139].

A rapid, selective and sensitive double-template imprinted polymer nanofilm-modified pencil graphite electrode has been fabricated for the simultaneous analysis of phosphorus-containing amino acid–type herbicides (glyphosate and glufosinate) in soil and human serum samples [145].

A new biomimetic sensor based on MIPs for the determination of herbicide hexazinone has been proposed by Taboada Sotomayor et al. [146]. The MIP, containing recognition sites for hexazinone, has been prepared with a functional monomer selected by means of molecular modelling. A schematic representation is reported in Fig. 5.

Some piezoelectric sensors using MIP have been developed for fast and on-site determination of pesticides in contaminated foods. Liu et al. described a highly selective and sensitive quartz crystal microbalance realized by mixing with polyvinyl chloride and MIP microspheres for rapid endosulfan detection. Detection of pesticide in water and milk samples has been observed with

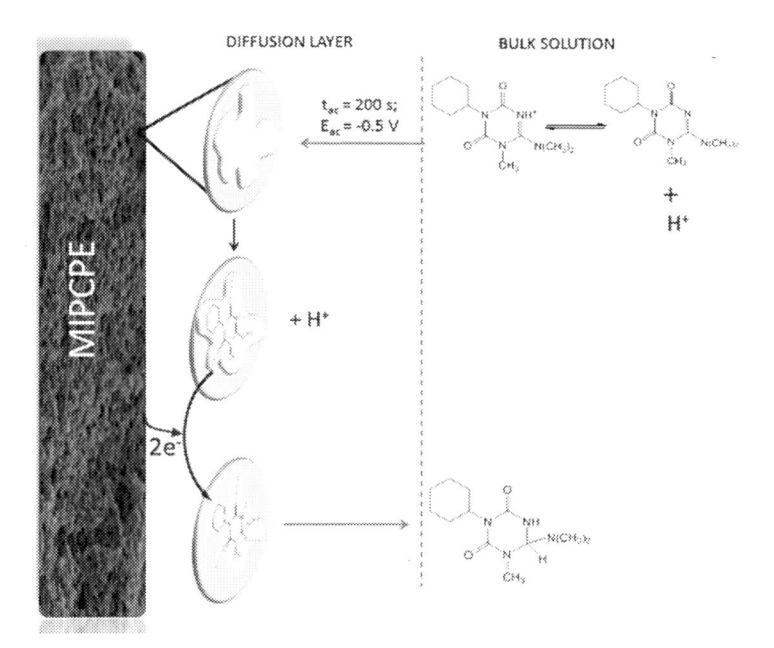

FIGURE 5 Schematic representation of the hexazinone (HXZ) sensor response. In the first step, an accumulation potential (-0.5 V) is used to preconcentrate HXZ within the MIP cavities of the paste, after which the applied potential resulted in quantitative reduction of HXZ (in an acidic medium) and its return to the solution. *Reprinted from M.J.U. Toro, L.D. Marestoni, M.D. Pilar, T. Sotomayor, A new biomimetic sensor based on molecularly imprinted polymers for highly sensitive and selective determination of hexazinone herbicide, Sensors and Actuators B: Chem. 208 (2015) 299–306, with permission from Elsevier.*

recoveries of 96.0–104.1% and 101.8–108.0%, respectively [147]. Özkütük et al. presented a sensor based on the modification of molecular imprinted film for the determination of paraoxon. The study also included the measurement of binding interaction quartz crystal microbalance sensor, selectivity experiments and analytical performance of quartz crystal microbalance [148].

Atrazine detection in wastewater has been reported by imprinted p(HEMA-phenol) film on gold surface of a QCM chip. The developed QCM sensor shows high sensitivity and selectivity for determination of trace ATR in wastewater.

A chemiluminescence sensor for the determination of glyphosate has been made up based on MIP [149].

Wang et al. developed a paper-based, multidisc plate grafted with MIPs for chemiluminescence detection of pesticides. They focused on the challenges, eg, low-cost, portable, fast and easy-to-set-up detection for public use, and proposed grafting MIPs onto cellulose fibres in paper disks. By this approach, they achieved the detection of target pesticide at the femtomolar level under optimized conditions [150].

4. CONCLUSIONS AND FUTURE PERSPECTIVES

The widespread application of pesticides in the last decades and the continuous increase of their use nowadays have brought in an enormous diffusion of these compounds into the environment. The variety of their chemical classes and the degradation processes, in which they are subjected in the environment, make their intrinsic toxicity even more dangerous, because not only the active molecules but also their metabolites can contaminate crops and food, reaching the human food chain. Their toxicity derives from their general stability, mobility, bioaccumulation and long-term effects on living organisms, including plants, in which pesticides can accumulate and reach mammalians. Crops represent the main food source for humans, and the quality of the related products is extremely important, to protect consumers' health. Safe food should contain less amount of possible dangerous residues from the cultivation, and for this reason, the control of pesticides and their residues for food safety remains one of the most crucial target for analytical chemistry. Monitoring should be constant during the entire food processing chain, from the production of raw ingredients to the sale; however, due to the difficulties related to pesticide detection (mainly caused because of their low concentration in food samples), continuous developments are necessary to obtain new and powerful analytical tools. In this way, biosensors represent a valid alternative to classical analytical techniques, due to their low cost, fast response and simplicity of use.

A summary of the recent developments in sensing of pesticide for food safety analysis has been provided. As described, in recent years there has been an intense growth of system based on bio- or biomimetic molecules for pesticide detection, using assays based on biological macromolecules such as enzymes or antibodies, synthetic biomolecules such as aptamers and artificial macromolecules such as MIPs. A great investigation has been made to improve the analytical parameters, such as sensitivity, selectivity and response, and the chemical and physical ones, such as the stability and the number of turnover, using innovative immobilization approaches, new materials (even nanostructured) and different transducers. Despite the good results achieved, sensing of pesticide via biosensors still shows some limits to become a widely accepted and applicable approach, especially concerning the application in real samples. Nevertheless, their potential for environmental analytical purposes has not been much explored, and this makes their development one of the most fascinating targets in analytical chemistry.

REFERENCES

[1] FAO/WHO, Joint Meeting on Pesticide Specifications (JMPS), Available only on the Internet, 2010, who.int/iris/handle/10665/44527.

[2] G.W. Ware, J. Ariz. Acad. Sci. 9 (1974) 61—65.

[3] N. Verma, A. Bhardwaj, Appl. Biochem. Biotechnol. 175 (2015) 3093—3119.

[4] T. Cairns, J. Sherma (Eds.), Emerging Strategies for Pesticide Analysis, CRC Press, 1992.

[5] J. Fenik, M. Tankiewicz, M. Biziuk, TrAC Trends Anal. Chem. 30 (2011) 814—826.

[6] M. Tankiewicz, J. Fenik, M. Biziuk, TrAC Trends Anal. Chem. 29 (2010) 1050—1063.

[7] Y. Song, Y. Ge, Y. Zhang, B. Liu, Y. Lu, T. Dong, S. Wang, Anal. Bioanal. Chem. 393 (2009) 2001—2008.

[8] J. Walker, S. Asher, Anal. Chem. 77 (2005) 1596—1600.

[9] J. Whiteaker, K. Prather, Anal. Chem. 75 (2003) 49—56.

[10] C.E. Handford, C.T. Elliott, K. Campbell, Integr. Environ. Assess. Manag. 11 (2015) 525—536.

[11] C. Lesueur, P. Knittl, M. Gartner, A. Mentler, M. Fuerhacker, Food Control 19 (2008) 906—914.

[12] C. Lesueur, M. Gartner, A. Mentler, M. Fuerhacker, Talanta 75 (2008) 284—293.

[13] S. Liu, Z. Zheng, X. Li, Anal. Bioanal. Chem. 405 (2013) 63—90.

[14] T.D. Sterling, A.V. Arundel, Scand. J. Work Environ. Health 12 (1986) 161—173.

[15] S.K. Hoar, A. Blair, F.F. Holmes, et al., JAMA 256 (1986) 1141—1147.

[16] Effects of chlorsulfuron on early embryo development in Norway spruce cell suspensions, in: Y. Blume, D.J. Durzan, P. Smertenko (Eds.), Cell Biology and Instrumentation: UV Radiation, Nitric Oxide, and Cell Death in Plants, IOS Press, 2006.

[17] P. Tchounwou, B. Wilson, A. Ishaque, R. Ransome, M.-J. Huang, J. Leszczynski, Int. J. Mol. Sci. 1 (2000) 63—74.

[18] M. Abdollahi, A. Ranjbar, S. Shadnia, S. Nikfar, A. Rezaiee, Med. Sci. Rev. 10 (2004) RA141—RA147.

[19] K. Mwila, M.H. Burton, J.S. Van Dyk, B.I. Pletschke, Environ. Monit. Assess. 185 (2013) 2315—2327.

[20] M.P. Longnecker, W.J. Rogan, G. Lucier, Annu. Rev. Public Health 18 (1997) 211—244.

[21] A. Ranjbar, P. Pasalar, M. Abdollahi, Hum. Exp. Toxicol. 21 (2002) 179—182.

[22] M.V. Kumar, T. Desiraju, Toxicology 75 (1992) 13—20.

[23] L.G. Costa, G. Giordano, M. Guizzetti, A. Vitalone, Front Biosci. 13 (2008) 1240−1249.

[24] S. Feng, Z. Kong, X. Wang, P. Peng, E.Y. Zeng, Ecotoxicol Environ. Saf. 61 (2005) 239−246.

[25] A.Y. Kocaman, M. Topaktas, Environ. Mol. Mutag. 48 (2007) 483−490.

[26] T. Pastoor, Toxicol. Sci. 86 (2005) 56−60.

[27] H.P.M. Vijverberg, J. vanden Bercken, Crit. Rev. Toxicol. 21 (1990) 105−126.

[28] G. Meco, V. Bonifati, N. Vanacore, E. Fabrizio, Scand. J. Work Environ. Health 20 (1994) 301−305.

[29] L. Gray, J. Ostby, J. Furr, C. Wolf, C. Lambright, L. Parks, D. Veeramachaneni, V. Wilson, M. Price, A. Hotchkiss, APMIS 109 (2001) S302−S319.

[30] C. Sanfeliu, J. Sebastia, R. Cristofol, E. Rodriguez-Farre, Neurotox Res. 5 (2003) 283−305.

[31] J. Pope, M. Skurky-Thomas, C. Rosen, Toxicity, Organochlorine Pestic. Emedicine (1994).

[32] C. De Jong, Toxicol. Lett. (1991) 1−206.

[33] Pesticides: classification and properties, in: J.L. Tadeo (Ed.), Analysis of Pesticides in Food and Environmental Samples, CRC Press, United States of America, 2008.

[34] K. Lányi, Z. Dinya, Microchem. J. 80 (2005) 79−87.

[35] J. Bethsass, A. Colangelo, Int. J. Occup. Environ. Health 12 (2006) 260−267.

[36] P. Jovanov, V. Guzsvany, M. Franko, S. Lazic, M. Sakac, B. Saric, V. Banjac, Talanta 111 (2013) 125−133.

[37] E. Caldas, M. Conceição, M.C.C. Miranda, L.C.K.R. de Souza, J. Lima, J. Agric. Food Chem. 49 (2001) 4521−4525.

[38] R. Noon (Ed.), New Developments in Fungicides: 2004 Edition, PJB Publications, 2004.

[39] F.P. Kaloyanova, M.A. El Batawi (Eds.), Human Toxicology of Pesticides, CRC press, 1991.

[40] Food and nutritional analysis. pesticide residues, in: P.W.T. Poole (Ed.), Encyclopedia of Analytical Science, second ed., Elsevier, Oxford, 2005.

[41] Food and nutritional analysis. sample preparation, in: P.W.T. Poole (Ed.), Encyclopedia of Analytical Science, second ed., Elsevier, Oxford, 2005.

[42] J. Pan, X.X. Xia, J. Liang, Ultrason. Sonochem. 15 (2008) 25−32.

[43] A. Juan-Garcia, J. Manes, G. Font, Y. Pico, J. Chromatogr. A 1050 (2004) 119−127.

[44] D. Stajnbaher, L. Zupancic-Kralj, J. Chromatogr. A 1015 (2003) 185−198.

[45] J.L. Martínez Vidal, P. Plaza-Bolaños, R. Romero-González, A. Garrido Frenich, J. Chromatogr. A 1216 (2009) 6767−6788.

[46] E. Sobhanzadeh, N.K.A. Bakar, M.R. Abas, K. Nemati, Mal. J. Fund. Appl. Sci. 5 (2014).

[47] F.J. Schenck, S.J. Lehotay, J. Chromatogr. A 868 (2000) 51−61.

[48] Z. Sharif, Y.B.C. Man, N.S.A. Hamid, C.C. Keat, J. Chromatogr. A 1127 (2006) 254−261.

[49] Consumer perceptions of food safety and their impact on food choice, in: G.G. Birch, G. Campbell-Platt (Eds.), Food Safety—the Challenge Ahead Intercept, Andover, 1993.

[50] A. Wilcock, M. Pun, J. Khanona, M. Aung, Trends Food Sci. Technol. 15 (2004) 56−66.

[51] A. Amine, H. Mohammadi, I. Bourais, G. Palleschi, Biosens. Bioelectron. 21 (2006) 1405−1423.

[52] J.S. Van Dyk, B. Pletschke, Chemosphere 82 (2011) 291−307.

[53] G. Jeanty, A. Wojciechowska, J.L. Marty, M. Trojanowicz, Anal. Bioanal. Chem. 373 (2002) 691−695.

[54] A.P. Periasamy, Y. Umasankar, S.M. Chen, Sensors (Basel) 9 (2009) 4034−4055.

[55] S. Andreescu, J.L. Marty, Biomol. Eng. 23 (2006) 1−15.

[56] J. Wang, R. Krause, K. Block, M. Musameh, A. Mulchandani, M.J. Schoning, Biosens. Bioelectron. 18 (2003) 255–260.

[57] P. Mulchandani, W. Chen, A. Mulchandani, J. Wang, L. Chen, Biosens. Bioelectron. 16 (2001) 433–437.

[58] A. Mulchandani, W. Chen, P. Mulchandani, J. Wang, K.R. Rogers, Biosens. Bioelectron. 16 (2001) 225–230.

[59] B.G. Choi, H. Park, T.J. Park, M.H. Yang, J.S. Kim, S.-Y. Jang, N.S. Heo, S.Y. Lee, J. Kong, W.H. Hong, ACS Nano 4 (2010) 2910–2918.

[60] J.H. Lee, J.Y. Park, K. Min, H.J. Cha, S.S. Choi, Y.J. Yoo, Biosens. Bioelectron. 25 (2010) 1566–1570.

[61] D. Du, W. Chen, W. Zhang, D. Liu, H. Li, Y. Lin, Biosens. Bioelectron. 25 (2010) 1370–1375.

[62] Y. Zhao, W. Zhang, Y. Lin, D. Du, Nanoscale 5 (2013) 1121–1126.

[63] A. Amine, F. Arduini, D. Moscone, G. Palleschi, Biosens. Bioelectron. 76 (2016) 180–194.

[64] V. Scognamiglio, F. Arduini, G. Palleschi, G. Rea, TrAC Trends Anal. Chem. 62 (2014) 1–10.

[65] W.Y. Zhang, A.M. Asiri, D.L. Liu, D. Du, Y.H. Lin, TrAC Trends Anal. Chem. 54 (2014) 1–10.

[66] G. Marrazza, Biosensors (Basel) 4 (2014) 301–317.

[67] Y. Liu, G. Wang, C. Li, Q. Zhou, M. Wang, L. Yang, Mater. Sci. Eng. C Mater. Biol. Appl. 35 (2014) 253–258.

[68] K.A. Hassall (Ed.), Biochemistry and Uses of Pesticides, Macmillan Press Ltd, 1990.

[69] A. Simonian, E. Efremenko, J. Wild, Anal. Chim. Acta 444 (2001) 179–186.

[70] F. Berti, L. Lozzi, I. Palchetti, S. Santucci, G. Marrazza, Electrochim. Acta 54 (2009) 5035–5041.

[71] F. Berti, C. Eisenkolbl, D. Minocci, P. Nieri, A.M. Rossi, M. Mascini, G. Marrazza, J. Electroanal. Chem. 656 (2011) 55–60.

[72] L. Rotariu, L.-G. Zamfir, C. Bala, Anal. Chim. Acta 748 (2012) 81–88.

[73] C. Zhai, X. Sun, W. Zhao, Z. Gong, X. Wang, Biosens. Bioelectron. 42 (2013) 124–130.

[74] D. Du, M. Wang, J. Cai, A. Zhang, Sensors Actuators B: Chem. 146 (2010) 337–341.

[75] D. Du, M. Wang, J. Cai, Y. Qin, A. Zhang, Sensors Actuators B: Chem. 143 (2010) 524–529.

[76] R.-S. Saberi, S. Shahrokhian, G. Marrazza, Electroanalysis 25 (2013) 1373–1380.

[77] Z. Taleat, A. Ravalli, M. Mazloum-Ardakani, G. Marrazza, Electroanalysis 25 (2013) 269–277.

[78] Z. Taleat, A. Khoshroo, M. Mazloum-Ardakani, Microchim. Acta 181 (2014) 865–891.

[79] D. Du, X. Ye, J. Cai, J. Liu, A. Zhang, Biosens. Bioelectron. 25 (2010) 2503–2508.

[80] Z.-J. Shang, Y.-L. Xu, Y. Wang, D.-X. Wei, L.-L. Zhan, Procedia Eng. 15 (2011) 4480–4485.

[81] Y. Li, Y. Bai, G. Han, M. Li, Sensors Actuators B: Chem. 185 (2013) 706–712.

[82] R. Xue, T.-F. Kang, L.-P. Lu, S.-Y. Cheng, Anal. Lett. 46 (2013) 131–141.

[83] L. Yang, G. Wang, Y. Liu, M. Wang, Talanta 113 (2013) 135–141.

[84] P. Nayak, B. Anbarasan, S. Ramaprabhu, J. Phys. Chem. C 117 (2013) 13202–13209.

[85] A. Ravalli, G.P. dos Santos, M. Ferroni, G. Faglia, H. Yamanaka, G. Marrazza, Sensors Actuators B: Chem. 179 (2013) 194–200.

[86] Y. Wang, S. Zhang, D. Du, Y. Shao, Z. Li, J. Wang, M.H. Engelhard, J. Li, Y. Lin, J. Mater. Chem. 21 (2011) 5319–5325.

[87] L. Zhang, L. Long, W. Zhang, D. Du, Y. Lin, Electroanalysis 24 (2012) 1745−1750.

[88] Y. Yang, A.M. Asiri, D. Du, Y. Lin, Analyst 139 (2014) 3055−3060.

[89] T. Liu, H. Su, X. Qu, P. Ju, L. Cui, S. Ai, Sensors Actuators B: Chem. 160 (2011) 1255−1261.

[90] J. Dong, T. Liu, X. Meng, J. Zhu, K. Shang, S. Ai, S. Cui, J. Solid State Electrochem. 16 (2012) 3783−3790.

[91] X. Meng, J. Wei, X. Ren, J. Ren, F. Tang, Biosens. Bioelectron. 47 (2013) 402−407.

[92] A. Hatefi-Mehrjardi, Electrochim. Acta 114 (2013) 394−402.

[93] T.M. Oliveira, M.F. Barroso, S. Morais, M. Araújo, C. Freire, P. de Lima-Neto, A.N. Correia, M.B. Oliveira, C. Delerue-Matos, Bioelectrochemistry 98 (2014) 20−29.

[94] C. Tortolini, P. Bollella, R. Antiochia, G. Favero, F. Mazzei, Sensors Actuators B: Chem. 224 (2016) 552−558.

[95] D. Ying, W. JiHua, H. Ping, M. Shuai, F. XiaoYuan, JFSQ 6 (2015) 2976−2980.

[96] G. Liu, W. Guo, D. Song, Biosens. Bioelectron. 52 (2014) 360−366.

[97] Impedimetric immunosensor for pesticide detection, in: R. Toonika (Ed.), State of the Art in Biosensors − Environmental and Medical Applications, InTech, 2013.

[98] H.V. Tran, R. Yougnia, S. Reisberg, B. Piro, N. Serradji, T.D. Nguyen, L.D. Tran, C.Z. Dong, M.C. Pham, Biosens. Bioelectron. 31 (2012) 62−68.

[99] N. Belkhamssa, C.I.L. Justino, P.S.M. Santos, S. Cardoso, I. Lopes, A.C. Duarte, T. Rocha-Santos, M. Ksibi, Talanta 146 (2016) 430−434.

[100] B. Van Dorst, J. Mehta, K. Bekaert, E. Rouah-Martin, W. De Coen, P. Dubruel, R. Blust, J. Robbens, Biosens. Bioelectron. 26 (2010) 1178−1194.

[101] W. Wei, X. Zong, X. Wang, L. Yin, Y. Pu, S. Liu, Food Chem. 135 (2012) 888−892.

[102] E. Valera, D. Muñiz, Á. Rodríguez, Microelectron. Eng. 87 (2010) 167−173.

[103] E. Valera, J. Ramón-Azcón, A. Barranco, B. Alfaro, F. Sánchez-Baeza, M.-P. Marco, Á. Rodríguez, Food Chem. 122 (2010) 888−894.

[104] R.E. Ionescu, C. Gondran, L. Bouffier, N. Jaffrezic-Renault, C. Martelet, S. Cosnier, Electrochim. Acta 55 (2010) 6228−6232.

[105] K. Jia, P.-M. Adam, R.E. Ionescu, Sensors Actuators B: Chem. 188 (2013) 400−404.

[106] M. Pan, L. Kong, B. Liu, K. Qian, G. Fang, S. Wang, Sensors Actuators B: Chem. 188 (2013) 949−956.

[107] J. Chiou, A.H.H. Leung, H.W. Lee, W.-T. Wong, J. Integr. Agric. 14 (2015) 2243−2264.

[108] M. Blažková, P. Rauch, L. Fukal, Biosens. Bioelectron. 25 (2010) 2122−2128.

[109] B. Holubová-Mičková, M. Blažková, L. Fukal, P. Rauch, Eur. Food Res. Technol. 231 (2010) 467−473.

[110] W. Zheng, GB 2763−2014: Translated English PDF of Chinese Standard GB2763-2014: National Food Safety Standard − Maximum Residue Limits for Pesticides in Food, 2015. chinesestandard.net.

[111] X. Yan, H. Li, Y. Yan, X. Su, Anal. Methods 6 (2014) 3543−3554.

[112] X. Hua, J. Yang, L. Wang, Q. Fang, G. Zhang, F. Liu, PLoS One 7 (2012) e53099.

[113] D.P. Nikolelis, A. Erdem, G.-P. Nikoleli, T. Varzakas (Eds.), Portable Biosensing of Food Toxicants and Environmental Pollutants, CRC Press, 2013.

[114] C. Tuerk, L. Gold, Science 249 (1990) 505−510.

[115] A.D. Ellington, J.W. Szostak, Nature 346 (1990) 818−822.

[116] J.H. Lee, M.V. Yigit, D. Mazumdar, Y. Lu, Adv. Drug Deliv. Rev. 62 (2010) 592−605.

[117] T. Tang, J. Deng, M. Zhang, G. Shi, T. Zhou, Talanta 146 (2016) 55−61.

[118] L. Fan, G. Zhao, H. Shi, M. Liu, Z. Li, Biosens. Bioelectron. 43 (2013) 12−18.

[119] S. Pang, T.P. Labuza, L. He, Analyst 139 (2014) 1895−1901.

[120] P. Weerathunge, R. Ramanathan, R. Shukla, T.K. Sharma, V. Bansal, Anal. Chem. 86 (2014) 11937—11941.

[121] T. Hermann, D.J. Patel, Science 287 (2000) 820—825.

[122] D.J. Patel, A.K. Suri, J. Biotechnol. 74 (2000) 39—60.

[123] R. Sharma, K. Ragavan, M. Thakur, K. Raghavarao, Biosens. Bioelectron. 74 (2015) 612—627.

[124] Y.S. Kwon, V.T. Nguyen, J.G. Park, M.B. Gu, Anal. Chim. Acta 868 (2015) 60—66.

[125] J. He, Y. Liu, M. Fan, X. Liu, J. Agric. Food Chem. 59 (2011) 1582—1586.

[126] R. Rapini, G. Marrazza, DNA technology for small molecules sensing: a new approach for acetamiprid detection, XVIII AISEM Annual Conference Trento (2015) 1—4.

[127] H. Shi, G. Zhao, M. Liu, L. Fan, T. Cao, J. Hazard Mater. 260 (2013) 754—761.

[128] R.V. Shutov, A. Guerreiro, E. Moczko, I.P. de Vargas-Sansalvador, I. Chianella, M.J. Whitcombe, S.A. Piletsky, Small 10 (2014) 1086—1089.

[129] T. Panasyuk, V. Campo Dall'Orto, G. Marrazza, A. El'skaya, S. Piletsky, I. Rezzano, M. Mascini, Anal. Lett. 31 (1998) 1809—1824.

[130] F. Berti, S. Todros, D. Lakshmi, M.J. Whitcombe, I. Chianella, M. Ferroni, S.A. Piletsky, A.P. Turner, G. Marrazza, Biosens. Bioelectron. 26 (2010) 497—503.

[131] A. Poma, A. Guerreiro, M.J. Whitcombe, E.V. Piletska, A.P. Turner, S.A. Piletsky, Adv. Funct. Mater. 23 (2013) 2821—2827.

[132] D. Lakshmi, M. Akbulut, P.K. Ivanova-Mitseva, M.J. Whitcombe, E.V. Piletska, K. Karim, O. Güven, S.A. Piletsky, Ind. Eng. Chem. Res. 52 (2013) 13910—13916.

[133] P.S. Sharma, F. D'Souza, W. Kutner, TrAC Trends Anal. Chem. 34 (2012) 59—77.

[134] Molecularly imprinted polymer-based biosensors portable biosensing of food toxicants and environmental pollutants, in: D.P. Nikolelis, A. Erdem, G.-P. Nikoleli, T. Varzakas (Eds.), Portable Biosensing of Food Toxicants and Environmental Pollutants, CRC Press, 2013.

[135] C. Xie, B. Liu, Z. Wang, D. Gao, G. Guan, Z. Zhang, Anal. Chem. 80 (2008) 437—443.

[136] X.L. Song, S.F. Xu, L.X. Chen, Y.Q. Wei, H. Xiong, J. Appl. Polym. Sci. 131 (2014).

[137] V. Suryanarayanan, C.T. Wu, K.C. Ho, Electroanalysis 22 (2010) 1795—1811.

[138] M.-F. Pan, G.-Z. Fang, B. Liu, K. Qian, S. Wang, Anal. Chim. Acta 690 (2011) 175—181.

[139] I. Bakas, A. Hayat, S. Piletsky, E. Piletska, M.M. Chehimi, T. Noguer, R. Rouillon, Talanta 130 (2014) 294—298.

[140] Y. Zhao, Y. Ma, H. Li, L. Wang, Anal. Chem. 84 (2012) 386—395.

[141] B.O. Najwa, B. Idriss, I. Georges, A.-I. Ihya, A.-A. Elhabib, R. Régis, N. Thierry, Science 2 (2014) 1—6.

[142] D. Zhang, D. Yu, W. Zhao, Q. Yang, H. Kajiura, Y. Li, T. Zhou, G. Shi, Analyst 137 (2012) 2629—2636.

[143] T.S. Anirudhan, S. Alexander, Biosens. Bioelectron. 64 (2015) 586—593.

[144] X. Tan, Q. Hu, J. Wu, X. Li, P. Li, H. Yu, X. Li, F. Lei, Sens. Actuators B 220 (2015) 216—221.

[145] B.B. Prasad, D. Jauhari, M.P. Tiwari, Biosens. Bioelectron. 59 (2014) 81—88.

[146] M.J.U. Toro, L.D. Marestoni, M.D.P.T. Sotomayor, Sensors Actuators B: Chem. 208 (2015) 299—306.

[147] N. Liu, J. Han, Z. Liu, L. Qu, Z. Gao, Anal. Methods 5 (2013) 4442—4447.

[148] E.B. Özkütük, S.E. Diltemiz, E. Özalp, R. Say, A. Ersöz, Mater. Sci. Eng. C 33 (2013) 938—942.

[149] P. Zhao, M. Yan, C. Zhang, R. Peng, D. Ma, J. Yu, Spectrochim. Acta Pt. A: Mol. Biomol. Spectrosc. 78 (2011) 1482—1486.

[150] S. Wang, L. Ge, L. Li, M. Yan, S. Ge, J. Yu, Biosens. Bioelectron. 50 (2013) 262—268.

Chapter 2

Trends on Biosensing Systems for Heavy Metal Detection

N. Verma* and G. Kaur
Punjabi University, Patiala, Punjab, India
Corresponding author: E-mail: neelam_verma2@rediffmail.com

Chapter Outline

1. INTRODUCTION

Group of metals and metalloids having atomic weight in the range of 63.5–200.6 are being regarded as heavy metals. These metals have specific gravity more than 5.0, five times higher compared to water [1]. Nonessential heavy metals include cadmium, chromium, mercury, lead, arsenic and antimony. Some heavy metals, such as cobalt, copper, iron, manganese, molybdenum, vanadium, strontium and zinc, are being used as essential elements in living systems when occurring in trace amounts [2]. Due to their high chemical reactivity, heavy metals can be poisonous even at very low concentration [3]. However, if the concentration of essential and nonessential heavy metals exceeds the optimum threshold level, it leads to various mental, genetic, morphological abnormalities in humans, plants and animals [4,5]. In addition, they are persistent in the environment, nonbiodegradable and can bioaccumulate causing serious health hazards [6]. Heavy metals are present in the earth crust and can enter the food chain [7] via biogeochemical cycles, becoming highly toxic or even carcinogenic [2], affecting the health of people who have consumed contaminated water and food products. For this reason, food safety is the foremost requirement to preserve the environment balance and proper health of human beings [8,9]. Environmental awareness is growing among consumers and industrialists, because of the recognition of toxic effects of heavy metals even at low concentration. Legal constraints have also become strict both at national and at international levels for the release of heavy metals from industries [10], hence there is a prime requirement of analytical techniques to detect heavy metals in agrifood sector.

2. INDUSTRIAL USES OF HEAVY METALS

In developing countries, various heavy metals are used in metal-plating process, mining operations, fertilisers, tanneries, paper, batteries, automobiles, pesticides and manufacturing industries [11]. Arsenic is mostly used in the preparation of pesticides, fungicides and metal smelters. Cadmium is used for welding, electroplating and cadmium-containing batteries. Lead is mainly found in paints, pesticides, smoke, automobile fuel and mining industries. Manganese is used in welding and ferromanganese production. Mercury occurs in batteries, pesticides and paper industries. Copper is used in chemical industry, metal piping and electric wires. Zinc is required in brass manufacture, refineries, batteries and metal plating [12].

3. RELEASE OF HEAVY METALS IN THE ENVIRONMENT AND EXPOSURE TO HUMANS

Heavy metals are naturally present in earth crust and rocks in the form of sulphides and oxide ores. These can be extracted as minerals from various ores such as sulphides of lead, iron, mercury, cadmium, arsenic or cobalt [7]. Leaching of heavy metals into lakes, rivers and oceans, due to weathering of rocks and volcanic eruptions and mining processes, can cause serious pollution by affecting its surrounding areas via acid rains. Several scientific data report that water, soil, vegetables, crops and dust in a close distance to the mining areas have been highly polluted by lead, arsenic, copper, chromium, zinc and cadmium. These heavy metals are the main toxicity-generating elements for living beings identified by World Health Organization (WHO) [13]. A large number of commercial industries, municipal sewerage and wastewater irrigation systems release their heavy metals containing effluents into the environment by inappropriate means [14]. The release of heavy metals leads to contamination of soil and water, further used for agricultural purposes resulting in increased accumulation of heavy metals in food crops and vegetable plants and affecting food security throughout the world [15]. It was observed that 90% of the total heavy metal uptake by human beings often occur by consuming vegetables grown in contaminated fields, and the remaining sources are contaminated through air inhalation or direct skin contacts [16]. It has been also reported that green leafy vegetables like lettuce are the major sources of heavy metal accumulation, without showing any morphological symptoms of toxicity by this accumulation [17,18]. Other sources of metals, leaching from household pipelines, cook wares or utensils also contaminate food and water used by humans in daily routine [10]. Milk and dairy products (main source of diet for infants and adults) are also affected via grazing of animals in fields contaminated by heavy metals, such as lead, cadmium, copper, chromium, arsenic and zinc [19]. Similarly, human beings are exposed to heavy metals by using industrial products, like electric batteries, paints, wires and pipes, to make up their routine needs. These industrially prepared products require heavy metals as a part of manufacturing [20]. The permissible limits of some toxic heavy metals declared by WHO and USEPA (US Environmental Protection Agency) are being described in Table 1 [21].

4. TOXICITY MECHANISMS AND HARMFUL EFFECTS OF HEAVY METALS TO LIVING BEINGS

Heavy metals usually interfere with the normal metabolism of the body by forming stable oxidation states, such as Zn^{2+}, Pb^{2+}, Cd^{2+}, As^{3+}, Hg^{2+} and

TABLE 1 Permissible Limit of Some Potential Toxic Heavy Metals Ions in Drinking Water by International Agencies

Heavy Metal	Permissible Limit (µg/L)	
	WHO	USEPA
Lead	10	05
Cadmium	03	05
Mercury	01	02
Arsenic	10	50
Chromium	50	100

Ag^+, which combine with various proteins and enzymes in the body by chemical bonding. The heavy metal poisoning causes the increased production of reactive oxygen species (ROS), interfering in normal electron transport system, and leading to ion leakage, oxidative stress, DNA damage, suppressed natural killer cells, distortion of structure and impaired functioning of biomolecules. The oxidation states of these heavy metals also form highly stable toxic compounds, which are persistent in the body and very difficult to degrade [7]. Essential heavy metals, like iron, cobalt, copper, zinc, manganese and molybdenum, are mostly required by the body. They play a very vital role in different metabolic activities up to certain level. Other nonessential metals, like arsenic, cadmium, lead and mercury, have not beneficial role in the body even at low concentration [4]. Most heavy metals replace the essential elements in the biochemical reactions involving various proteins, enzymes and hormones by acting as inhibitor. As an example, lead can replace the calcium ion present in the body, disturbs calcium homeostasis and restricts vitamin D absorption, resulting in impaired cell growth, restricted tooth and bones development [22]. Cadmium ions can replace iron, zinc and copper (acting as a cofactor) inside the body, affecting the oxidative phosphorylation and causing many harmful effects [19]. Different heavy metals are known to have unique features to produce toxic effects in the body; they have been described in Table 2.

5. HEAVY METAL ION DETECTION BY CONVENTIONAL METHODS

Improvement of life quality is one of the most important objectives of research worldwide, and it is naturally related to control of diseases as well as to assess food quality and safety of our environment. These issues require continuous, fast, sensitive monitoring systems to detect heavy metals in soil, water and food samples, due to the extreme toxic effects produced by these compounds to environment and living organisms. Conventional methods, like volumetric, gravimetric and calorimetric assays to detect heavy metals, are rarely practised because of their low sensitivity. Other techniques include atomic absorption spectrophotometry (AAS) (based on absorption of light by free atoms of the element generated after exposure to flame, detection limit − 25 μg/L), electro thermal AAS (used for the detection of elements at even low concentration than by flame AAS, detection limit −5.0 μg/L), differential pulse polarography (measures the redox potential of certain elements, detection limit − 100 μg/L), inductively coupled plasma mass spectrometry (ICP-MS) (based on determining number of ions at particular mass of the element, detection limit − 0.025 μg/L), electrochemical metal analyser (measures the change in current by oxidising or reducing the metal, detection limit 0−500 μg/L) [23], anodic stripping voltammetry (ASV) (accumulates the trace metals on to the working electrode and stripping current is measured, detection limit − 1 μg/L) [28].

TABLE 2 Harmful Effects of Toxic Heavy Metal Ions to the Living System

Heavy Metal	Harmful Effects	References
Lead (Pb)	Lead is found to be carcinogen and produces toxic effects related to neurology, haematology and nephrology of humans like mental retardation, paralysis, neural deafness.	[23]
Cadmium (Cd)	Cadmium mainly causes damage to kidneys, liver, lungs, brain and bones. It also causes the renal dysfunction and demineralisation of bones, hypertension, cancer and immune disorders in humans.	[24,19]
Mercury (Hg)	Mercury harms the nervous system, poisoning of protoplasm, acrodynia and psychological problems to the humans.	[12]
Nickel (Ni)	Nickel toxicity interferers the functioning of T-cell system by inhibiting the natural killer cells activity. Its higher concentration even causes cancer.	[25]
Arsenic (As)	Arsenic causes mainly skin and respiratory problems	[12]
Manganese (Mn)	It damages central nervous system severely by contact or by inhaling its dust. It also weakens muscles and disturbs speech.	[12,22]
Chromium (Cr)	It is a kind of carcinogen to the human beings causing tumours in lungs, dermatitis and allergies.	[21]
Zinc (Zn)	Excess of zinc in the body can cause anaemia, arteriosclerosis, stomach cramps, dermatitis problems, nausea	[26]
Copper (Cu)	Higher amount of copper leads to Wilson's disease which mainly affects liver. It also causes kidney problems, stomach aches, diarrhoea, vomiting	[27]

6. WHY BIOSENSOR IS CONSIDERED AS A PROMISING TOOL TO DETECT HEAVY METAL IONS OVER CONVENTIONAL METHODS

Although the above-mentioned analytical techniques are sensitive and reliable to detect heavy metal ions, these techniques have a number of shortcomings. ICP-MS, for example, is affected by spectral and nonspectral interference; AAS, ICP-MS and ASV are quite expensive, time-consuming and require well-trained and skilled laboratory personnel.

If compared with these conventional techniques, biosensors are considered to be very promising tools for the detection of heavy metal ions because of their advantages in terms of specificity, low cost, portability, real-time monitoring, fast response time, easy handling, compactness, sensitivity, user friendly and reliability and sample does not require much preprocessing. In addition, biosensors are also able to measure biotoxicity level of heavy metal ions towards living beings [25].

Biosensing technology is a synergic combination of nano/biotechnology and microelectronics. Biosensors are analytical devices (Fig. 1) that mainly relies on a sensing element (enzyme, metal-binding proteins, antibodies, whole cell, DNA, recombinant cells, aptamers, as shown in Fig. 2) able to recognise the target analyte and the signal of this interaction is transduced by the transduction element (electrochemical, optical, piezoelectric and thermal, as

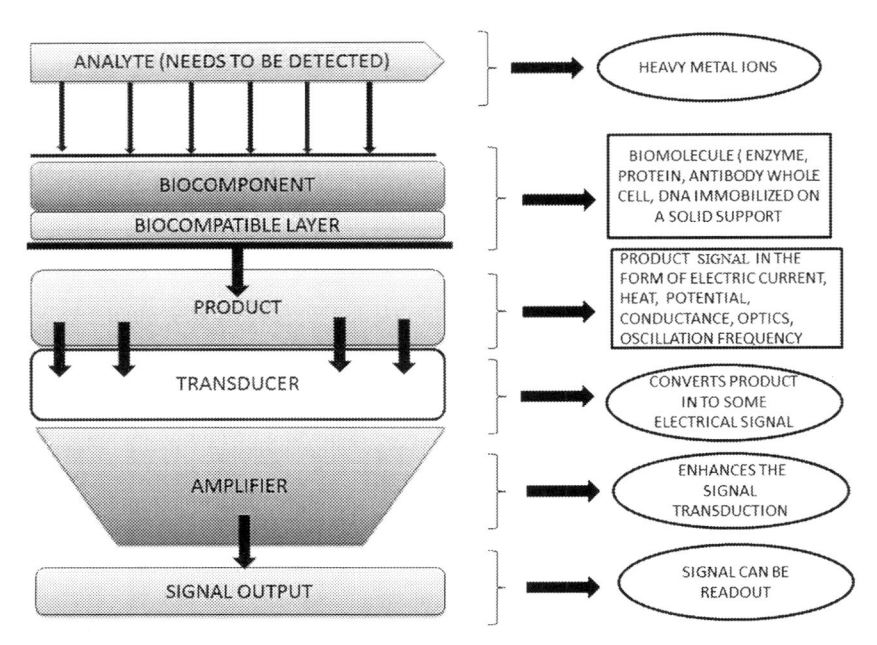

FIGURE 1 General outline of biosensing system involving different parts in series.

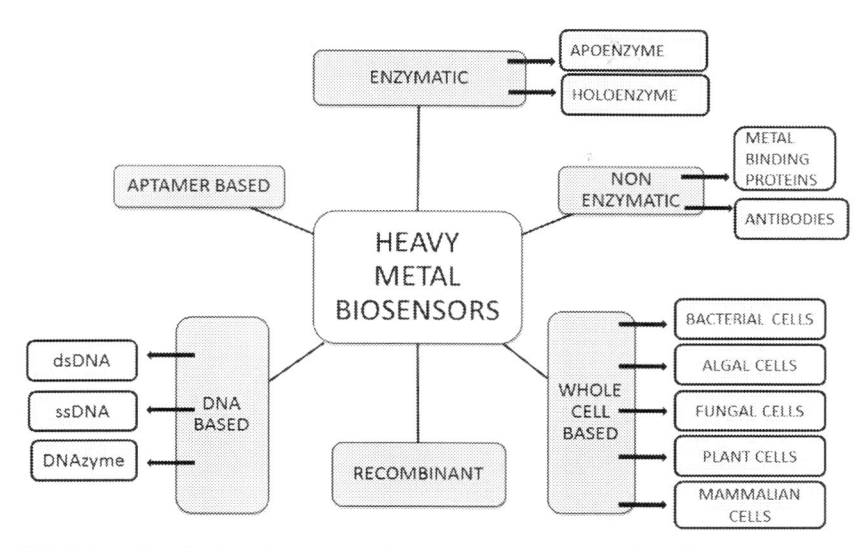

FIGURE 2 Classification of heavy metal biosensors based on the choice of biocomponent.

shown in Fig. 3). Different types of immobilisation procedures have been adopted, according to the suitability of the biological element and transducing system (as shown in Fig. 4). These techniques involve physical methods, like entrapment, encapsulation and adsorption, or chemical methods like covalent bonding and cross linking. Physical methods, like entrapment or encapsulation, are mainly used to immobilise whole cells or cellular organelles. Thickness and porosity of membranes used for encapsulation and entrapment affect the functioning of biocomponents by providing extra diffusional barriers, resulting in lower sensitivity and delayed response time. To overcome this drawback, chemical methods can be adopted, which include covalent bond formation or cross linking with functional groups of enzymes, proteins or whole cells [29].

The choice of immobilisation method affects the stability and activity of the biocomponent. Matrix used for the immobilisation of biocomponent also acts as a barrier and controls the rate of diffusion of substrate to provide maximum proximity between the analyte and biorecognition element [30,31]. The main advantages and disadvantages of various immobilisation strategies are discussed in Table 3.

7. DIFFERENT STRATEGIES FOR BROAD RANGE AND SPECIFIC HEAVY METAL ION BIOSENSORS

For the construction of a successful biosensing system, the first main step is the selection of the biocomponent. Some biosensors are specific for a

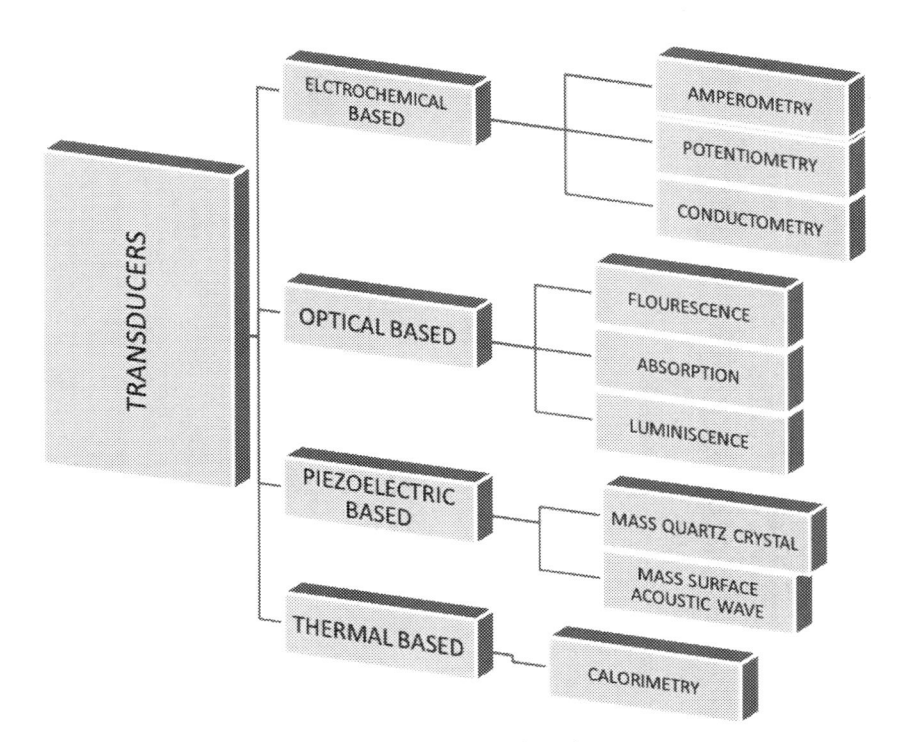

FIGURE 3 Types of transducers involved in various biosensing systems.

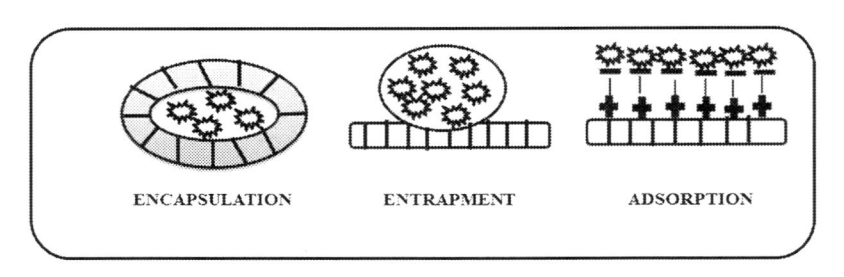

PHYSICAL METHODS

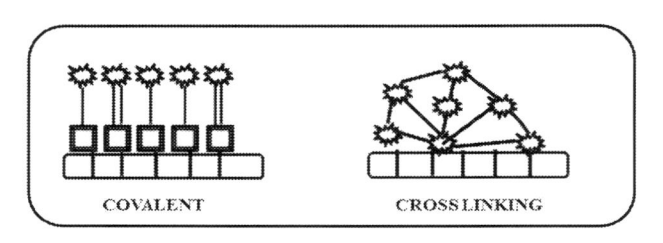

CHEMICAL METHODS

FIGURE 4 Overview of different immobilization techniques used in biosensing systems.

TABLE 3 Immobilisation Strategies Used in Construction of a Biosensor (Advantages and Disadvantages)

Type of Immobilisation Technique	Advantages	Disadvantages
Physical methods • Entrapment • Encapsulation • Adsorption	• Cause proper retention of cells/enzymes inside a protective membrane. • The specificity, thermal and chemical stability of biocomponent is retained. • Native structure of enzyme/cell is not changed in adsorption as there is no covalent bond formation.	• Additional resistance due to the membrane used for encapsulation or entrapment cause poor diffusion of substrate and product. • It leads to lower sensitivity and delayed response. • There is poor long time stability due to noncovalent interactions like ionic, hydrophobic, polar and hydrogen bonding.
Chemical methods • Covalent bonding • Crosslinking	• Form stable covalent bonds with the functional groups of enzymes/cell walls of microbial cells, cause longer stability. • Crosslinking involves bridge formation with different functional groups of a biocomponent by forming stable network.	• Biological activity of biocomponent may be affected in harsh chemical environment due to absence of protective cover. • Native structure of biocomponent is changed due to covalent bond formation with functional groups, which may change its activity. • Crosslinking reagent may affect the cell viability.

particular metal ion and others may involve broad range of detection with the different biocomponents. The choice and suitability of a particular biocomponent depend on the type of the analyte to be detected. Various metal ions inhibit or enhance the activity of different biological enzymes, while other heavy metal ions distort the structure of particular proteins, antibodies and whole cells resulting in the altered functioning inside the body. Depending upon the metal ion to be detected, the suitability of biocomponents in response to particular metal ion can be studied, to generate the quantifiable and detectable signal in accordance with the reliable transduction system.

7.1 Enzymatic Biosensors

Enzymes are biocatalysts involved in various biochemical reactions in living systems. Some metal ions act as a cofactor in the functioning of various enzymes, while others may act as inhibitors distorting the structure and function of enzymes. Depending on the metal ion to be detected enzymatic, biosensors can be classified into apoenzyme- and holoenzyme-based biosensors.

7.1.1 Apoenzyme-Based Biosensors

In apoenzyme-based biosensors, heavy metal ions act as cofactors for the activation of enzymes, binding the active site of apoenzymes. The main examples of heavy metal ions (Cu^{2+} and Zn^{2+}) acting as cofactors for various apoenzymes have been described in Fig. 5 [32,33]. As an example, apoform of

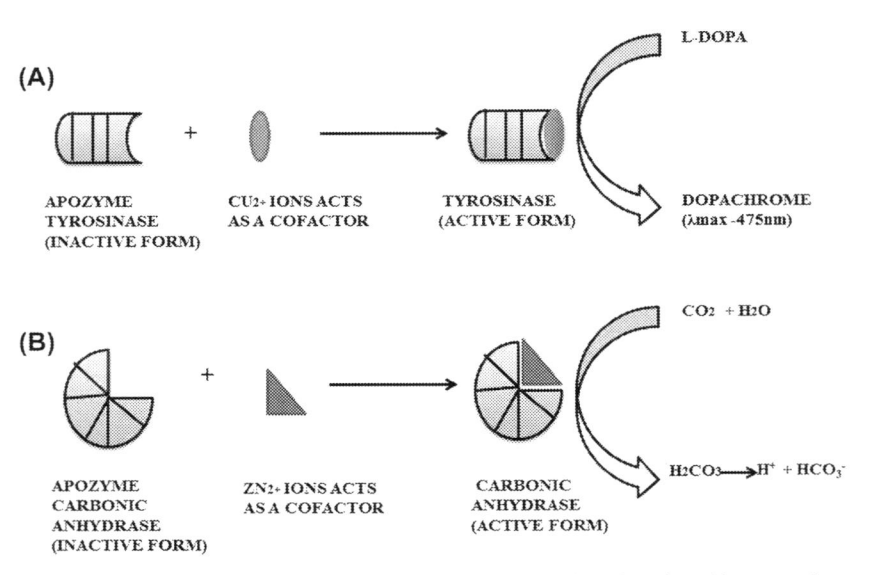

FIGURE 5 Apoenzyme-based biosensor approach: (A) apotyrosianse is activated by copper ions; (B) apocarbonic anhydrase is activated by zinc ions.

tyrosinase is converted into active form in the presence of Cu^{2+} ions for the conversion of L-DOPA (dihydroxyphenylalanine) into dopachrome, which is an orange-coloured product and can be detected spectrophotometrically [34]. Similarly, Zn^{2+} ions act as microactivators for the activity of carbonic anhydrase and alkaline phosphatase. These enzymes are used in various chemical reactions. The mechanism is reversible and thus apoenzyme biosensors can be reused after chelating the already present metal ions by using ethylenediaminetetraacetic acid (EDTA) complexes.

7.1.2 Holoenzyme-Based Biosensors

In holoenzyme-based biosensors, heavy metal ions mostly inhibit the activity of holoenzyme reversibly or irreversibly. Heavy metal ions detection is thus provided by an inhibition-based mechanism, which affect the overall activity of the enzyme and can be competitive, uncompetitive or noncompetitive. Heavy metal ions acting as inhibitors compete with the substrate, binding to the enzyme-active site or substrate-binding site by noncovalent interactions in case of competitive inhibition. The activity of enzymes can be regained after the removal of inhibitors and the increase in substrate concentration, resulting in reversible inhibition. In case of uncompetitive inhibition, the inhibitor binds reversibly to the enzyme—substrate complex (other than the active site), altering the formation of the product. During noncompetitive inhibition, inhibitors irreversibly bind to either the enzyme or the enzyme—substrate complex, but not at the active site, leading to the impaired structure and function of the enzyme by deforming its active centre. Substrate binding is not altered but enzyme—substrate inhibitor (ESI) complex is unable to form products. Substrate and inhibitor do not compete with each other in case of noncompetitive inhibition [35,29]. The inhibition-based enzymatic assay has been described in Fig. 6. As an example, urease enzyme is used for the conversion of urea into ammonia and carbon dioxide, leading to an increase in pH of the solution. The activity of urease is altered in the presence of Hg^{2+} ions, resulting in a decrease in ammonia production and a consequent decrease in potential response, which can be sensed by a potentiometer [36].

In case of inhibition-based biosensors, the strength of inhibition helps us to find the limit of detection for heavy metal ions present in the sample. Level of inhibition depends on many factors like concentration of inhibitor or reaction conditions (temperature, pH and incubation time). In case of reversible inhibition, the biosensor can be regenerated by different methods, where the activity of enzyme can be regained after removing the inhibitor and increasing the substrate concentration. In irreversible inhibition-based biosensors, regeneration is not possible because of the complete distortion of enzyme activity. This could affect the applicability of similar biosensors in commercial applications.

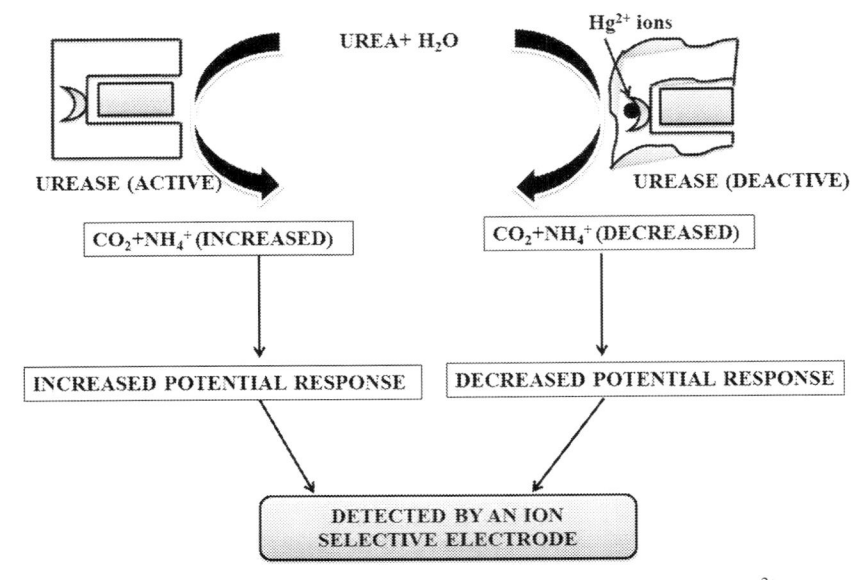

FIGURE 6 Holoenzyme inhibition-based approach showing urease inhibition by Hg^{2+} ions.

A huge number of different enzymes, including tyrosinase, carbonic anhydrase, acetylcholine esterase (AChE), urease, invertase, alkaline phosphatase, peroxidase, lactate dehydrogenase, nitrate reductase, can be used for the detection of heavy metal ions by exploiting either activation-based or inhibition-based reaction mechanisms [29]. Zn and Cu ions detection mainly involves apoenzyme-based reaction mechanism. These biosensors are highly specific with limits of detection in the range from picomolars to nanomolars [33,34]. Other nonessential heavy metal ions, like lead, cadmium, mercury and arsenic, are generally detected by holoenzyme-based inhibition mechanism and mostly developed biosensors are broad range with detection limits in the range of micro molars [36,42,44,48]. Biosensors based on the apoenzyme activation or holoenzyme inhibition are summarised in Table 4. As an example, glucose oxidase-reversible inhibition-based enzyme biosensor has been reported for the detection of Hg^{2+} ions. The enzyme is immobilised on electropolymerised aniline membrane along with glutaraldehyde by cross-linking on Pt electrode. The presence of Hg^{2+} ions in the sample inhibit glucose oxidase activity, resulting in reduction in current on GOx electrode, sensed by electrochemical system. The generated biosensor can detect as low as 2.4 nM of Hg^{2+} ions in compost extracts. The 70% activity of this system can be retained up to one month [39]. In a similar work, GOx has been immobilised in polyphenylenediamine with a lowest response time of 2 min and detection limit 2.5 μM for Hg^{2+} and Cu^{2+} ions both, based on ampero-metric studies [35]. Another similar inhibitory effects on GOx has been

TABLE 4 Enzyme Based Biosensors for Detection of Heavy Metal Ions

Enzyme (Bio Component)	Mechanism Involved	Immobilisation Technique	Transducer	Heavy Metal Ion Detected	Detection Limit	References
Horseradish peroxidase	Holoenzyme inhibition	Adsorption on carbon paste electrode	Cyclic voltammetry	Zn^{2+}	1.14×10^{-7} M	[26]
Alcohol oxidase	Holoenzyme inhibition	Quantum dots (QDs) and enzyme hybrids	QDs fluorescence	Cu^{2+}	2.7×10^{-8} M	[37]
Carbonic anhydrase	Apoenzyme activation	—	Optical (fluorescence)	Zn^{2+}	2.0×10^{-12} M	[33]
Nitrate reductase	Holoenzyme inhibition	Crosslinking on neflon	Conductometry	$Cu^{2+}, Cd^{2+}, Pb^{2+}, Zn^{2+}$	0.05, 0.1, 1, 0.05 (µM)	[38]
Glucose oxidase	Holoenzyme inhibition	Crosslinking using glutaraldehyde and electropolymerisation on Pt electrode	Amperometry	Hg^{2+}	0.49 (µg/L)	[39]
Alkaline phosphatase	Holoenzyme inhibition	Entrapment in sol gel	Amperometry	$Hg^{2+}, Cu^{2+}, Cd^{2+}, Zn^{2+}, Ag^{+}$	mg/L	[40]
Invertase and glucose oxidase	Holoenzyme inhibition	Entrapment in agarose guasgum	Amperometry	$Hg^{2+}, Ag^{2+}, Pb^{2+}, Cd^{2+}$	3×10^{-8} M	[41]
Glucose oxidase	Holoenzyme inhibition	Electropolymerised on Pt electrode	Amperometry	$Hg^{2+}, Ag^{+}, Cu^{2+}, Cd^{2+}, Pb^{2+}, Cr^{3+}, Fe^{3+}, Co^{2+}, Ni^{2+}, Zn^{2+}, Mn^{2+}$	10^{-6} M	[42]
Urease	Holoenzyme inhibition	Electro polymerised	Amperometry	Hg^{2+}	7.4 µM	[43]
Tyrosinase	Apoenzyme activation	Crosslinking on Pt electrode	Conductometry	Cu^{2+}	1.57×10^{-8} M	[32]

Continued

TABLE 4 Enzyme Based Biosensors for Detection of Heavy Metal Ions—cont'd

Enzyme (Bio Component)	Mechanism Involved	Immobilisation Technique	Transducer	Heavy Metal Ion Detected	Detection Limit	References
Urease	Holoenzyme inhibition	Adsorption on polyvinyl chloride	Potentiometry	Hg^{2+}	0.05 μM	[36]
Glucose oxidase	Holoenzyme inhibition	Electro polymerisation	Amperometry	Hg^{2+}, Cu^{2+}	2.5 μmol/L, 2.5 μmol/L	[35]
Urease	Holoenzyme inhibition	Entrapment in sol—gel	Optical	Hg^{2+}, Cu^{2+}, Cd^{2+}	10 nM, 50 μM, 500 μM	[44]
Invertase	Holoenzyme inhibition	Crosslinking on glutaraldehyde	Amperometry	$HgCl_2$, methyl mercury, phenyl mercury	0.27, 0.34, 0.12 (ppm)	[45]
Tyrosinase	Apoenzyme Activation	Covalent binding using gelatin and glutaraldehyde	Amperometry	Cu^{2+}	10^{-6} M	[46]
Urease/ glutamate dehydrogenase	Holoenzyme inhibition	Electropolymerisation on screen-printed electrode	Amperometry	Hg^{2+}, Cu^{2+}, Cd^{2+}, Zn^{2+}	7.2 μg/L, 8.5 μg/L, 0.3 mg/L, 0.2 mg/L	[47]
Urease	Holoenzyme inhibition	Encapsulation in ultrabind membrane	Optical fibre	Hg^{2+}, Ag^+, Cu^{2+}, Ni^{2+}, Zn^{2+}, Co^{2+}, Pb^{2+}	mM	[48]
Alkaline phosphatae	Holoenzyme inhibition	Sol—gel matrix	Fluorescence	Ag^+	10.1 μM	[49]
Urease	Holoenzyme inhibition	Entrapment in sol—gel	Conductometry	Cu^{2+}, Cd^{2+}, Pb^{2+}, Hg^{2+}	0.02, 0.1, 0.9, 0.05 (mM)	[50]
Glucose oxidase	Holoenzyme inhibition	Entrapment in polyacrylamide gel	Amperometry	Cu^{2+}	1.9×10^{-7} M	[51]

reported for the range of heavy metal ions like Hg^{2+}, Ag^+, Cu^{2+}, Cd^{2+}, Pb^{2+}, Cr^{3-}, Fe^{3+}, Co^{2+} and Ni^{2+} by determining the hydrogen peroxide decomposition detected by an amperometer. Ag^+ has shown the maximum inhibitory effect than other heavy metal ions up to the concentration of micromolars [42]. By immobilising glucose oxidase (GOx) enzyme in polyacrylamide gel, an amperometric biosensor was developed, for the determination of copper ions in food samples. The response time for the reaction was optimised to 2.5 min. Linear range for the detection of Cu^{2+} was found in the range 1.9×10^{-7} M to 1.5×10^{-6} M [51]. Electrochemical biosensor has been constructed by assembling carbon film electrode supported with glucose oxidase enzyme that was immobilised with glutaraldehyde by crosslinking on topmost of a film [poly(neutral red)] acting as redox mediator. The detection limits for Cd^{2+}, Cu^{2+}, Pb^{2+} and Zn^{2+} were found to be 8.9, 94.3, 14.4 and 137 nM, respectively. Copper-induced inhibition of glucose oxidase was maximum followed by cadmium, lead and zinc. EDTA was used to regenerate the activity of glucose oxidase. The suitability of biosensor was checked in various food samples and beverages, which contains trace amount of these metal ions [52].

Amperometric inhibition–based urease biosensor has been reported for the detection of Hg^{2+} ions. Urease enzyme electrode has been prepared by using polyvinylferrocinium film. The occurrence of Hg^{2+} ions in the sample inhibit urease activity, resulting in decrease in oxidation current. The detection limit is found to be 7.4 μM for Hg^{2+} ions [43]. In another work potentiometric determination of Hg^{2+} ions has been developed based on inhibition of urease, immobilised on self-assembled gold nanoparticles. Gold nanoparticles are further adsorbed on polyvinylchloride (PVC)–NH_2 matrix on pH electrode. The response is based on change in pH after inhibition of urease activity, with a detection limit of 0.05 μM Hg^{2+} ions [36]. An optical biosensor based on inhibition of urease has also been reported with highest sensitivity for Cu^{2+} and Cd^{2+} ions up to range of 10 μM enzyme is entrapped in sol gel along with a fluorescent dye that can sense the change in pH due to decrease in activity of urease by the presence of heavy metal ions. The developed biosensor has also been applied to various biological and environmental samples [44]. In a similar work, fibre-optic biosensor is developed for the detection of various heavy metal ions like Hg^{2+}, Ag^+, Cu^{2+}, Ni^{2+}, Zn^{2+}, Co^{2+} and Pb^{2+}. The urease enzyme is immobilised on ultrabind membrane. The inhibition of urease due to the presence of heavy metal ions in the sample has been checked by measuring the wavelength at 615 nm using pH indicator strip. Hg^{2+} ions have shown maximum inhibitory effect, with a detection limit as low as 1 nM [48]. Enzyme biosensor based on the inhibition of urease and acetyl cholinesterase by Pb^{2+} ions was reported and applied in drinking water samples. A detection limit of 4.83×10^{-9} M was achieved, by using the method of total reflection at the interface between the Si_3N_4 core and complex polyelectrolyte self-assembled (PESA) membranes containing cyclotetrachromotropylene (CTCT) as an indicator [53]. An optical biosensor by immobilising urease and a pH indicator chlorophenol red in a PVC—sol—gel matrix for monitoring heavy metals in water samples was also developed [54]. For the ultrasensitive determination of Cu^{2+} ions, alcohol oxidase (AO) enzyme fluorescent transducer in combination with Cd/Te quantum dots (QDs) was successfully constructed. Methanol was oxidised to hydrogen peroxide (H_2O_2), was catalysed by AO, further induce the quenching of QDs fluorescence. The activity of AO was inhibited in the presence of copper ions, resulting in decreased quenching of QDs fluorescence. Other metal ions even at 10 or 100 times conc. of Cu^{2+} ions showed no severe inhibition to the AO activity, leading to highly selectable fluorescent sensor for copper. QDs-enzyme hybrid system showed the detection limit for Cu^{2+} ions as low as 2.7×10^{-8} M due to the superior fluorescence character of QDs [37]. An alkaline phosphatase conductometric biosensor consisting of interdigitated gold electrodes for the assessment of Pb in water was developed. A detection limit of 1.93×10^{-4} M was achieved and the storage stability of the biosensor in buffer solution at $4°C$ was more than one month [55].

7.2 Nonenzymatic Biosensors

Nonenzymatic-based biosensors include various naturally occurring proteins, such as metalloproteins or antibodies, as biocomponent. Nonenzymatic based can be classified further into two main categories.

7.2.1 Metal-Binding Proteins

Different types of nonenzymatic proteins, like metalloproteins, require metal ions as central element for the formation of metal—chelate complexes. These metal-binding proteins can be used as a better source of biocomponent for the detection of heavy metal ions. These metalloproteins can be derived from natural sources or can be synthesised by arranging a peptide chain that binds to specific metal ion for their proper 3-D folding and functioning. As an example, zinc fingers are synthesised using 20—25 amino acids, which require zinc ions for the folding of zinc finger peptides. Cysteine and histidine residues act as a ligand to the zinc ions whose coordination is essential to stabilise the tertiary structure of the peptide [56]. Several nonenzymatic proteins, like metallothionein (MT) [57], bovine serum albumin (BSA) [58], human angiotensinogen I [59], phytochelatin [60], human carbonic anhydrase II [61], cytochrome C3 [62], have been exploited for heavy metals detection, in combination with optical or electrochemical transducers showing detection limits in the picomolar range [58—60] have been summarised in Table 5. Pb^{2+} ions have been monitored by using BSA-based biosensor, immobilised on piezoelectric quartz crystal (PQC) which is Au modified. The change in piezoelectric quartz crystal impedance (PQCI) is being measured in response to the presence of Pb^{2+} ions in the sample, which is influenced by change in pH of the reaction buffer. The lowest detection limit is found to be

TABLE 5 Nonenzymatic Protein and Antibody-Based Biosensors for Heavy Metal Ions

Biocomponent	Immobilisation Technique	Transducer Used	Metal Ion Detected	Detection Limit	References
Metallothionein	Carboxymethylated dextran matrix	Optical (surface plasmon resonance)	Cd^{2+}, Zn^{2+} and Ni^{2+}	2.0×10^{-5} M	[57]
BSA (bovine serum albumin)	Au-modified piezoelectric quartz crystal	Piezoelectric	Pb^{2+}	1.0×10^{-9} M	[58]
Human angiotensinogen I	Gold electrode (self-assembly by thiocitic acid)	Electrochemical	Pb^{2+}	1.0×10^{-9} M	[59]
Phytochelatin	Gold nanoparticles	Optical	Cd^{2+} and Zn^{2+}	1.42×10^{-9} M	[60]
Mer R	Gold electrode (self-assembly by thiocitic acid)	Capacitance	Hg^{2+}, Cu^{2+}, Cd^{2+}, Zn^{2+}, Pb^{2+}	10^{-15} M	[63]
Cue R	Au-modified piezoelectric quartz crystal	Piezoelectric (quartz crystal)	Cu^{2+}	10^{-21} M	[64]
Human carbonic anhydrase II variant	Adsorption on an optical fibre	Fibre optic	Cu^{2+}	1.0×10^{-13} M	[61]
Cytochrome c3 from desulphomicrobium norvegicum/ GCE	Glassy carbon electrode	Amperometric	Cr^{6+}	0.2 mg/L	[62]
Antibody (2A81G5)	Covalent (streptavidin–biotin) coupling	Optical (absorption)	Cd^{2+}	0.1 µM	[65]

Continued

TABLE 5 Nonenzymatic Protein and Antibody-Based Biosensors for Heavy Metal Ions—cont'd

Biocomponent	Immobilisation Technique	Transducer Used	Metal Ion Detected	Detection Limit	References
Antibody (ISB4)	Covalent (streptavidin—biotin) coupling	Optical (colorimeter)	Cd^{2+}, Co^{2+}, Pb^{2+}, U^{6+}	0.001 μM	[66]
Antibody (12F6)	Covalent (streptavidin—biotin) coupling	Optical (fluorescence)	U^{6+}	0.12 nM	[67]

1.0×10^{-9} M for Pb^{2+} ions. The biosensing chip can be regenerated almost eight times by using EDTA [58]. In another approach mammalian MT, immobilised on carboxymethylated dextran matrix in combination with surface plasmon resonance—based transducer has been reported. The MT-based sensor is sensitive mainly for $Cd^{2+} > Zn^{2+} > Ni^{2+}$ up to the range of micromolars [57]. A synthetic phytochelatin (Glu—Cys) 20 Gly (EC20) fusion with maltose-binding domain has been explored for the detection of $Zn^{2+} > Cu^{2+} > Hg^{2+} > Cd^{2+}$ with lowest detection limit of 100 fM by measuring the change in capacitance. The sensor be regenerated by using EDTA and stable for 15 days [63].

7.2.2 Antibodies-Based Biosensors

Antibody—antigen—based biosensors can be developed by exploiting the high selectivity and sensitivity of immunological reactions [68]. These immunosensors in combination with different transducers (eg, electrochemical, piezoelectric or optical) can be used to detect the amount of total toxicity caused by specific heavy metal ions. The metal chelate complex acting as an antigen in response to a specific antibody generates the detectable signal after forming a metal—chelate antibody complex, which helps to determine the amount of toxicity caused by the heavy metal ions. Antibodies are preferred in comparison to enzymes because of their high specificity to a particular metal ion [69]. Many monoclonal antibodies have been exploited, being highly specific to a particular metal—EDTA (ethylenediamine tetraacetic acid), metal—DTPA (diethylene triamine pentaacetic acid) or metal—CHXDTPA (cyclohexyl diethylene triamine pentaacetic acid) complex of mercury, lead, cadmium, cobalt, nickel, copper and silver. Also, a number of monoclonal antibodies have been fabricated for the complexes of Cd—EDTA, Co—DTPA and Pb—CHXDTPA with high sensitivity [66]. Fig. 7 reports the basic outline of the steps involving monoclonal antibody mechanism in response to a particular labelled metal chelate complex, which generate a direct fluorescent signal after binding with the specific antibody. Some examples for antibody-based biosensors for the detection of heavy metal ions have been summarised in Table 5. Antibodies 2A81G5, ISB4, 12F6 have been also used as biocomponents by covalent coupling with streptavidin—biotin along with optical transducers for cadmium, cobalt, uranium, and lead ions, with detection limits in the nanomolar range [65—67]. As an example, for monitoring the Cd^{2+} ions in aqueous samples, a monoclonal antibody has been generated, which only bound to Cd^{2+}—EDTA complex, but not with a metal-free EDTA. An immunoassay is performed by incubating the antibody along with the Cd^{2+}—EDTA conjugate. This assay is completely sensitive for Cd^{2+} ions up to the range of micromolars [65]. Other researchers have explored competitive immunoassay for the detection of various heavy metal ions like Cd^{2+}, Co^{2+}, Pb^{2+} and U^{6+}. Monoclonal antibodies specific for four different metal chelate complexes of Cd^{2+}—EDTA, Co^{2+}—DTPA, Pb^{2+}—CHXDTPA and U^{6+}-2,9-dicarboxyl-1,10-phenanthroline

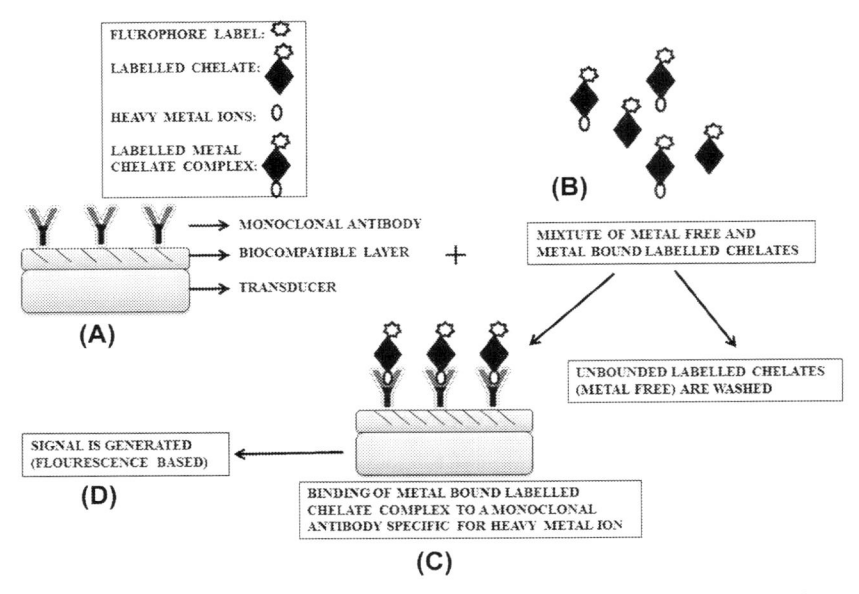

FIGURE 7 Schematic representation for the construction of monoclonal antibodies-based heavy metal biosensors (immunosensors).

have been generated. The antibodies and soluble metal−chelate conjugates were incubated to perform competitive microwell assay and immunosensor format KinExA (kinetic exclusion assay). For all the four metals, the KinExA format has shown a 10 to 1000-fold greater sensitivity due to the difference in monoclonal antibodies affinity toward soluble and immobilised metal−chelate conjugate. The sensor is also applied to spiked ground water samples for the detection of Cd^{2+} ions showing detection limit of 0.001 μM [66]. In a similar work, monoclonal antibody for uranium−dicarboxyphenanthroline (DCP) complex has been generated to follow kinetic exclusion immunoassay format for field base and inline monitoring of uranium-containing ground water samples. The field portable sensor (FPS) has shown a response time of 5−10 min with a lowest detection limit of 0.33 nM for single uranium-containing sample, in comparison to inline sensor, which can analyse multiple samples simultaneously with lowest detection limit of 0.12 nM [67].

7.3 Whole Cell−Based Biosensors

Whole cell−based biosensors involve bacteria, algae, fungi, plants or mammal cells as biocomponents in intimate contact with a transducer. The change in cellular response in comparison to the heavy metal toxicity can be diagnosed [70]. Whole cells have many advantages over enzymes or antibodies, being easily growth in bulk in inexpensive optimum media and easily adapted to alterations in reaction conditions [71]. Whole cell–based biosensors are relatively cheap, because cells need not to be isolated or purified and are relatively more stable and adaptive to the nutrient media. The main drawback of using whole cells as biocomponents involves low selectivity, sensitivity and in some cases delayed response time, which could further be improved by recombinant DNA technology, involving insertion of foreign gene in to a particular host which responds to the presence of specific heavy metal ion [29]. As shown in Fig. 8, bacterial cell *Bacillus badius* has been used for the detection of Cd^{2+} ions in milk samples. Cells were immobilised on to a plastic discs using sol–gel approach. The presence of Cd^{2+} ions inhibits the activity of urease present inside the bacterial cell by attaching itself to the enzyme site other than the catalytic site. This results in the decreased production of ammonium ions and in a consequent decrease in absorbance, sensed by using optical transducers [19]. The main examples of whole-cell biosensors have been summarised in Table 6. For an example, urease enzyme activity can be inhibited by many heavy metal ions like mercury, nickel, cadmium and lead. Based on this approach, whole-cell *Bacillus sphaericus*, biosensor has been explored for the detection of Ni^{2+} ions in food and wheat flour samples. This microbial biosensor relies on the inhibition of urease enzyme present inside the cell, in response to the presence of nickel ions in the sample. The transducers used here is potentiometer and ammonium ion selective electrode which detects the presence of ammonium ion formed after urea hydrolysis. It

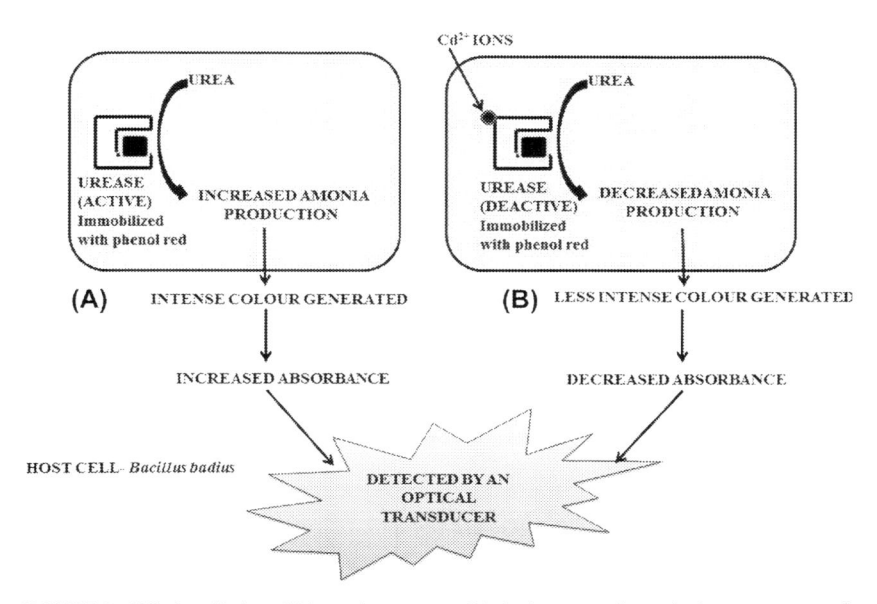

FIGURE 8   Whole cell–based biosensing system. (A) Active urease is producing more ammonia in the absence of Cd^{2+} ions. (B) Deactive urease is producing less ammonia when Cd^{2+} attaches itself to the enzyme inside the cell.

TABLE 6 Whole Cell–Based Biosensors for Detection of Heavy Metal Ions

Whole Cells	Biocomponent	Immobilisation Technique	Transducer	Metal Ion Detected	Detection Limit	References
	Escherichia coli	Deposition with polyelectrolyte multilayer	Piezoelectric (acoustic wave)	Cd^{2+} and Hg^{2+}	10^{-12} M	[72]
Bacterial cells	Bacillus sphericus	Carbon paste electrode	Cyclic voltammetry	Pb^{2+}	2.5×10^{-12} M	[23]
	Bacillus badius	Hydrosol gel, adsorption on nylon membrane	Potentiometer	Cd^{2+}	8.8×10^{-11} M	[24]
	B. badius	Sol–gel matrix, adsorption on plastic discs	Optical	Cd^{2+}	8.8×10^{-10} M	[19]
	Vibrio fisheri	In culture medium	Optical (luminescence)	Hg^{2+}, Cu^{2+}, Cd^{2+}, Pb^{2+}, Zn^{2+}, Fe^{2+}	0.001 mg/L	[73]
	Sulphur oxidising bacteria	In culture medium	Electrical conductivity	Cr^{6+}	5 ppb	[74]
	E. coli	Nanobeads on indium tin oxide electrode	Electrochemical impedance spectroscopy	Hg^{2+}	10^{-12} M	[75]
	Bacillus sphaericus	Adsorption on Whatman filter paper 1	Potentiometer	Ni^{2+}	0.03×10^{-9} M	[25]
	Acidithiobacillus ferrooxidans	Whatman glass-fibre felt and carbon felt	Amperometry	Cr^{3+}	2×10^{-5} M	[76]
	E. coli	Screen printing electrode	Conductometry	Hg^{2+}	1 ppm	[77]

Algal cells	*Chlorella* sp.	Glassy carbon electrode	Conductometry	Hg^{2+}	10^{-14} M	[78]
	Chlorella vulgaris	Sol gel, silica matrix	Spectrofluorimetry	Cd^{2+}	1.42×10^{-7} M	[79]
	C. vulgaris Bienzymatic approach (alkaline phosphatase and acetylcholine transferase)	Entrapment in bovine serum albumin membrane, crosslinking, glutaraldehyde	Conductometry	Cd^{2+}, Zn^{2+}	8.8×10^{-8} M	[80]
	Tetraselmis chuii (*Prasinophyceae*)	Carbon paste electrode	Amperometry	Cu^{2+}	4.6×10^{-10} M	[81]
	Phormidium sp.	Carbon paste electrode	Cyclic voltammetry	Pb^{2+}	2.5×10^{-8} M	[82]
Yeast cells	*Rhodotorula mucilaginosa*	Carbon paste electrode	Cyclic voltammetry	Cu^{2+}	10^{-7} M	[83]
Fungal cells	*Rhizopus arrhizus*	Carbon paste electrode	Cyclic voltammetry	Pb^{2+}	0.5×10^{-8} M	[84]
Plant cells	*Arabidopsis thaliana, Populus tremula*	–	Fluorescence resonance energy transfer	Zn^{2+}	1.0×10^{-3} M	[85]
Mammalian cells	Cardiac cells	–	Potentiometer	Hg^{2+}, Cu^{2+}, Cd^{2+}, Pb^{2+}, Zn^{2+}, Fe^{2+}	10 μM	[86]

has quite low detection limit up to 0.03 nM with a very quick response time of 1.5 min. It can fairly substitute the current conventional methods of monitoring [25]. A microbial whole-cell *B. sphaericus* biosensor based on urease inhibition for the determination of copper in electroplating industrial effluents has also been developed. It has range of detection from 0.3 to 30 pM for copper ions by using ion-selective electrode [87]. In recent approach of whole-cell lead biosensor, electrochemical transducer along with urease-producing *B. sphaericus* isolate MTCC 5100 has been developed. Carbon paste electrode has been used to immobilise the whole cells. The limit of detection achieved was 2.5×10^{-12} M with a response time of 5 min. This biosensor was also applied on milk samples by the researchers [23]. In another approach *B. badius* whole cells have been used as a biocomponent for the detection of cadmium ions in milk, based on optical studies. The limit of detection achieved was 8.8×10^{-10} M with 10 μL of sample volume. Storage stability of biocomponent was found at 4°C in 10% glycerol for more than 90 days [19]. Other research involves ammonium ion-sensing electrode used as a transducer. In this study, *B. badius* cells containing urease enzyme in hydro sol gel were immobilised on nylon membrane and used as a biocomponent. Inhibition of enzyme in the presence of cadmium ions was achieved within the range of 8.8×10^{-8} to 8.8×10^{-6} M with a lowest detection limit of 8.8×10^{-9} M of cadmium. The biocomponent was stable for 65 days at 4°C in 10% glycerol. The applicability of this biosensor was checked in spiked milk samples [24]. The *Escherichia. coli* microbial sensors have also been reported in response to the presence of mercuric chloride in the sample, resulting in change in metabolic activity of cells. The generated biosensor is single use, disposable in conjugation with an amperometric-based transduction system with a lowest detection limit up to micromolars [77]. Algal species like *Chlorella vulgaris* have also been explored for the detection of various heavy metal ions by many researchers. As an example, *Chlorella* species immobilised on glassy carbon electrode has been reported for the detection of mercury, with lowest detection limit of 10^{-14} M. The sensor can measure the change in conductance of algal cells (stable for 14 days) in response to the selective presence of Hg^{2+} ions [78]. In a similar approach, whole-cell *C. vulgaris* microalgae-containing alkaline phosphatase (AP), immobilised in BSA membranes (crosslinked by glutaraldehyde vapours), in addition to the interdigitated conductometric transducers is experimented for the detection of Cd^{2+} ions in aquatic samples. The change in conductance due to the inhibition of alkaline phosphatase (AP) activity, in response to the presence of Cd^{2+} ions can be measured achieving the detection limit of 8.8×10^{-8} M. The generated biosensor is highly stable in comparison to enzymatic biosensors, having APs on walls on algal cells [80]. Some researchers have also explored cells of *C. vulgaris*, immobilised in translucent silica matrix for the detection of heavy metal ions like Cd^{2+} and various herbicides like atrazine, diuron and paraquat in water. These toxic compounds cause quenching of fluorescence generated by algal cells by affecting their photosynthetic activity. The synchronous-scan spectrofluo-rometry-based biosensor can measure up to 1.42×10^{-7} M of Cd^{2+} ions in various water samples [79].

In the literature, most of the whole cell—based biosensors studies for the detection of heavy metal ions has been done on the bacterial species like *B. badius* or *E.coli* and algal species of *C. vulgaris*, entrapped in sol—gel matrix, beads or on electrodes in combination with the optical (luminescent) or con-ductometry-based transducers [19,77—79]. Other species, like yeast cells, fungal cells, plant cells, mammalian cells, have been rarely exploited for the generation of biosensors [83—86]. The detection limit has been attained up to picomolars for the lead, cadmium and mercury ions [23,24,75].

7.3.1 Recombinant Biosensors

Recombinant whole cell—based biosensors encompass genetic engineering techniques in which the promoter gene specific for particular heavy metal ions is constructed and added upstream to the reporter gene present in the host cell. In Fig. 9, a bacterial biosensor (*E. Coli* DH5α) able to determine Pb^{2+} ions by using green fluorescent proteins (GFPs) has been reported. A regulatory protein gene (PbrR) acts as a biorecognition element along with PbO/P operator/promoter of the operon for lead resistance from plasmid (pMOL30) regulates the expression of *gfp* reporter gene. This gene construct can sense the presence of Pb^{2+} ions, leads to enhanced expression of *gfp* genes resulting in an increase in fluorescent signal. The interference of other heavy metal ions like Cd^{2+}, Zn^{2+} and Hg^{2+} were also checked, only Zn^{2+} showed the mild induction in fluorescence at highest concentration, but 8.5 lower fluorescent

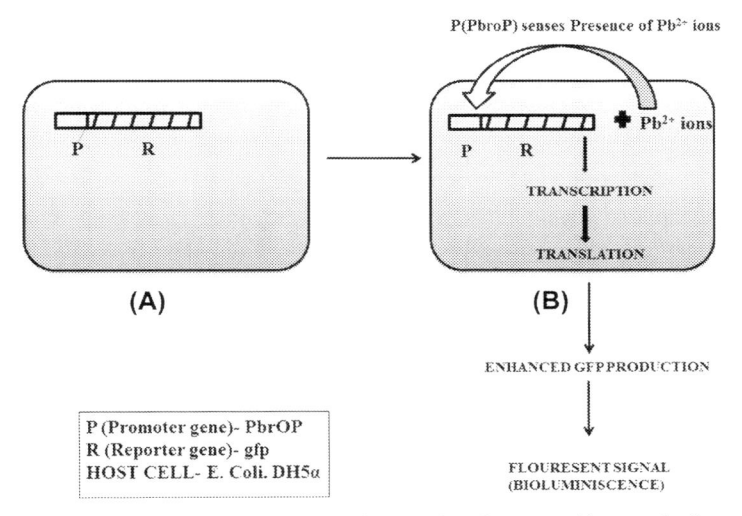

FIGURE 9   Genetically modified whole-cell biosensor based on recombinant technology.

signal than that of Pb^{2+} ions. The contribution of recombinant technology makes the biosensor specific for Pb^{2+} ions in the range of detection from 50 to 400 μM [88]. Various examples of recombinant whole cell—based biosensors have been summarised in Table 7. A recombinant whole-cell biosensor based on yellow fluorescent protein (YFP) has been reported for the detection of arsenic compounds in various environmental samples. *E. coli* DH5α has been used as a host cell for the expression of arsenic-resistant promoter and regulatory gene (arsR) along with phiYFP as a reporter gene. The WCB-11 showed a change in yellow fluorescence in response to As^{3+} and As^{5+} ions with lowest detection limit of 100 μM [90]. In a similar work, *E. coli* DH5a-based recombinant biosensor is explored for monitoring Cd^{2+}, Pb^{2+} and Sb^{2+} ions in contaminated soil samples with a detection limit of 0.1, 10 and 0.1 nM, respectively. The expression of green fluorescent protein (GFP) gene is regulated by cad promoter and cadC gene of the plasmid p1258 obtained from *Staphylococcus aureus*. The change in fluorescence is measured in response to the bioavailability of heavy metal ions with greater sensitivity and response time of 2 h [95]. The wild-type red fluorescent protein (DsRed) has also been reported for monitoring both Cu^+ and Cu^{2+} ions with a detection limit of nanomolars. The DsRed is highly sensitive and selective for copper ions without the interference of other heavy metals. This wild-type red fluorescent protein is seven times more sensitive toward Cu^{2+} ions than wild-type GFPs [96]. In a different approach, *Pseudomonas putida* X4 is used as a host cell for the expression of czcR3 promoter, enhanced green fluorescent protein (EGFP) gene and lacZ gene fusions in response to the presence of Zn^{2+} ions, which affects the β-glycosidase activity. The constructed biosensor (pczcR3GFP) is highly selective for measuring zinc toxicity in soil samples up to micromolars of concentration [89]. In a related approach, host cell of *P. putida* X4 and *E. aerogenes* NTG-01 host strain has been reported (pmerRluxCDABE-Kan) and compared for the construction of two Hg^{2+}-specific sensing systems in a similar reaction conditions. The X4 strain has shown maximum bioluminescence at midexponential phase (at 4th hour), but NTG-10 strain found to be highest luminescent at late exponential phase (at 5th hour) of incubation time. The lowest detection limit is found to be 100 pM of Hg^{2+}ions for both the strains, without the interference of other heavy metal ions in various contaminated samples [101].

7.4 DNA-Based Biosensors

In recent years, a number biosensing system have been designed using DNA, since many heavy metal ions show affinity for the metal-binding sites and catalytic property for DNA. Detection of heavy metal ions is based on their ability to bind with certain bases in the oligonucleotide chain resulting in stable (metal-dependent) DNA duplex [105]. The guanine and adenine bases are highly affected causing DNA oxidative damage due to interaction with the

TABLE 7 Biosensors Based on Recombinant Technology for Heavy Metal Ions

Host Cell	Biocomponent				
	Promoter/Reporter Gene	Transducer Used	Metal Ion Detected	Detection Limit	References
Pseudomonas putida X4	pczcR3GFP/lacZ	Optical (fluorescence)	Zn^{2+}	5 μmol/L	[89]
E. coli DH5α	arsR/phiYFP	Optical (fluorescence)	As^{3+}, As^{5+}	100 μmol/L	[90]
E. coli (MC1061)	merR/luxFF	Optical (luminescence)	Hg^{2+}	nmol/L	[91]
E. coli (K12), *Caenorhabditis elegans*	–/eGFP205C	Optical (fluorescence)	Hg^{2+}	nM	[92]
Burkholderia sp. (RASC c2)	–/lux	Optical (luminescence)	Zn^{2+}, Cu^{2+}	1.7 μg/mL, 0.09 μg/mL	[93]
Saccharomyces cerevisiae	CUP 1/lacZ	Amperometer	Cu^{2+}	16.0 mg/L	[94]
E. coli (DH5α)	Cad/rs-GFP	Optical (fluorescence)	Cd^{2+}, Pb^{2+}, Sb^{2+}	10^{-10} M	[95]
E. coli (DH5α)	–/DsRed-GFP	Optical (fluorescence)	Cu^{2+}	4.5×10^{-8} M	[96]
E. coli (S30)	merR/luxFF	Optical (bioluminescence)	Hg^{2+}	1×10^{-7} M	[97]

Continued

TABLE 7 Biosensors Based on Recombinant Technology for Heavy Metal Ions—cont'd

Biocomponent					
Host Cell	Promoter/Reporter Gene	Transducer Used	Metal Ion Detected	Detection Limit	References
E. coli (W3110)	copA/lux	Optical (bioluminescence)	Cu^{2+}, Ag^+, $Au3^+$	0.01, 0.2, 0.30 μM	[98]
Alcaligenes eutrophus (AE1239)	pMOL 90+Tn4431/luxCDABE	Optical (bioluminescence)	Cu^{2+}	10^{-6} M	[99]
Caulobacter crescentus	urcA/gfp	Optical (fluorescence)	Uranium	0.1 μM	[100]
P. putida X4	pmerR:luxCDABE-Kan	Optical (bioluminescence)	Hg^{2+}	100 pM	[101]
P. putida	PmerT/EGFP	Optical (bioluminescence)	Hg^{2+}	200 nM	[102]
E. coli	cadR30/gfp	Optical (fluorescence)	Cd^{2+}, Pb^{2+}, Hg^{2+}	—	[103]
Deinococcus radiodurans	DR_0659/lacZ	Optical (colorimeter)	Cd^{2+}	5.0×10^{-8} M	[104]

lead ions. Cadmium and nickel cause the stress on the DNA confirmation and affects its stability, which is sensed by electrochemical transduction [106,107]. DNA can be used as a biocomponent in its double-standard form (dsDNA) having catalytic property known as DNAzymes or molecular beacons. The DNAzyme biosensors are very popular for the detection of lead ions in various samples. These DNAzymes can be synthesised involving oligonucleotide strands much shorter than the fully extended DNA. The DNAzyme consists of two oligonucleotide strands, where one strand acts as a substrate and the other one has catalytic properties and it is known as the enzymatic strand. In the presence of heavy metal ions, the enzymatic strand acts on the ribonucleotide cleavage site present on the substrate strand causing its cleavage, which can be sensed using a fluorophore and a quencher. In Fig. 10, DNAzyme-based Pb^{2+} biosensor has been shown. The DNAzyme is attached with a fluorophore at $5'$ end of substrate strand and quencher at $3'$ end of enzyme strand, which results in decreased fluorescent signal due to quenching of fluorophore by the quencher, but in the presence of Pb^{2+} ions enzyme strand cleaves the substrate strand at ribonucleotide cleavage site, resulting in decreased quenching and increased fluorescent signal. The DNAzyme and fluorophore are assembled at gold substrate by thiolation, showing detection limit of 1.0×10^{-8} M for Pb^{2+} ions [108]. Researchers have experimented with so many different ssDNA (single-stranded DNA), dsDNA (double-stranded DNA) and DNAzyme-based

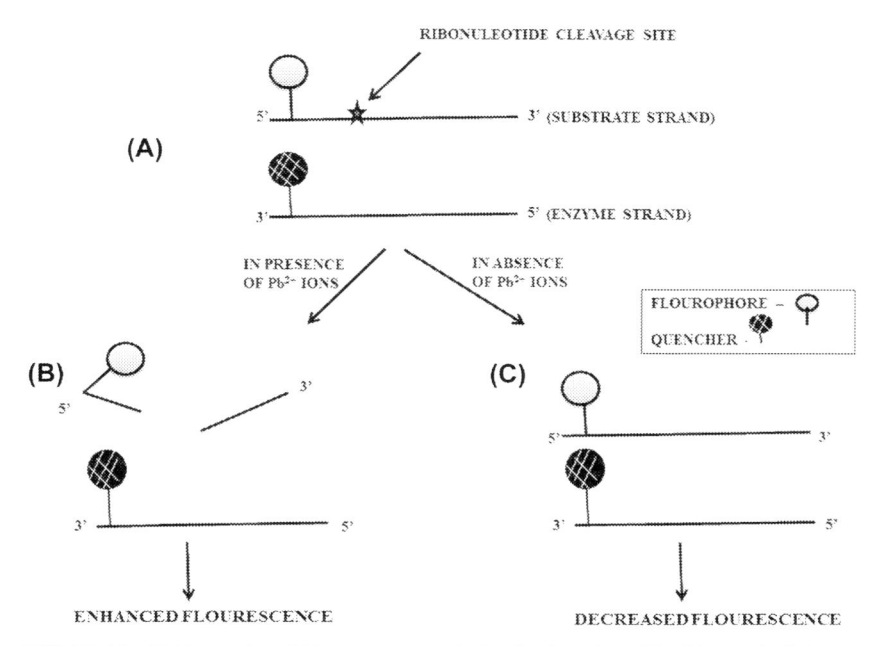

FIGURE 10   DNAzyme-based biosensor approach for the detection of lead ions. (A) Substrate strand attached with a flourophore, enzyme strand attached with a quencher; (B) cleavage of substrate strand at ribonucleotide cleave site by enzyme strand in the presence of lead ions leads to enhanced fluorescent because of no quenching of the flourophore by quencher; (C) decreased fluorescence due to no cleavage and quenching of the flourophore in the absence of lead ions.

biosensors for the heavy metal ion detection, which are being summarised in Table 8. An amperometric ssDNA-based biosensor has been constructed for the study of heavy metal ions and its complexing with DNA, so that these can be determined in blood serum, water and food samples. The method of determination of lead is based on the preconcentration of metal on the biosensor monitored by the destruction of DNA–metal complex by treating with EDTA and taking voltammogram recording of metal–EDTA complex. The lower limit of detection for Pb^{2+} was found to be 1.0×10^{-10} M [109]. In another approach, dsDNA (immobilised on glassy carbon electrode) interaction with Pb^{2+}, Cd^{2+} and Ni^{2+} ions has been reported electrochemically. The change in oxidation peaks of adenine and guanosine bases of dsDNA, causing DNA-oxidative damage and conformationalchange due to interaction with the heavy metal ions is monitored [107]. An electrogenerated chemiluminescence (ECL) biosensor based on the use of ruthenium complex tagged 5′- amino -17 E′ as an ECL probe has been developed. The modified Ru 1-17 E′ and substrate strand were covalently coupled on a graphite electrode modified with 4-aminobenzoic acid. Pb^{2+} ions induced cleavage of substrate strand leads to dissociation of dsDNA complex and hence high ECL signals. The detection limit achieved by the workers is 1.4×10^{-12} M [124].

7.5 Aptamer-Based Biosensor

Aptamers are synthetic oligonucleotides (DNA or RNA) whose specific binding affinity can be exploited for the detection a particular metal ion. Aptamers are constructed *in vitro* using selection evolution of ligands by exponential enrichment (SELEX) technique which makes them specific to bind a particular heavy metal ion [125]. Fig. 11 shows the main steps involved in the construction of aptamers by using SELEX approach. Aptamers are more stable in comparison with enzymes, whole cells, recombinant cells and DNA, having long shelf life, high specificity and affinity for a wide range of targets [69]. The aptamer-based biosensing systems are successful in combination with a suitable transducer. Aptamer-based biosensors have been discussed in Table 9. As an example, for the detection of arsenic compounds in Vietnamese ground water, ssDNA aptamer–based biosensor has been reported. The specific aptamer has been generated from random library in response to the presence of arsenic heavy metal ions. The *in vitro* selection of aptamers is done by using arsenic aptamer affinity column. Out of Ars-1 to Ars-8 selected aptamers, Ars-3 has shown greatest affinity for As^{5+} and As^{3+} ions without the interference of other heavy metals. The detection limit is found to be in the range of micromolars for arsenic compounds with response time of 5 min [126]. In another work, colorimetric detection of Hg^{2+} ions has been done by using anti-Hg^{2+} aptamers and bare gold nanoparticles as colorimetric probe. The generated aptamer (ssDNA) is rich in thymine (T) bases and resulting in T–Hg^{2+}–T complex (dsDNA) in response to the presence of Hg^{2+} ions in the

TABLE 8 DNA-Based Biosensors for Heavy Metal Ion Detection

Bio Component	Immobilisation Technique	Transducer Used	Heavy Metal Ion Detected	Detection Limit	References
ssDNA	Inclusion in cellulose nitrate membrane	Electrochemical (amperometer)	Pb^{2+}, Cd^{2+}	10^{-10} M (Pb) 10^{-9} M (Cd)	[109]
dsDNA	Electrodepositing on glassy carbon electrode	Electrochemical (amperometer)	Pb^{2+}	—	[107]
dsDNA	Assembly at gold substrate	Electrochemical	Cd^{2+}	10 pM	[106]
dsDNA	Electrodepositing on screen-printing electrode	Electrochemical	Sn^{2+}, As^{3+}	—	[110]
DNAzyme+TMR fluorophore	Assembly at gold substrate by thiolation	Optical (fluorescence)	Pb^{2+}	1.0×10^{-8} M	[108]
Intermolecular quencher	Assembly at gold substrate by thiolation	Optical (fluorescence)	Pb^{2+}	1.0×10^{-9} M	[111]
DNAzyme+13 nm GNP	Deposition on gold nanoparticles	Optical (colorimeter)	Pb^{2+}	1.0×10^{-9} M	[112]
DNAzyme+fluorophore	Nanofluid thiolated track	Optical (fluorescence)	Pb^{2+}	1.1×10^{-8} M	[113]
DNAzyme Au coated	Au-coated assembly	Optical (fluorescence)	Pb^{2+}	1.7×10^{-8} M	[114]
DNAzyme complex	Adsorption on gold electrode by thiolation	Electrochemical	Pb^{2+}	3.0×10^{-7} M	[115]

Continued

TABLE 8 DNA-Based Biosensors for Heavy Metal Ion Detection—cont'd

Bio Component	Immobilisation Technique	Transducer Used	Heavy Metal Ion Detected	Detection Limit	References
DNAzyme in sol gel	Entrapment in sol gel	Optical (fluorescence)	Pb^{2+}	–	[116]
ssDNAzyme 17 E+GNP	Adsorption on gold nanoparticles	Optical (Colorimeter)	Pb^{2+}	5.0×10^{-7} M	[117]
HRP mimicking DNA	Gold nanoparticles	Optical (chemiluminescence)	Pb^{2+}	1.0×10^{-8} M	[118]
DNAzyme as catalytic and molecular beacon	–	Optical (fluorescence)	Pb^{2+}	6.0×10^{-10} M	[119]
Gquadruplex DNAzyme PS2.M	–	Optical (chemiluminescence)	Pb^{2+}	3.2×10^{-9} M	[120]
T30695 DNAzyme	–	Optical (fluorescence)	Pb^{2+}	2.0×10^{-8} M	[121]
G4-AGR0100 hemin complex	–	Optical (fluorescence)	Pb^{2+}	4.0×10^{-10} M	[122]
GNP-labelled DNAzyme	Gold substrate assembly	Optical	Pb^{2+}	2.0×10^{-8} M	[123]
Ru1-17EdsDNA complex	Graphite electrode	Optical (chemiluminescence)	Pb^{2+}	1.4×10^{-12} M	[124]

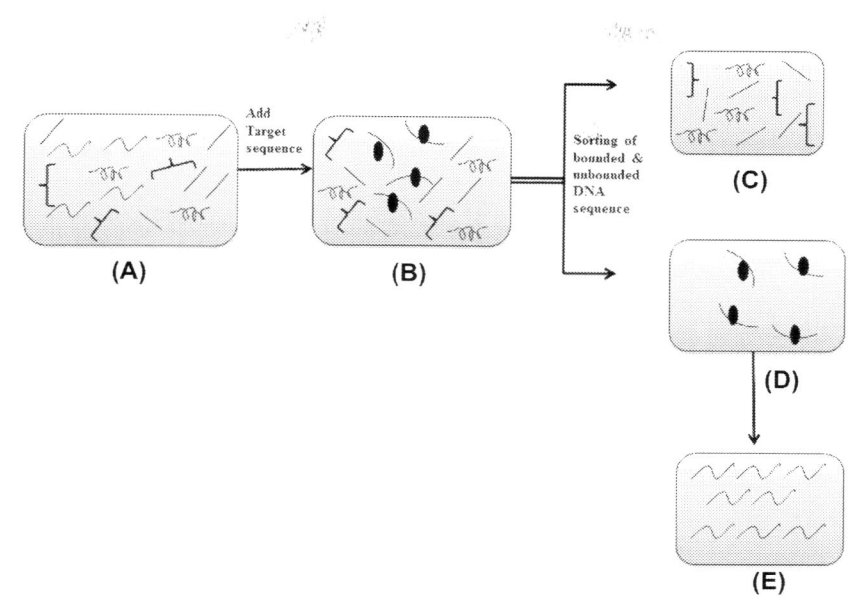

FIGURE 11   Schematic representation of construction of an aptamer. (A) DNA oligonucleotide sequences from DNA library. (B) Incubation with a target sequence: (C) unbounded sequences (D) bounded sequences with a particular target. (E) Amplification of the selected bounded sequences by using PCR.

sample. Hg^{2+} ion-induced change in the confirmation of aptamer, measured by spectrometric method, as bare gold responds differently to ssDNA and dsDNA. The sensor has shown detection limit of 0.6 nM of Hg^{2+} ions with highest selectivity over other metals [127]. Similar research has been reported for the detection of Hg^{2+} ions by using nanogold-aptamer resonance scattering (RS) probe (NGssDNA). The thymine$-Hg^{2+}-$thymine form stable mismatches, resulting in the aggregation of unbounded gold nanoparticles. The more is the concentration of Hg^{2+} ions, lesser is the resonance scattering intensity showing detection limit up to 5 fM [128].

There has been a tremendous growth in the field of specific biosensors for the detection of different chemical compounds like heavy metal ions, pesticides, antibiotics, pathogens and other organic pollutants. The reliability and success of a biosensor depends upon its applicability and commercial value, along with its detection limit, stability and sensitivity. Many researchers have applied biosensors in different water, soil and food samples for the detection of heavy metal ions which are toxic to human beings. Most of the biosensors are broad range, which are able to detect number of potential toxic heavy metal ions, while others are specific for a particular metal. A number of different biosensors reported in literature for the detection of heavy metals ions are discussed.

Although, many researchers have applied their developed biosensors in different water, soil and food samples, but these methods are designed and

TABLE 9 Aptamer (Synthetic Oligonucleotide-Based) Biosensors for Heavy Metal Ions

Biocomponent	Immobilisation Technique	Transducer Used	Heavy Metal Ion Detected	Detection Limit	References
Oligonucleotide (DNA aptamer)	–	Optical (surface plasmon resonance)	As^{3+}	µM	[126]
Oligonucleotide (DNA aptamer)	Gold nanoparticles	Optical (colorimetric)	Hg^{2+}	0.6 nM	[127]
Oligonucleotide (DNA aptamer)	Gold nanoparticles	Optical	Hg^{2+}	5 fM	[128]
Oligonucleotide (DNA aptamer)	–	Optical (fluorescence)	Hg^{2+}	20 µM	[129]
8–17 DNAzyme and adenosine aptamer	–	Optical (fluorescence)	Pb^{2+}	4 nM	[130]
DNA (aptamer)	Grapheme oxide sheets	Optical (fluorescence)	Hg^{2+}	0.92 nM	[131]
DNA aptamer	Gold substrate	Electrochemical	Pb^{2+}, Cd^{2+}	10^{-18} M	[132]

practised mainly at a laboratory scale. The main factors affecting the commercialisation of developed biosensors involve the portability of instrumentation for the infield application. Other factors like stability of the biocomponent are also delaying its applicability at commercial level. There requires a comprehensive study regarding the interface between biocomponent and the transducer, to make these biosensors more popular and effective commercially.

8. CONCLUSION AND SCOPE

In the last few years, there has been a major growth in the development of biosensors to detect various chemical contaminants present in water, soil and food, with a huge consideration toward heavy metal monitoring. Biosensors are very fast, reliable, specific, sensitive, highly selective, compact and sensitive tools to monitor these chemical pollutants, which are hazardous to living systems by entering through the food chain. A large number of biosensors based on enzymes, proteins, antibodies, whole cells, DNA and recombinant systems have been designed and developed in research till date. Some are specific for particular metal ions, while others have a broad range of applicability for different heavy metals. Every biosensing system has its particular property depending upon the suitability with a transducer. Enzyme biosensors are less stable and involve additional cost for purification and isolation of the enzyme, which can be improved by using different immobilisation approaches. Whole-cell biosensors are nonspecific and there is an additional cell membrane barrier for the proper diffusion of substrate inside the cell, which in turns shows slow response. Recombinant-based biosensors are highly specific but involve additional cost and time for the design of a particular reporter gene. Different types of toxins like pesticides and other organic compounds can also interfere with the proper detection of a specific heavy metal ion in the sample. It is being reported that number of enzymes are inhibited by several compounds, as an example AChE activity is affected by Cu^{2+} ions along with organophosphorus pesticides, toxins like oxalic acids and anatoxins. Other compounds like trichloroform, nitric oxide affect the activity of horse radish peroxidase (HRP) along with mercury and methyl mercury compounds, resulting in interference for detecting the exact cause of inhibition [133]. DNAzyme- and aptamer-based biosensors are highly specific for a particular metal ion, but they affect economic value of the system due to the extra preparation steps and selection from the DNA library. Despite the main drawbacks described, biosensors are still considered promising tools for recognising various pollutants and several research efforts are constantly accomplished to enhance their performance with the help of latest trends in nanotechnology, biomimetic chemistry and material science.

9. FUTURE PERSPECTIVES

There is a lot of debate for the absolute real-time monitoring of heavy metal ions at a commercial level, and skilled research is still needed to be done for the success of biosensors over the conventional techniques. In future, broad-range biosensors can be used for primary screening of heavy metals in food samples. Among them DNA- or aptamer-based biosensors at a nanoscale level are going to be the excellent choice for the total detection of these environmental and agrifood pollutants at a very large scale. Hence, the main hope relies on biosensors to monitor these compounds, whose toxic effects can be avoided by determining their presence timely in the food chain and their accumulation in the ecosystem.

ACKNOWLEDGEMENTS

The authors are highly thankful to Department of Science and Technology, New Delhi, India, and Department of Biotechnology, Punjabi University, Patiala, India, for their overall financial support.

REFERENCES

[1] N. Shrivastva, C. Majumder, J. Hazard. Mater. 151 (2008) 1−8.
[2] A. Khan, S. Khan, M. Khan, Z. Qamar, M. Waqas, Environ. Sci. Pollut. Res. 22 (2015) 13772−13799.
[3] I. Saxena, G. Shekhawat, Nitric Oxide 32 (2013) 13−20.
[4] Q. Li, S. Cai, C. Mo, B. Chu, L. Peng, F. Yang, Ecotoxicol. Environ. Saf. 73 (2010) 84−88.
[5] C. Luo, C. Liu, Y. Wang, X. Liu, F. Li, G. Zhang, X. Li, J. Hazard. Mater. 186 (2011) 481−490.
[6] G. Nabulo, C. Black, S. Young, Environ. Pollut. 159 (2011) 368−376.
[7] J. Abuduwaili, Z. Zhang, F. Jiang, PLoS One 10 (2015) 3.
[8] J.P.F. D'Mello, Food Safety: Contaminants and Toxins, Wallingford, 2003.
[9] A. Singh, N. Verma, Curr. Biotechnol. 3 (2014) 127−132.
[10] N. Verma, M. Singh, BioMetals 18 (2005) 121−129.
[11] F. Fu, Q. Wang, J. Environ. Manage. 92 (2011) 407−418.
[12] R. Singh, N. Gautam, A. Mishra, R. Gupta, Indian J. Pharmacol. 43 (2011) 246−253.
[13] D. Song, D. Zhuang, D. Jiang, J. Fu, Q. Wang, Int. J. Environ. Res. Public Health 12 (2015) 7100−7117.
[14] K. Ajah, J. Ademiluyi, C. Nnaji, J. Environ. Health Sci. Eng. 13 (2015) 15.
[15] K. Zhao, W. Fu, Z. Ye, C. Zhang, Int. J. Environ. Res. Public Health 12 (2015) 1577−1594.
[16] I. Martorell, G. Perelló, R. Martí-Cid, J.M. Llobet, V. Castell, J.L. Domingo, Biol. Trace Elem. Res. 142 (2011) 309−322.
[17] M. Intawongse, J.R. Dean, Food Addit. Contam. 23 (2006) 36−48.
[18] I. Ramos, E. Esteban, J.J. Lucena, A. Gárate, Plant Sci. 162 (2002) 761−767.
[19] N. Verma, S. Kumar, H. Kaur, J. Biosens. Bioelectron. 1 (2010) 102.
[20] J. Song, X. Yang, X. Zhang, Y. Long, Y. Zhang, T. Zhang, Int. J. Environ. Res. Public Health 12 (2015) 8243−8262.
[21] D. Sud, G. Mahajan, M.P. Kaur, Bioresour. Technol. 99 (2008) 6017−6027.

[22] B. Gutti, B. Zubairu, D. Buba, Int. J. Sci. Environ. Tech. 3 (2014) 2120−2126.

[23] N. Verma, H. Kaur, S. Kumar, Biotechnology 10 (2011) 259−266.

[24] N. Verma, S. Kumar, H. Kaur, Adv. Appl. Sci. Res. 2 (2011) 354−363.

[25] N. Verma, M. Singh, J. Automated Methods Manage. Chem. 1−4 (2006) 2006.

[26] M. Moyo, Open J. Appl. Biosensor 3 (2014) 1−7.

[27] A.C. Onkowska, G. Gromadzka, J. Buttner, G. Chabik, Arch. Gynecol. Obstet. 281 (2010) 129−134.

[28] N. Verma, A. Singh, P. Kaur, J. Anal. Chem. 70 (2015) 1111−1115.

[29] G.L. Turdean, Int. J. Electrochem. 1-15 (2011) 2011.

[30] J. Wang, J. Pharm. Biomed. Anal. 19 (1999) 47−53.

[31] A. Leung, P.M. Shankar, R. Mutharasan, Sens. Actuators B 125 (2007) 688−703.

[32] T. Anh, S. Dzyadevych, N. Prieur, Mater. Sci. Eng. C 26 (2006) 453−456.

[33] D. Wang, T. Hurst, R. Thompson, C. Fierke, Biomed. Opt. 16 (087011) (2011).

[34] K. Zhang, Cross Sections 2 (2006) 195−206.

[35] C. Malitesta, M. Guascito, Biosens. Bioelectron. 20 (2005) 1643−1647.

[36] Y. Yang, Z. Wang, M. Yang, Sens. Actuators B 114 (2006) 1−8.

[37] C. Guo, J. Wang, J. Chenq, Z. Dai, Biosens. Bioelectron. 36 (2012) 69−74.

[38] X. Wang, S. Xia, J. Zhao, H. Zhao, N. Jaffrezic, Chem. Res. Chin. Univ. 25 (2009) 443−445.

[39] J. Liu, X. Xu, L. Tang, G. Zeng, Trans. Nonferrous Metals Soc. China (English Edition) 19 (2009) 235−240.

[40] L. Shyuan, L. Heng, M. Ahmad, S. Aziz, Z. Ishak, Asian J. Biochem. 3 (2008) 359−365.

[41] D. Bagal-Kestwal, M. Karve, B. Kakade, V. Pillai, Biosens. Bioelectron. 24 (2008) 657−664.

[42] M. Guascito, C. Malitesta, E. Mazzotta, A. Turco, Sens. Actuators B 131 (2008) 394−402.

[43] F. Kuralay, H. Zyoruk, A. Yildiz, Enz. Microb. Tech. 40 (2007) 1156−1159.

[44] H. Tsai, R. Doong, T. Chiang, K. Chen, Anal. Chim. Acta 481 (2003) 75.

[45] H. Mohammadi, A. Amine, S. Cosnier, C. Mousty, Anal. Chim. Acta 543 (2005) 143−149.

[46] E. Akyilmaz, E. Yorganci, E. Asav, Biochemistry 78 (2010) 155−160.

[47] B. Rodriguez, J. Bolbot, I. Tothill, Anal. Bioanal. Chem. 380 (2004) 284−292.

[48] B. Kuswandi, Anal. Bioanal. Chem. 376 (2003) 1104−1110.

[49] F. Garcia, A. Navas, M. Ramos, C. Belledone, Anal. Chim. Acta 484 (2003) 45−51.

[50] S. Lee, W. Lee, Korean Chem. Soc. 23 (8) (2002) 1169.

[51] J. Singh, N. Verma, Portugaliae Electrochim. Acta 26 (2008) 527−532.

[52] M. Ghica, C. Brett, Microchim. Acta 163 (2008) 185−193.

[53] S. Haron, A. Ray, Med. Eng. Phys. 28 (2006) 978−981.

[54] A. Gani, M. Ashari, B. Kuswandi, Sensor Lett. 8 (2010) 320−327.

[55] A. Berezhetsky, O. Sosovska, C. Durrieu, J. Chovelon, S. Dzyzderzych, C. Tran- Mink, IRBM 29 (2008) 136−140.

[56] Y. Berezovskaya, C. Armstrong, A. Boyle, M. Porrini, D. Woolfson, P. Barran, Chem. Comm. 47 (2011) 412−414.

[57] C. Wu, L. Lin, Biosens. Bioelectron. 20 (2004) 864−871.

[58] J. Yin, W. Wei, X. Liu, B. Kong, L. Wu, S. Gong, Anal. Biochem. 360 (2007) 99−104.

[59] E. Chow, D. Hibbert, J. Gooding, Anal. Chim. Acta 53 (2005) 167−176.

[60] T. Lin, M. Chung, Biosens. Bioelectron. 24 (2009) 1213−1218.

[61] H. Zeng, R. Thompson, B. Maliwal, G. Fones, J. Mosffet, C. Fierke, Anal Chem. 75 (2003) 6807−6812.

[62] C. Michel, A. Ouerd, F. Battaglia-Brunet, Biosens. Bioelectron. 22 (2006) 285−290.

[63] I. Bontidean, J. Ahlqvist, A. Mulchandani, W. Chen, W. Bae, R. Mehra, A. Mortari, E. Csoregi, Biosens. Bioelectron. 18 (2003) 547−553.

[64] A. Changela, K. Chen, Y. Xue, J. Holschen, C. Outten, T. Halloran, A. Mondragon, Science 301 (2003) 1383−1386.

[65] M. Khosraviani, A. Pavlov, G. Flowers, D. Blake, Environ. Sci. Tech. 32 (1998) 137−142.

[66] D. Blake, R. Jones, R. Blake, A. Pavlov, I. Darwish, H. Yu, Biosens. Bioelectron. 16 (2001) 799−809.

[67] S. Melton, H. Yu, K. Williams, S. Morris, P. Long, D. Blake, Environ. Sci. Technol. 43 (2009) 6703−6709.

[68] M. Farre, L. Kantiani, S. Perez, D. Barcelo, Trends Anal. Chem. 28 (2009) 170−185.

[69] N. Verma, A. Bhardwaj, Biosens. Technol. Pestic. (2015). http://dx.doi.org/10.1007/s12010-015-1489-2.

[70] S. Rodriguez, M. Marco, M. Lopez, D. Barcelo, Pure Appl. Chem. 76 (2004) 723−752.

[71] K. Yagi, Appl. Microbiol. Biotechnol. 73 (2007) 1251−1258.

[72] I. Gammoudi, H. Tarbague, J. Lachaud, S. Destor, A. Othmane, D. Moynet, R. Kalfat, D. Rebiere, C. Dejous, Sensor Lett. 9 (2011) 816−819.

[73] I. Tsybulskii, M. Sazykina, Appl. Biochem. Microbiol. 46 (5) (2010) 505−510.

[74] S. Oh, S. Hassan, S. Ginkel, Sens. Actuators B: Chem. 154 (2011) 17−21.

[75] M. Souiri, I. Gammoudi, H. Quada, Procedia Chem. 1 (2009) 1027−1030.

[76] R. Zlatev, J. Magnin, P. Ozil, M. Stoytcheva, Biosens. Bioelectron. 21 (2006) 1501−1506.

[77] R. Bhatia, J. Dilleen, A. Atkinson, D. Rawson, Biosens. Bioelectron. 18 (2003) 667−674.

[78] J. Singh, S. Mittal, Sens. Actuators B: Chem. 165 (2012) 48−52.

[79] H. Nguyen, C. Durrieu, C. Tran, Ecotoxicol. Environ. Saf. 72 (2009) 316−320.

[80] C. Chouteau, S. Dzyadevych, J. Chovelon, C. Durrieu, Biosens. Bioelectron. 19 (2004) 1089−1096.

[81] S. Alpat, S. Alpat, B. Kutlu, O. Zbayrak, H. Buyukisik, Sens. Actuators B 128 (2007) 273−278.

[82] M. Yuce, H. Nazir, G. Donmez, Biosens. Bioelectron. 26 (2010) 321−326.

[83] M. Yuce, H. Nazir, G. Donmez, Bioelectrochemistry 79 (2010) 66−70.

[84] M. Yuce, H. Nazir, G. Donmez, Biores. Technol. 101 (2010) 7551−7555.

[85] J. Adams, A. Adeli, C. Hsu, R. Harkness, G. Page, C. Depamphilis, E. Schultz, C. Yuceer, Plant Biotechnol. J. 10 (2011) 207−216.

[86] Q. Liua, H. Cai, Y. Xua, L. Xiaoa, M. Yangc, P. Wang, Biosens. Bioelectron. 22 (2007) 3224−3229.

[87] N. Verma, M. Singh, A Rapid and Disposable Microbial Biosensor for Determination of Copper in Electroplating Industrial Effluents, Indian Patent No 197519, 2006.

[88] T. Chakraborty, P. Babu, A. Alam, A. Chaudhari, Curr. Sci. 94 (2008) 800−805.

[89] P. Liu, Q. Huanga, W. Chen, Environ. Pollut. 164 (2012) 66−72.

[90] Q. Hu, L. Li, Y. Wang, W. Zhao, H. Qi, G. Zhuang, J. Environ. Sci. 22 (2010) 1469−1474.

[91] P. Barrocas, W. Landing, J. Hudson, J. Environ. Sci. 22 (2010) 1137−1143.

[92] R. Chapleau, M. Sagermann, Toxicology 261 (2009) 136−142.

[93] F. Chinalia, G. Paton, K. Killham, Biores. Tech. 99 (2008) 714−721.

[94] K. Tag, K. Riedel, H. Bauer, G. Hanke, K. Baronian, G. Kunze, Sens. Actuators B 122 (2007) 403−409.

[95] V. Liao, M. Chien, Y. Tseng, K. Ou, Environ. Pollut. 142 (2006) 17−23.

[96] J. Sumner, N. Westerberg, A. Stoddard, Biosens. Bioelectron. 21 (2006) 1302−1308.

[97] T. Pellinen, T. Huovinen, M. Karp, Anal. Biochem. 330 (2004) 52−57.

[98] J. Stoyanov, D. Magnani, M. Solioz, FEBS Lett. 546 (2003) 391−394.

[99] S. Leth, S. Maltoni, R. Simkus, Electroanalysis 14 (2002) 35−42.

[100] N. Hillson, P. Hu, G. Andersen, L. Shapiro, Appl. Environ. Microbiol. 73 (2007) 7615−7621.

[101] Y. Fu, C. Wen-Li, H. Qiao-Yun, Microbiol. Biotechnol. 79 (2008) 363−370.

[102] H. Wei, H. Cheng, T. Mao, W. Zhong, X. Lin, Microbiol. Biotechnol. 87 (2010) 981−989.

[103] C. Raja, G. Selvam, Int. J. Environ. Sci. Tech. 8 (2011) 793−798.

[104] M. Joe, K. Lee, S. Lim, S. Im, H. Song, I. Lee, D. Kim, Bioprocess Biosyst. Eng. 35 (2012) 265−272.

[105] M. Tencaliec, S. Laschi, V. Magearu, M. Mascini, Talanta 69 (2006) 365−369.

[106] E. Wong, E. Chow, J. Gooding, Electrochem. Comm. 9 (2007) 845−849.

[107] S. Oliveira, O. Corduneanu, A. Oliveira-Brett, Bioelectrochemistry 72 (2008) 53−58.

[108] J. Liu, Y. Lu, J. Am. Chem. Soc. 122 (2000) 10466−10467.

[109] S. Babkina, N. Ulakhovich, Bioelectrochemistry 63 (2004) 261−265.

[110] A. Ferancova, M. Adamovski, P. Grundler, Bioelectrochemistry 71 (2007) 33−37.

[111] J. Liu, Y. Lu, Anal. Chem. 75 (2003) 6666−6672.

[112] J. Liu, Y. Lu, J. Am. Chem. Soc. 125 (2003) 6642−6643.

[113] I. Chang, J. Tulock, J. Liu, W. Kim, D. Cannon, Y. Lu, P. Bohn, J. Sweedler, D. Cropek, Environ. Sci. Technol. 39 (2005) 3756−3761.

[114] D. Wernette, C. Swearingen, D. Cropek, Y. Lu, J. Sweedler, P. Bohn, Analyst 131 (2006) 41−47.

[115] Y. Xiao, A. Rowe, K. Plaxco, J. Am. Chem. Soc. 129 (2006) 262−263.

[116] Y. Shen, G. Mackey, N. Rupcich, D. Gloster, W. Chiuman, Y. Li, J. Brennan, Anal. Chem. 79 (2007) 3494−3503.

[117] H. Wei, B. Li, J. Li, S. Dong, E. Wang, Nanotechnology 19 (2008) 95501−95505.

[118] J. Elbaz, B. Shlyahovsky, I. Willner, Chem. Commun. (2008) 1569−1571.

[119] X. Zhang, Z. Wang, H. Xing, Y. Xiang, Y. Lu, Anal. Chem. 82 (2010) 5005−5011.

[120] T. Li, E. Wang, S. Dong, Anal. Chem. 82 (2010) 1515−1520.

[121] T. Li, S. Dong, E. Wang, J. Am. Chem. Soc. 132 (2010) 13156−13157.

[122] C. Li, K. Liu, Y. Lin, H. Chang, Anal. Chem. 83 (2011) 225−230.

[123] Y. Wang, J. Irudayaraj, Chem. Commun. 47 (2011) 4394−4396.

[124] F. Ma, B. Sun, H. Qi, H. Zang, Q. Gao, C. Zang, Anal. Chim. Acta 683 (2011) 234−241.

[125] A. Sassolas, B. Prieto, J. Marty, Am. J. Anal. Chem. 3 (2012) 210−232.

[126] M. Kim, H. Um, S. Bang, S. Lee, S. Oh, J. Han, Y. Kim, Environ. Sci. Technol. 43 (2009) 9335−9340.

[127] L. Li, B. Li, Y. Qi, Y. Jin, Anal. Bioanal. Chem. 393 (2009) 2051−2057.

[128] Z. Jiang, G. Wen, Y. Fan, C. Jiang, Q. Liu, Z. Huang, A. Liang, Talanta 80 (2010) 1287−1291.

[129] J. Xu, Z. Song, Y. Fang, J. Mei, L. Jia, A. Qin, B. Tang, Analyst 135 (2010) 3002−3007.

[130] Y. Xiang, A. Tong, Y. Lu, J. Am. Chem. Soc. 131 (2009) 15352−15357.

[131] C. Lia, W. Zhoub, S. Guob, N. Wua, Biosens. Bioelectron. 41 (2013) 889−893.

[132] J. Hasen, J. Wang, A. Kawde, Y. Xiang, K. Gothelf, G. Collins, J. Am. Chem. Soc. 128 (2006) 2228−2229.

[133] A. Amine, H. Mohammadi, I. Bourais, G. Palleschi, Biosens. Bioelectron. 21 (2006) 1405−1423.

Chapter 3

Sensing of Biological Contaminants: Pathogens and Toxins

N.M. Saucedo and A. Mulchandani*
University of California, Riverside, CA, United States
*Corresponding author: E-mail: adani@engr.ucr.edu

Chapter Outline

1. INTRODUCTION

1.1 Prevalence of Pathogen Contamination

Food safety regulations set by the hazard analysis critical control point (HACCP) stipulate that food products must be tested for pathogens [1].

Contamination of food products with pathogens such as bacteria, pathogen products (toxins), viruses and fungi is a serious matter costing upwards of 10 billion dollars in medical costs and productivity losses for those infected [1]. Different from other contaminants such as heavy metals and pesticides, bacterial contamination can occur throughout the handling processes of food products and therefore, there exists more opportunity of exposure and subsequent contamination with bacteria and their toxins. For the purpose of brevity and because bacterial pathogens constitute 80% of hospitalizations and 77% of deaths attributed to confirmed and reported pathogenic infections [2], their detection will be the focus of this chapter. The most common foodborne bacteria are *Salmonella*, *Clostridium*, *Escherichia coli* such as O157:H7, *Campylobacter*, *Staphylococcus aureus*, *Shigella*, *Listeria* spp., *Listeria monocytogenes* and *Yersinia enterocolitica*, while the most common toxins are *E. coli* Shiga toxin, *S. aureus* enterotoxin and *Vibrio cholera* choleragen/cholera toxin [3–7]. Contamination of food products can occur anywhere in their handling from harvesting to packaging or storage and distribution. Therefore, detection methods are strategically implemented throughout the handling of the products and before distribution as HACCP regulates [7,8].

1.2 Current Methods for Their Detection in Food Industry

Current detection methods include enzyme-linked immunosorbent assay (ELISA) and polymerase chain reaction (PCR), the positive results of which are verified by culturing and plating [8]. Table 1 shows the sensitivities of these established methods along with methods not addressed in this chapter. Briefly, ELISA-based methods typically use antibodies to capture the analyte onto a surface. A secondary antibody, which is labelled with an enzyme, is then introduced to bind onto another site/epitope of the captured target, forming a 'sandwich'. The corresponding substrate is added to react with the enzyme and produce a colour change. The intensity of the colour change is semiquantitative in its relation to the concentration of bacteria present in the sample. This method is favoured because of its high specificity and sensitivity and ease of interpretation. For PCR, samples must first be subjected to a DNA extraction process (improved PCR methods include automation of DNA extraction step), followed by a series of DNA denaturing and copying cycles. The DNA amplification is visualized through gel electrophoresis. This method also has high sensitivity and selectivity. The efficiencies of these methods are increased when the identification of the bacteria is known, ie, a specific bacteria is targeted for detection [9]. This allows technicians to properly choose the antibodies in the case of ELISA and primers (short DNA sequences) used for selective DNA amplification in PCR. Highly effective and selective antibodies have been isolated for common foodborne pathogens.

TABLE 1 Sensitivities of Established Methods for Detection of Bacteria and Toxins

Assay	Bacteria (cells/g)	Toxins (ng/mL)
Culture	10^7-10^8	NA[a]
Adenosine triphosphate (ATP)	10^4	NA
Latex agglutination (LA)	10^7	NA
Reverse passive LA (RPLA)	NA	0.5–4.0
Enzyme-linked immunosorbent assay (ELISA)	10^4-10^7	0.01–1.0
Immunomagnetic separation (IMS)	$<10^3$	NA
Immunodiffusion (1–2 test)	10^5-10^6	5–100
Immunoprecipitation (Ab-ppt)	10^4-10^8	NA
DNA probe	10^4-10^6	NA
Phage (*lux* or *ina*)	10^1-10^2	NA
Polymerase chain reaction (PCR)	10^1-10^2	NA
Biosensor	10^1-10^2	NA

[a]NA, information not available or applicable. Shows the best reported or achieved detection limits for common methods for detection of bacteria and their toxins.
Table reprinted from P.C.H. Feng, in: V.K. Juneja, J.P. Cherry, M.H. Tunick (Eds.), Advances in Microbial Food Safety, ACS Symposium Series 931, American Chemical Society, Washington, DC, 2006, pp 14–37 (Chapter 2). Reference J. Sofos (Ed.), Advances in Microbial Food Safety, UK, Cambridge, 2006 in this work.

1.3 Difficulties of Pathogen Detection in Food Samples

In addition to the inherent limitations of these methods, such as long assay times and complexity, detection of bacteria and their toxins in food samples possess additional challenges. Preparation of food samples often results in turbid solutions due to the sample matrix, interference from sample matrix and low cell count. Therefore, sample preparation often requires separation of bacteria and/or their toxins from the matrix. However, the variety of matrices, liquids, fats, oils and proteins affects the efficiency of separation and concentration. These challenges also affect the performance of ELISAs and PCR.

As for culturing and plating, there are limitations in using culturing as the 'final word' for verifying positive ELISA or PCR results. For example, exposure to low temperatures or acidic conditions may stress the bacteria, resulting in the need for higher initial concentrations of bacteria and extending growth times from a few hours to days to obtain results [9]. There are also fastidious bacteria which cannot be easily grown in vitro and those designated as 'viable but nonculturable'. A method which can ascertain the viability of the cells accurately and reliably in a shorter time period would improve the overall flow of production as it will result in shorter waiting times in production halts due to pathogen-positive results. As well, methods that are not affected by sample turbidity and complex matrices would increase method robustness.

2. COMMON BIOSENSORS AND BIORECEPTORS

Biosensors can achieve faster and comparable sensitivities to current methods. They are robust and their signal is easy to interpret making them well suited for routine use in food industry for determining bacterial contamination. Fig. 1 shows a schematic for biosensing. The two major components of a biosensor, the bioreceptor, a biomolecule which interacts selectively with the target analyte, and the transducer which converts the interaction event into a measurable signal, help achieve this process [10]. One of the methods of categorizing biosensors is by transduction mechanism. The most common biosensors are mass based, optical and electrochemical [11]. Table 2 provides a comparison of the various biosensors for pathogen sensing.

Many bioreceptors have been used to introduce selectivity into biosensors. Common bioreceptors include antibodies [12], DNA [13] and enzymes [14]. These bioreceptors are highly specific and sensitive to their target. One benefit of using antibodies or DNA is their high specificity towards a particular target even in a mixture of similar targets. However, if these bioreceptors were designed to have broader specificities, their affinities would decrease and may result in sensitivities below the desired limit of detection (LOD) [15]. Though comparatively costly, antibodies continue to be the most popular bioreceptors. They are used preferentially in separation and immunoassays such as immunomagnetic separation (IMS) and latex agglutination, respectively [8].

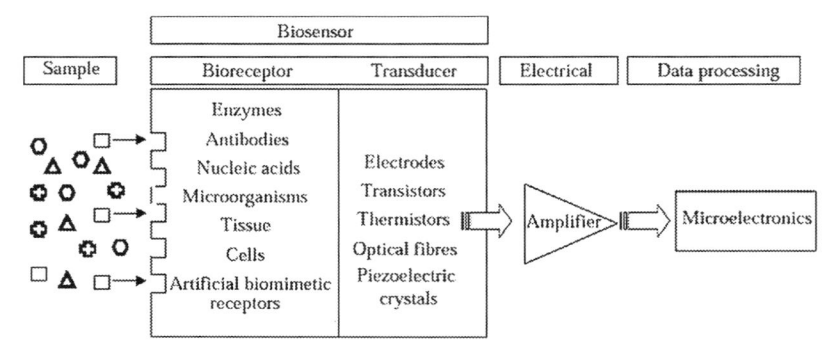

FIGURE 1   Biosensor schematic. *Used with permission from M.N. Velasco-Garcia, T. Mottram, Biosyst. Eng. 84 (1) (2003) 1–12, Fig. 1.*

TABLE 2 Comparison of Various Biosensor Performance for Common Pathogens

Target	Detection Method	Receptor	Matrix	Sensitivity	Assay Time	References
Salmonella	QCM	Aptamer	Milk	100 cfu/mL	Pre-enrichment ~30 min Detection ~few minutes	[21]
Campylobacter	SPR	Antibody	Meat	10^3 cfu/mL	45 min	[31]
Escherichia coli such as *O157:H7*	QCM	AMP	Water	400 cfu/mL	10 min	[22]
Listeria monocytogenes	EIS	AMP	Milk	10^3 cfu/mL	13 min	[44]
Staphylococcus aureus	EIS	DNA	N/A	100 fM	Incubation 30 min Detection ~few minutes	[53]
Rotavirus	FET	Antibody	Goose feces	10^3 pfu/mL	17 min	[54]
AFB1	QCM	Antibody	Groundnuts	0.1 ng/mL	Sample preparation ~1.5 h Detection ~few minutes	[24]

AMP, antimicrobial peptide; *EIS*, electrochemical impedance spectroscopy; *FET*, field-effect transistor; *QCM*, quartz crystal microbalance; *SPR*, surface plasmon resonance.

Aptamers [16] and lectins [17] are emerging as new bioreceptors for pathogen sensing. Aptamers are short synthetic oligonucleotides, RNA or DNA, which bind targets with high affinities and specificity comparable to antibodies [18]. Primary advantages of aptamers over antibodies include elimination of animal host or expensive hybridoma culture; selection using modern combinatorial chemistry tools and better stability. Lectins are proteins which bind mono- and oligosaccharides selectively and reversibly [19]. These bioreceptors can probe the surface of the target for saccharide composition. They have been used to study binding of pathogens in the body and distinguish between cancerous and normal tissue cells [20]. The detection mechanisms for antibodies, DNA, aptamers and lectins rely on affinity binding. These biological interaction events can be detected by a biosensor transducer to generate an electrical signal.

2.1 Mass-Based Biosensors

2.1.1 Quartz Crystal Microbalance

Quartz crystal microbalance (QCM) is an example of piezoelectric devices where mass accumulation on the surface results in attenuation of crystal oscillation frequency, which can be translated to electric current. The change in frequency can be used to quantify the concentration of the analyte in a sample. These devices have been used to detect bacteria in water and milk.

Ozalp et al. [21] demonstrated detection of *Salmonella* in milk samples using an aptamer. To achieve a detection of 100 cfu/mL milk, a pre-enrichment step utilizing a *Salmonella*-specific aptamer as the biorecognition element on QCM transducer and bioreceptor in IMS was employed. The beads and cells were allowed to interact for 30 min with shaking. The cells were then released from the beads into 1 mL phosphate buffer saline (PBS) using a 20 nM NaOH regeneration solution. This high pH disrupts the binding of the aptamer to the target. Using this method, *Salmonella* can be detected selectively as shown by comparison of frequency changes when the same beads were used in *E. coli* spiked milk samples (Fig. 2). However, detection of bacteria through direct injection of spiked milk samples (no separation step) resulted in much higher changes in frequency due to the adhesion of milk components onto the QCM (Fig. 2). Therefore, the separation step is necessary.

Dong et al. [22] demonstrated the use of an antimicrobial peptide (AMP) as bioreceptor for the detection of *E. coli* in water samples. AMPs are typically 10−40 residues in length and positively charged. They bind bacteria via the negatively charged head groups on the bacterial lipids. Injection and incubation of spiked water samples (microlitre volumes) onto a QCM device achieved a LOD of 400 cfu/mL in 10 min.

QCM can also be used to detect toxins produced by pathogens. Microbial toxins are proteins of smaller sizes. This poses a challenge for their detection.

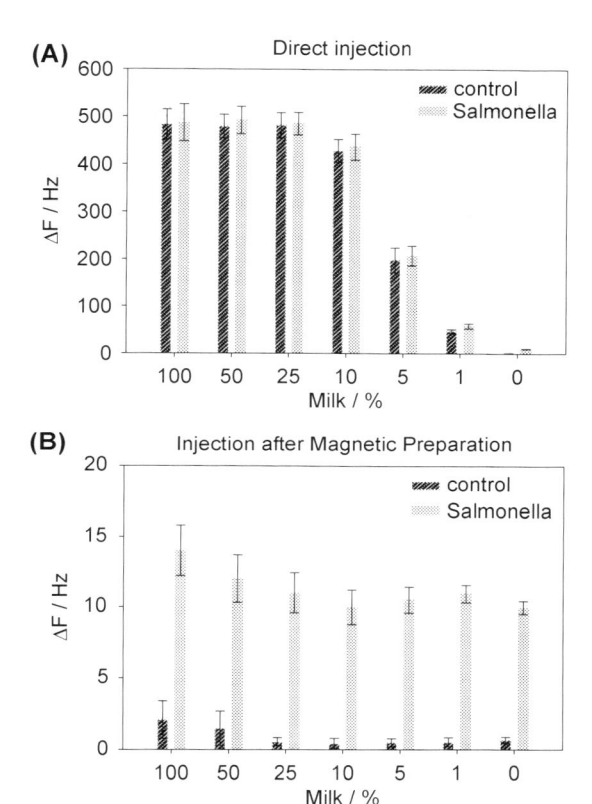

FIGURE 2 Frequency changes for samples of varying percentage concentrations of milk spiked with 10^4 cfu/mL *Salmonella* bacterial cells. (A) Spiked milk samples directly incubated onto quartz crystal microbalance (QCM) surface. (B) Spiked milk samples subjected to immuno-magnetic separation to separate *Salmonella* cells prior to exposure to QCM surface. *Used with permission from V.C. Ozalp, G. Bayramoglu, Z. Erdem, M.Y. Arica, Anal. Chim. Acta 853 (2015) 533–540, Fig. 7.*

A group of toxins of particular concern are aflatoxins which are produced as metabolites by *Aspergillus flavus* and are classified as fungal mycotoxins. They are highly toxic, mutagenic and carcinogenic [23]. Typically, chromatography or immunoassays such as ELISAs are used for their detection. Chauhan et al. [24] demonstrated the detection of aflatoxin B1 (AFB1) in groundnuts. Ten grams of contaminated groundnuts were ground and sonicated in a methanol/water/hexane and sodium chloride mixture for 30 min. The sample was centrifuged briefly at 10,000 rpm. The AFB1 was extracted from the supernatant via a column. It was dried and subsequently redissolved in acetonitrile/water mixture and exposed to antibody-functionalized QCM.

Using a 0.1-M glycine-HCl solution the biosensor surface was regenerated and reused. The lower concentrations successfully detected were 0.1–4 ng/mL, which are within the European Commission regulated limit of 2 µg/kg.

2.1.2 Cantilevers

Cantilevers are a newer type of mass-based sensors/transducers. Cantilevers are small planks or levers protruding from a larger base with typical dimensions in the micrometre range. Cantilever-based biosensors are operated in either static or dynamic mode. In the former, which has been the most widely used for molecular affinities identification, the addition of mass induces a deflection of the cantilever from a stable baseline. While highly sensitive, this arrangement is not preferred as it is more complex requiring a sophisticated laser setup. In the latter, the mass addition is correlated to the shift in the resonance frequency of the cantilever. Fig. 3 illustrates a schematic of cantilever setup and detection using change in frequency.

There has been a growing interest in lectin proteins as bioreceptors for the capture and detection of pathogens. Lectins bind selectively to saccharides. This lectin–carbohydrate interaction can be used to target the saccharide structures on bacteria and virus capsids/envelopes. Mader et al. [25] used this interaction in a clever way by functionalizing the surface of cantilevers with saccharides to bind bacteria through their surface lectins. In this way, the detection becomes more selective as only some bacteria have lectins, and each lectin binds saccharides in different affinity. Thus by changing the saccharide on the cantilever, other lectins and therefore bacterial cells can be specifically targeted. Mader et al. demonstrate the capture of *E. coli* bacterial cells through mannose-functionalized cantilevers. This interaction is well known and reportedly occurs through the Fimbriae H lectin (Fim H) located on the pili of the cells [26]. For controls, the pili were truncated at various lengths to see its effect on binding. It was found that as the pili were truncated, the specificity for the cell decreased. However, full truncation led to an increase in unspecific binding of cells due to their attraction to the cantilever surface. Cantilevers were also functionalized with galactose (Fig. 3) to show the specificity of Fim H, as it is reported that Fim H does not bind galactose [25]. A consistent LOD of 8×10^3 cfu/mL was achieved.

2.1.3 Summary of Mass-Based Biosensor Sensitivities

The signals of mass-based detection are improved upon by the miniaturization of these devices. However, smaller devices mean decreased interaction with target and may be difficult to use when bacterial concentrations are low [7]. US federal regulations dictate a maximum acceptable concentration of 10 cfu/mL *E. coli* in milk samples [27] and a zero tolerance for *E. coli* in beef samples. Similarly, the maximum concentration of 5 cfu/mL *E. coli* for milk samples [28] required by European Commission regulation cannot be met with

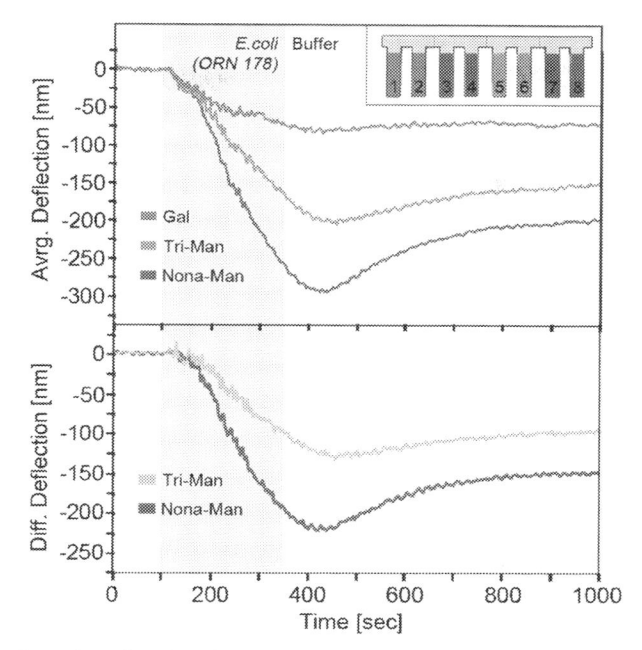

FIGURE 3 Detection of *Escherichia coli* strain ORN 178 with glycan cantilever arrays. Upper panel: Averaged deflections (Avrg. Deflection) for galactose, trimannose and nonamannose-functionalized cantilevers of an array against time. Each graph represents an average signal of 2−4 identically functionalized cantilevers. Upon *E. coli* sample injection (OD = 0.5) the mannose cantilevers react significantly stronger (two to three times) than the galactose reference due to increased sample recognition. Inset: Scheme of a cantilever array functionalized with different carbohydrates. In this example, cantilevers 1 and 2 are coated with the internal reference galactose; cantilevers 3, 4, 7 and 8 with nonamannose and cantilevers 5 and 6 with trimannose, respectively. Lower panel: Differential deflections (Diff. Deflection) representing the specific binding events for the trimannose and nonamannose sensors derived by subtracting the galactose reference. *Used with permission from A. Mader, K. Gruber, R. Castelli, B.A. Hermann, P.H. Seeberger, J.O. Rädler, M. Leisner, Nano Lett. 12 (1) (2011) 420−423, Fig. 1.*

the reported detection limits. Mass-based biosensors must be improved to be a practical method for pathogen detection in the food industry. Furthermore, there is also the challenge of small volumes (few microlitres) represent a much larger portion of food, as bacteria is often localized and not present uniformly throughout the food sample. Therefore, the determined concentration may not be a true representation of the bacterial concentration present in the food product as a whole. This also speaks to the challenges of developing methods for consistent sampling and sample preparation for food quality assessment.

2.2 Optical Biosensors

2.2.1 Surface Plasmon Resonance

Surface plasmon resonance (SPR) has been used for a wide variety of applications such as food analysis and drug discovery. It is an optical detection method which due to its reliability and reproducibility has become a well-established technique for use in biomolecular interactions studies and low-concentration analytes. Surface plasmons arise from free electrons at the interface of two materials. Metal, because of its high electron conductivity, is the material of choice for exploiting this phenomenon and is coated as a thin film on one side of a glass prism. The plasmons are excited by light, as can be seen in the simple SPR configuration illustrated in Fig. 4A [29]. Here, the light travelling through a prism is aligned to reach an angle, called the SPR angle, where it resonates with the oscillating electrons. The intensity of the light reflected reaches a minimum at this point (Fig. 4B). Either the change in SPR angle or the intensity of light at the initial SPR angle can be monitored for change. The 'top' of the SPR sensor is designated as the side with the gold film which is used as the surface for functionalizing bioreceptors and capturing the target. The incident light can be reflected through the prism ('back') or through the medium ('top'). The configuration where light is applied through the back is called the Kretschmann configuration and is preferred due to its higher signal stability [30]. The surface plasmon occurs at both interfaces (prism/metal and metal/water) but is enhanced on the side with the material of lower refractive index (water). The penetration depth of the evanescent field depends on variables such as wavelength of incident light and refractive indices of materials (which for liquids are also temperature dependent) and are usually in the 200−400 nm range. Any changes in the refractive index within this penetration depth will cause a shift in the SPR angle. The shift in SPR angle is proportional to the concentration of absorbed analyte. This shift can be monitored in real time to produce a signal (Fig. 4C), called a sensorgram.

Dong et al. [31] demonstrated the detection of *Campylobacter jejuni* in meat samples. Fifty grams of chicken was washed in 200 mL of PBS. Four millilitres of that wash were taken out and further allocated to prepare blanks and spiked samples. The samples, whose spiked bacterial concentrations ranged from 0 to 10^7 cfu/mL, were introduced by a liquid flow system and incubated on the antibody-functionalized SPR surface for 45 min followed by a rinsing step. A LOD of 10^3 cfu/mL was achieved. The reported time of 45 min is one of the longest times reported for SPR detection. Typically, SPR signals stabilize within 15−20 min given that samples are clean and contain minimal matrix [32]. Leonard et al. [33] were able to improve upon the detection of *L. monocytogenes* via SPR by first exposing rabbit anti-*L. monocytogenes* antibodies to a sample, collecting the bacterial cell−antibody complex, then introducing the complex onto an anti-rabbit Fab antibody functionalized SPR surface. In this way, the research group maximized on obtaining mass, namely the antibodies, within the penetration depth of the evanescent field. They reported a LOD of 10^5 cfu/mL. As can be seen, LODs heavily depend on the target.

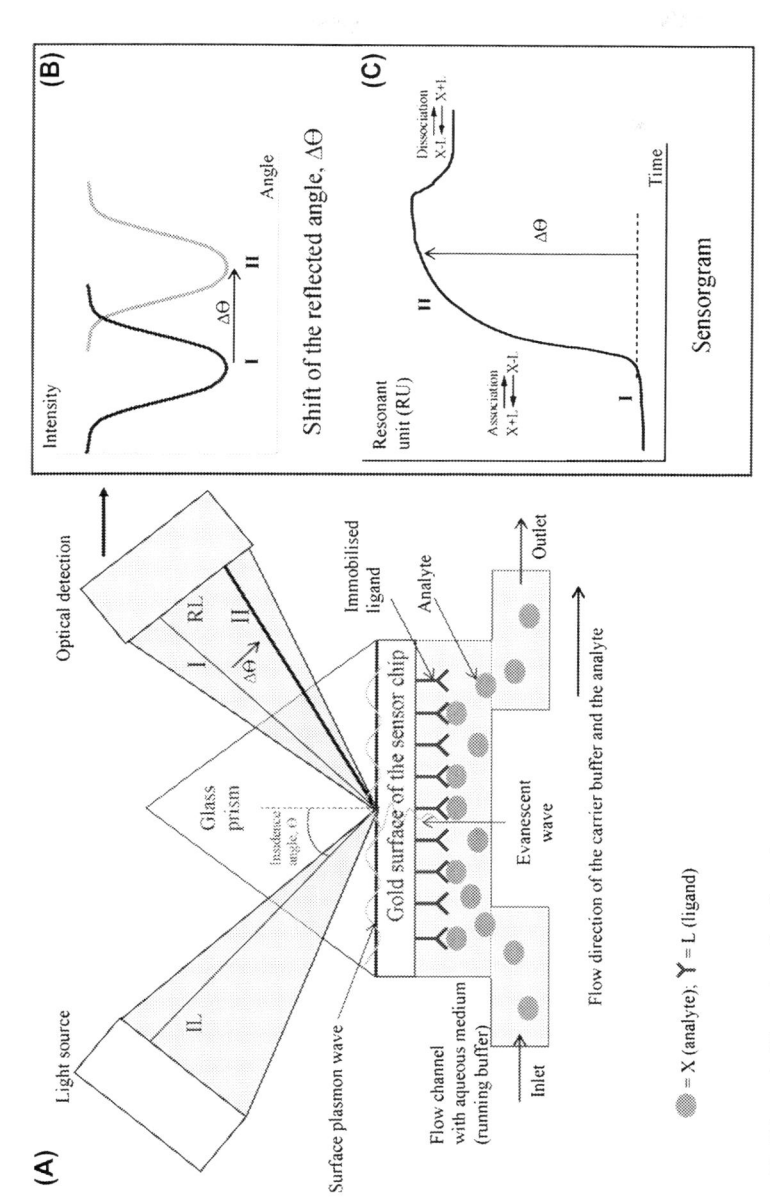

FIGURE 4 (A) Schematic illustration of surface plasmon resonance (SPR) mechanism and setup in Kretschmann configuration. Incident light (IL) is directed through the prism which is coated with a thin layer of gold at an angle at an angle (θ). Part of the reflected light (RL) energy is transferred to the surface plasmons. (B) SPR curve shows intensity minimum and angle shift (Δθ). Minimum intensity shifts from I to II when the composition (and thus the refractive index) near the surface of the gold film changes. (C) Sensorgram-real-time monitoring of SPR angle shift. The analyte (X) is introduced to the SPR surface immobilized with ligand (L) via a microfluidic system. As the analyte binds with its ligand (association X+L), the angle of reflection increases. Subsequently, some dissociation will occur after the surface is rinsed and equilibrium is reached. *Used with permission from G. Safina, Anal. Chim. Acta 712 (2012) 9–29, Fig. 1.*

2.2.2 Improvements Upon SPR

Instrumentation and components for this direct, label-free technique are commercially available adding to its widespread use which now includes its infancy in food industry. It has been said that SPR has reached its limits for sensitivity [34] but many have been working to cleverly circumvent the challenge of using a small evanescent field to detect large bacterial cells, a combination which seems incompatible. Such efforts include the development of gratings to enhance surface plasmons [35]. Grating or patterning the gold instead of depositing a continuous film on the glass provides 'hot spots' located between the patterns allowing for mass to accumulate well within the 400-nm height cutoff point. Others [36–38] have replaced the prism with an optical fibre. Srivastava et al. [36] demonstrated their setup as having the optical fibre coated with a thin film of gold which was placed inside a glass chamber having inlet and outlet channels. The light was focused to enter at one end of the fibre while the transmitted power was detected at the other end for various incident light angles. In regards to the complexity of setting up a light source, self-contained SPR devices have been developed to simplify SPR use [39]. Though mass-based and optical biosensors are gaining more acceptance for use in the food industry, positive results are still accompanied by verification through traditional methods [35]. This is reflective of food safety regulations and their diligence to human health.

2.3 Electrochemical Biosensors

Because of low cost, ease of use, portability and construction simplicity, biosensors based on electrochemical transducer have found widespread applications. There are four major types of electrochemical transducers: amperometric, potentiometric, conductometric and electrochemical impedance spectroscopy (EIS). The availability of these different configurations for electrochemical detection makes these methods ideal for use in the food industry where sample matrices and targeted analyte are inherently diverse. Electrochemical methods have demonstrated great performance in detection of pathogens achieved in shorter times and with increased sensitivity (depending on sample matrix and bacterial strain) when compared to other biosensors. A good example is EIS.

2.3.1 Electrochemical Impedance Spectroscopy

EIS is a characterization technique used for many material systems and applications. In electrochemical sensors, EIS is used to monitor modulation in electrical properties resulting from biorecognition events at the electrode surface. In an immunosensor, EIS is used for monitoring binding between antibody and antigen at the electrode surface that are proportional to the concentration of measured species [40–43]. Work by Etayash et al. [44]

showed that EIS can be used along with AMPs to effectively and selectively detect Gram-positive bacterial cells including *S. aureus*, *Listeria innocua*, *Enterococcus faecalis* and *L. monocytogenes*. Even so, the signals from each respective Gram-positive species resulted in unique binding kinetics, which can be used to distinguish between them as can be seen from Fig. 5. Pure milk

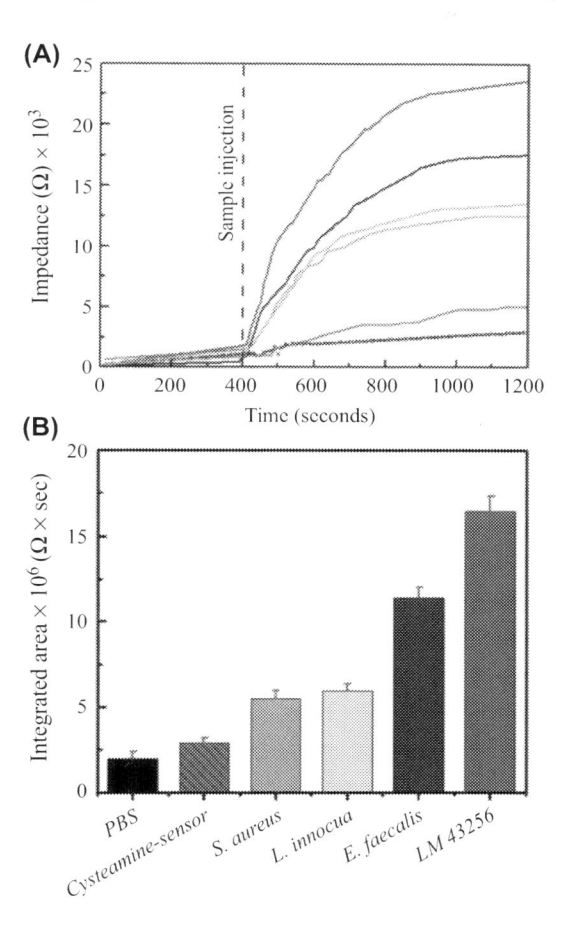

FIGURE 5 (A) Real-time impedimetric response of the peptide sensor to various bacterial species (10^3 cfu/mL) at 100 Hz. The red (darker grey in print versions), blue (darkest grey in print versions), green (light grey in print versions) and orange (grey in print versions) signals correspond to the peptide sensor response to *Listeria monocytogenes* (LM 43256), *Enterococcus faecalis*, *Listeria innocua* and *Staphylococcus aureus*, respectively. The pink (dark grey in print versions)-coloured signal is the control sensor response to *L. monocytogenes*, and the *dashed black line* is the peptide sensor signal against the phosphate buffer saline (PBS) solution. (B) The binding curve parameter (integrated area under the curve) for impedance sensor responses to the corresponding strains. Note that each impedance response is an average calculation of five replicates and *error bars* indicate corresponding standard deviations. *Used with permission H. Etayash, K. Jiang, T. Thundat, K. Kaur, Anal. Chem. 86 (3) (2014) 1693—1700, Fig. 4.*

samples showed interference through nonspecific binding of milk components. Systematic dilution of pure milk down to 10% proved to be optimal for minimizing interference. Spiking of diluted milk samples with 10^3 cfu/mL *L. monocytogenes* showed good differentiation from blank and determined as the LOD. Regeneration of the device was achieved by treatment with 70% ethanol which proved to be too harsh and damaged the electronics of the device after its second use (signal dropped to 50% of original by third trial).

2.3.2 Potentiometric Biosensors

In the electrochemical technique of potentiometry, the potential difference between the reference electrode and the working electrode arising from charge accumulation as a result of biorecognition is correlated to the analyte concentration. Zelada-Guillen et al. [16] used aptamer-functionalized carbon nanotubes (CNTs) on glassy carbon electrode (GCE) to detect *E. coli* in milk. The aptamers, upon exposure to their targets, change conformation as they wrap around their target. This alters the charge density near the electrode surface and thus results in a modulation in potential difference between the modified GCE and the Ag/AgCl reference electrode. In this work, bacterial cells were filtered out from spiked milk and apple juice samples via sterile cellulose acetate filters of 0.45-μm pore size. The concentrations used were 1 cfu/5 mL to 10^6 cfu/mL. The lower concentrations were reported using the final volume of the solution in the cell. The collected cells were washed with PBS and then dislodged by addition of PBS from the back side of the filter and collected in a fixed volume. The cells were then exposed to the aptamer-functionalized working electrode by introducing them into the solution. This method is unconventional as current methods first incubate the working electrode with the sample then rinse and reintroduce into the electrochemical cell. This allows for more interaction between the target and the bioreceptor and therefore a higher possibility of target binding. However, a calibration curve for increasing *E. coli* concentration in PBS resulted in an average response of 1.87 mV per log of bacteria concentration. The lowest *E. coli* concentration detected in real samples was 12 cfu isolated from 2 mL of spiked milk and 26 cfu isolated from 1 mL of apple juice. The specificity of the sensor was demonstrated by testing against nontarget bacteria, which included other strains of *E. coli* and *Lactobacillus casei* at a concentration of 10^3 cfu/mL (Fig. 6), which showed minimal binding.

2.4 Field-Effect Transistor/Chemiresistor Biosensors

In 1972 Bergveld introduced a device that combined a field-effect transistor (FET) and a glass electrode and named it ion selective field-effect transistor. In a biosensor, the ion selective membrane is modified with or replaced by a biological recognition element. In an FET immunosensor the binding event

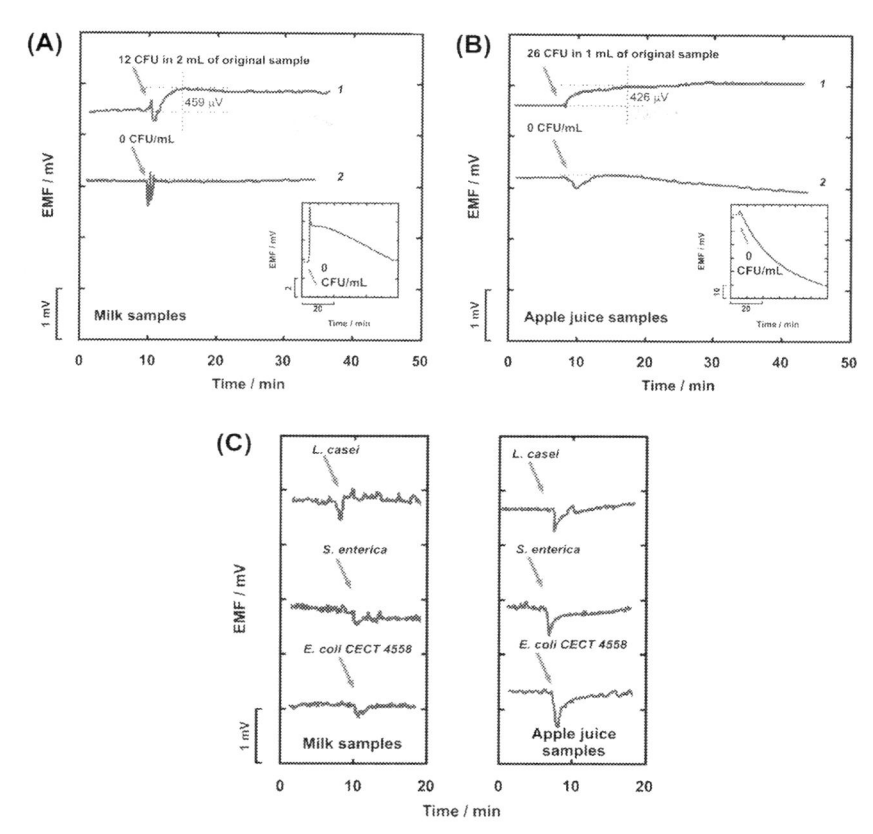

FIGURE 6 Potentiometric detection of microorganisms using aptasensors exposed to real samples (pretreatment steps previously carried out to remove original matrix). (A) Sample of milk: original matrix containing inoculated *Escherichia coli* CECT 675 12 CFU in 2 mL of sample (1) and without inoculated bacteria (2). (B) Sample of apple juice: original matrix containing inoculated *E. coli* CECT 675 26 CFU in 1 mL of sample (1) and without inoculated bacteria (2). Insets in (A) and (B), respectively, show the potentiometric response for milk and apple juice samples if no preprocessing steps are carried out in a sample without target bacteria. (C) Selectivity assays in real samples (milk in left-hand chart, and apple juice in right-hand chart): original matrix containing 10^3 CFU/mL of the bacteria *Lactobacillus casei*, *Salmonella enterica*, and *E. coli* CECT 4558. The bacteria were counted using the standard plate count method in triplicate. *Used with permission from G.A. Zelada-Guillén, S.V. Bhosale, J. Riu, F.X. Rius, Anal. Chem. 82 (22) (2010) 9254–9260, Fig. 3.*

between the antibody and antigen influences the presence of accumulated charge carriers at the gate surface in proportion to analyte concentration. Chemiresistive devices are a simplified version of FET containing only the source and gate electrodes bridged by a semiconductor channel without a gate electrode. Recently, one-dimensional (CNTs, silicon and conducting polymer nanowires) and two-dimensional (graphene, transition metal dichalcogenides) nanostructure-based FETs have gained significant attention as a powerful sensing platform for label-free immunosensing [45−47].

The methods discussed so far have exemplified detection of bacteria via their physical capture onto a bioreceptor-modified surface. However, chemiresistive devices are proving to possess the sensitivity, robustness and simplicity needed to probe cell viability via their metabolism [48]. In this way, a positive signal would indicate the detection of live bacteria. This distinction is important in the food industry to determine the survival of bacterial cells through critical points such as refrigeration and preservation in low-pH environments. For example, *Listeria* can grow in low-temperature environments, and many more bacteria can withstand low-pH environments, most notably *S. aureus* and *Salmonella* [36].

Huang et al. [49] constructed a graphene-based chemiresistive device and determined the sensitivity of the material to changes in pH. It was concluded that the electronic properties, such as conductance, of graphene were dependent on the pH of the environment (Fig. 7). To exploit this sensitivity, anti-*E. coli* antibodies were used to capture cells onto its surface selectively and then glucose was added to induce metabolism. The excreted organic acids produced a change in conductance which was recorded by monitoring the current traversing the channel. Glucose is a neutral molecule therefore its addition to the solution does not result in a change of pH and does not interfere with detection. An LOD of 10 cfu/mL was achieved using this method which supersedes sensitivities of PCR and ELISA and other biosensors.

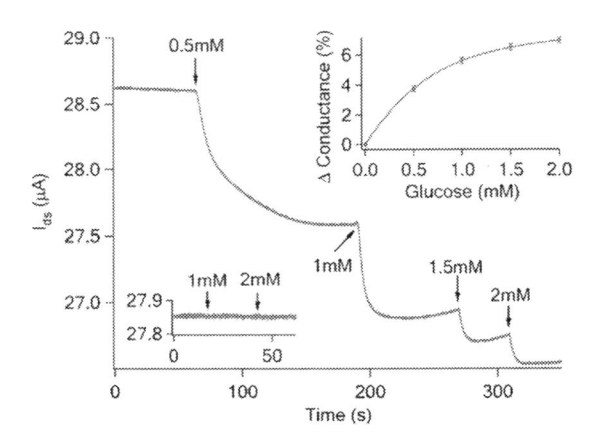

FIGURE 7 Real-time current recording (V_{ds} 100 mV and V_g 0 V) of a graphene device immobilized with *Escherichia coli* bacteria (incubated with 10^5 cfu/mL *E. coli* for 30 min followed by rinsing), with application of glucose to the phosphate buffer saline (PBS) recording buffer at the indicated time points to reach the final concentrations of 0.5, 1, 1.5 and 2 mM. Lower inset: Bacteria-free graphene sensor was not responsive to glucose. Upper inset: Percentage change in graphene conductance versus glucose concentration. Each data point is the average from three devices. The *error bars* indicate the standard errors. *Used with permission from Y. Huang, X. Dong, Y. Liu, L.J. Li, P. Chen, J. Mater. Chem. 21 (33) (2011) 12358−12362, Fig. 6.*

3. CONCLUSIONS AND FUTURE PERSPECTIVES

Biosensors are capable of meeting the demands of food safety assessment in regards to detection of pathogen contamination. There are a variety of platforms which facilitate the development of customizable biosensors to perform well under very particular environments and situations, such as low pH or temperature and high turbidity. The examples presented in this chapter allude to the directions and improvements towards which many groups are working. Bioreceptor arrays [17,25,50] are the current solutions for moving away from single-target systems. Bioreceptor arrays have individually addressable electrodes each functionalized with a different bioreceptor. The bioreceptors would bind targets, such as bacteria, with different affinities resulting in unique binding profiles. Therefore, these arrays could be used for multiple targets and thus minimize cost of inventory as they would decrease the use of single-target biosensors. Electrode responses are analysed with a pattern recognition tool making simultaneous detection possible [51]. The transduction material is also very important to the performance of a biosensor. Discovery and development of new materials and the exploitation of their electronic properties can provide increased sensitivity [46,49,52] and achieve LODs at or below regulation values. Such success is necessary for their application in food safety assessment in part due to the loss of target (bacterial cells) that results from separation and sample cleanup techniques. Electrochemical biosensors have shown the most promise with their ability to distinguish between live and dead cells, creating a possible alternative to or reducing the need for culturing and plating. Within the last 10 years biosensors have made good advancement towards the sensitive detection of foodborne pathogens and toxins. Their robustness and cost-effectiveness afford them a continually bright future.

ACKNOWLEDGEMENTS

We acknowledge the financial support of the National Science Foundation (1144635, 1265044 and 1307671), U.S. Department of Agriculture (2014-67021-21589) and W. Ruel Johnson Chair in Environmental Engineering.

REFERENCES

[1] FSIS, USDA, Pathogen reduction; Hazard analysis and critical control point (HACCP) systems, preliminary regulatory impact assessment for Docket No. 93-016P, Fed. Regist. 60 (23) (1995) 6774—6889.

[2] L.H. Gould, K.A. Walsh, A.R. Vieira, K. Herman, I.T. Williams, A.J. Hall, D. Cole, MMWR Surveill. Summ. 62 (2) (2013) 1—34.

[3] Bacterial Foodborne Disease: Medical Costs and Productivity Losses, Agricultural Economic Reports, 741, 2005.

[4] P.A. Meyer, P.W. Yoon, R.B. Kaufmann, Centers for Disease Control and Prevention (CDC), MMWR Surveill. Summ. 62 (3) (2013) 3—5.

[5] Centers for Disease Control and Prevention (CDC), Surveillance for Foodborne Disease Outbreaks, United States, 2013, Annual Report, US Department of Health and Human Services, CDC, Atlanta, Georgia, 2015.

[6] S. Escott-Stump, Nutrition and Diagnosis-Related Care, United States of America, Philadelphia, 2012.

[7] L. Serna-Cock, J.G. Perenguez-Verdugo, Biosensors applications in agri-food industry, in: V. Somerset (Ed.), Environmental Biosensors, InTech, 2011, pp. 43−64.

[8] J. Sofos (Ed.), Advances in Microbial Food Safety, 2006. UK, Cambridge.

[9] I. Safarik, M. Safarikova, S.J. Forsythe, J. Appl. Bacteriol. 78 (1995), 575−585.

[10] M.N. Velasco-Garcia, T. Mottram, Biosyst. Eng. 84 (1) (2003) 1−12.

[11] A. Mortari, L. Lorenzelli, Biosens. Bioelectron. 60 (2014) 8−21.

[12] Y. Wan, Z. Lin, D. Zhang, Y. Wang, B. Hou, Biosens. Bioelectron. 26 (5) (2011) 1959−1964.

[13] H. Peng, C. Soeller, N. Vigar, P.A. Kilmartin, M.B. Cannell, G.A. Bowmaker, R.P. Cooney, J. Travas-Sejdic, Biosens. Bioelectron. 20 (9) (2005) 1821−1828.

[14] J.S. Caygill, S.D. Collyer, J.L. Holmes, F. Davis, S.P. Higson, Analyst 138 (1) (2013) 346−352.

[15] P. Ramnani, Y. Gao, M. Ozsoz, A. Mulchandani, Anal. Chem. 85 (17) (2013) 8061−8064.

[16] G.A. Zelada-Guillén, S.V. Bhosale, J. Riu, F.X. Rius, Anal. Chem. 82 (22) (2010) 9254−9260.

[17] E. Peter, S.R. Mikkelsen, Anal. Chem. 73 (17) (2001) 4241−4248.

[18] A. Geiger, P. Burgstaller, H. von der Eltz, A. Roeder, M. Famulok, Nucleic Acids Res. 24 (6) (1996) 1029−1036.

[19] L. Halina, N. Sharon, Chem. Rev. 98 (2) (1998) 637−674.

[20] L.R.A. Lima, M.F. Bezerra, S.M.V. Almeida, L.P.B.G. Silva, E.I.C. Beltrão, L.B. Carvalho Júnior, Dis. Markers 35 (3) (2013) 149−154.

[21] V.C. Ozalp, G. Bayramoglu, Z. Erdem, M.Y. Arica, Anal. Chim. Acta 853 (2015) 533−540.

[22] Z.M. Dong, G.C. Zhao, Talanta 137 (2015) 55−61.

[23] International Agency for Research on Cancer, Monograph on the Evaluation of Carcinogenic Risk to Humans, 56: 245, Lyon, 1993.

[24] R. Chauhan, P.R. Solanki, J. Singh, I. Mukherjee, T. Basu, B.D. Malhotra, Food Control 52 (2015) 60−70.

[25] A. Mader, K. Gruber, R. Castelli, B.A. Hermann, P.H. Seeberger, J.O. Rädler, M. Leisner, Nano Lett. 12 (1) (2011) 420−423.

[26] B. Mukhopadhyay, M.B. Martins, R. Karamanska, D.A. Russell, R.A. Field, Tetrahedron Lett. 50 (8) (2009) 886−889.

[27] Food and Drug Administration, US Public Health Service, Grade "A" Pasteurized Milk Ordinance, 2011.

[28] C. Regulation, 2073/2005 of 15 November 2005 on microbiological criteria for foodstuffs, Off. J. Eur. Union 338 (2005) 1−29.

[29] G. Safina, Anal. Chim. Acta 712 (2012) 9−29.

[30] B.K. Lavine, D.J. Westover, L. Oxenford, N. Mirjankar, N. Kaval, Microchem. J. 86 (2007) 147−155.

[31] W. Dong, O.A. Oyarzabal, T.S. Huang, S. Balasubramanian, S. Sista, A.L. Simonian, J. Microbiol. Methods 69 (1) (2007) 78−85.

[32] P.A. Van Der Merwe, Surface plasmon resonance, in: S. Harding, P.Z. Chowdhry (Eds.), Protein-Ligand Interactions: Hydrodynamics and Calorimetry, Oxford University Press, Oxford, 2001, pp. 137−170.

[33] P. Leonard, S. Hearty, J. Quinn, R. O'Kennedy, Biosens. Bioelectron. 19 (10) (2004) 1331–1335.
[34] M. Piliarik, J. Homola, Opt. Express 17 (19) (2009) 16505–16517.
[35] S. Moon, Y. Oh, D. Kim, H. Lee, H.C. Kim, K. Lee, SPIE NanoScience + Engineering, International Society for Optics and Photonics, 2011, pp. 809914–809917.
[36] K.S. Srivastava, R. Verma, B.D. Gupta, Sens. Actuators B 153 (1) (2011) 194–198.
[37] T.B. Tims, D.V. Lim, J. Microbiol. Methods 55 (1) (2003) 141–147.
[38] T. Geng, M.T. Morgan, A.K. Bhunia, Appl. Environ. Microbiol. 70 (10) (2004) 6138–6146.
[39] A.M. Sesay, D.C. Cullen, Environ. Monit. Assess. 70 (1–2) (2001) 83–92.
[40] X. Muñoz-Berbel, N. Godino, O. Laczka, E. Baldrich, F.X. Muñoz, F.J. Del Campo, Impedance-based biosensors for pathogen detection, in: M. Zouroub, S. Elwary, A.P.F. Turner (Eds.), Principles of Bacterial Detection: Biosensors, Recognition Receptors and Microsystems, Springer-Verlag, New York, 2008, pp. 341–376.
[41] D. Wenchao, W. Tang, G. Zhao, Electrochim. Acta 97 (2013) 79–85.
[42] N. Zainudin, A.R.M. Hairul, M.M. Yusoff, L.L. Tan, K.F. Chong, Anal. Methods 6 (19) (2014) 7935–7941.
[43] F. Xi, J. Gao, J. Wang, Z. Wang, J. Electroanal. Chem. 656 (1) (2011) 252–257.
[44] H. Etayash, K. Jiang, T. Thundat, K. Kaur, Anal. Chem. 86 (3) (2014) 1693–1700.
[45] P. Ramnani, N.M. Saucedo, A. Mulchandani, Chemosphere 143 (2016) 85–98.
[46] A. Mulchandani, N.V. Myung, Curr. Opin. Biotechnol. 22 (2011) 502–508.
[47] T.T. Tran, A. Mulchandani, TrAC Trends in Analytical Chemistry (2015). http://dx.doi.org/10.1016/j.trac.2015.12.002.
[48] Y. Huang, H.G. Sudibya, P. Chen, Biosens. Bioelectron. 26 (10) (2011) 4257–4261.
[49] Y. Huang, X. Dong, Y. Liu, L.J. Li, P. Chen, J. Mater. Chem. 21 (33) (2011) 12358–12362.
[50] J. Gao, D. Liu, Z. Wang, Anal. Chem. 82 (22) (2010) 9240–9247.
[51] S. Haswell (Ed.), Practical Guide to Chemometrics, CRC Press, 1992.
[52] J. Chang, S. Mao, Y. Zhang, S. Cui, G. Zhou, X. Wu, C.H. Yang, J. Chen, Nanoscale 5 (9) (2013) 3620–3626.
[53] Z. Wang, J. Zhang, P. Chen, X. Zhou, Y. Yang, S. Wu, L. Niu, Y. Han, L. Wang, P. Chen, F. Boey, Q. Zhang, B. Liedberg, H. Zhang, Biosens. Bioelectron. 26 (9) (2011) 3881–3886.
[54] F. Liu, Y.H. Kim, D.S. Cheon, T.S. Seo, Sens. Actuators B 186 (2013) 252–257.

Chapter 4

Biosensor-Based Technologies for the Detection of Pathogens and Toxins

K.L.M. Moran,[a] J. Fitzgerald,[a] D.A. McPartlin, J.H. Loftus and R. O'Kennedy*

Dublin City University, Dublin, Ireland
Corresponding author: E-mail: richard.okennedy@dcu.ie

Chapter Outline

---

[a] These authors contributed equally to this work.

Comprehensive Analytical Chemistry, Vol. 74. http://dx.doi.org/10.1016/bs.coac.2016.04.002

1. INTRODUCTION

Microbial pathogens include bacteria, viruses and protozoa, and failure to detect them can have severe impacts on public health and safety. In the food or water services industries, legislation developed by the appropriate associated regulatory bodies to monitor and control the presence of these microorganisms is vital. Rapid and cost-efficient detection methods, with high-throughput capacity, are essential to implement effective monitoring systems to protect human health. Such an approach would act as an early warning system for the presence of contamination, particularly in food manufacturing plants, or in potable water, to concurrently confirm the effectiveness of sanitary procedures and personnel practices, and to ensure production and quality methods used are effective.

Exposure to pathogenic organisms in water can occur during recreational swimming, through drinking water or also through consumption of shellfish/fish from pathogen-impaired estuarine waters. In 2012, the Environmental Protection Agency (EPA) released new Recreational Water Quality Criteria recommendations for protecting human health in waters designated for primary contact recreation. The bacterial indicators monitored under these criteria for recreational waters are *Escherichia coli* and Enterococci. Further important waterborne bacteria include *E. coli* 0157, Campylobacter and Salmonella. *Clostridium perfringens* is often used as a bacterial faecal indicator in water. Protozoa routinely detected in water include *Cryptosprydium parvum* and Giardia. Viruses potentially infective to humans present in animal waste include hepatitis, E virus, reoviruses, rotaviruses, caliciviruses (noroviruses), adenoviruses, enteroviruses and retroviruses [1].

Food monitoring programmes for pathogenic microorganism are routinely carried out by the European Food Safety Authority (EFSA). EFSA monitors and analyses the situation in relation to zoonoses, zoonotic microorganisms, antimicrobial resistance, microbiological contaminants and foodborne outbreaks across Europe. EU-wide baseline surveys are carried out to obtain fully comparable figures of the prevalence of zoonotic microorganisms in food. EFSA analyses and publishes the results of these surveys covering the prevalence of Salmonella, Campylobacter and Staphylococcus. In the United States, the National Antimicrobial Resistance Monitoring System for Enteric Bacteria (NARMS) was established in 1996 providing collaboration between the state/local public health departments, the Centre for Disease Control and Prevention (CDC), the US Food and Drug Administration (FDA) and the US Department of Agriculture (USDA). Acting as a national public health surveillance system, NARMS tracks changes in the antimicrobial susceptibility of certain enteric (intestinal) bacteria found in ill people (CDC), retail meats (FDA) and food animals (USDA). These include Salmonella, Campylobacter, Shigella, *E. coli* O157, Enterococci and Vibrio, and the program aims to provide information regarding emerging bacterial resistance, the spread of resistance and how resistant infection differs from susceptible infection.

To perform large-scale monitoring programs on a routine-basis, rapid methods for the detection of pathogens need to be established. Conventional detection methods are primarily culture-based as they are reliable, sensitive to the target organism and can be applied to a wide range of matrices. However, these methods are laborious, require specially trained personnel, can have a long turnaround time and are not suitable for use at the 'point-of-need'. Polymerase chain reaction (PCR)-based methods are gaining attention as they are highly specific, sensitive, have rapid turnaround time and can be automated. However, they also require trained personnel and laboratory equipment [2]. In recent years, biosensors have become prominent, notably for their ability to perform sample-to-answer analyses, their high-throughput capacity, speed of operation and portability.

2. BIOSENSORS

Biosensors are analytic tools consisting of a biological recognition element coupled to a physical or chemical signalling system referred to as a transducer and a read-out system (Fig. 1). A biological recognition element is selected and immobilized within the biosensor. It is capable of binding the analyte (contaminant of interest) which the user seeks to identify and quantify. Interaction of the target and biorecognition element generates a signal facilitated by the transducer, which is interpreted and delivered to the user by a read-out device. A simple result is produced that represents the extent of the binding event that originated between the immobilized recognition element

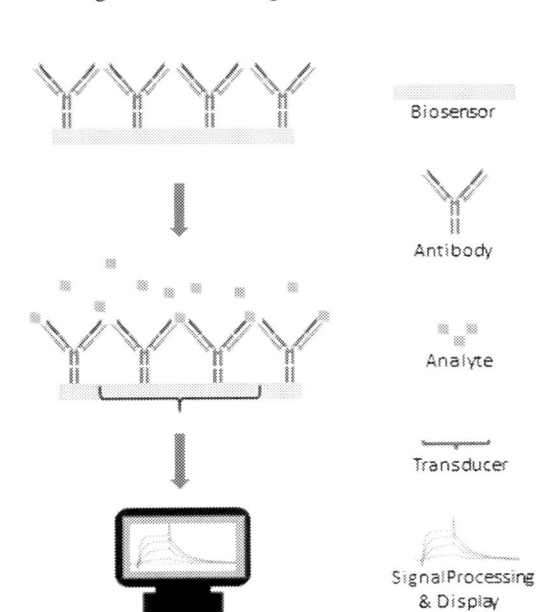

FIGURE 1 Biosensor Components. The selected biological recognition element (antibodies shown) is immobilized on the surface of the biosensor. The antibody binds the target analyte producing a signal based on physical or chemical changes, which are transduced and interpreted by a read-out system into a simple output.

and the compound of interest [3—7]. Biosensors, therefore, offer an excellent resource for converting biological activity into a quantifiable signal. This process permits sensitive detection of the presence (or absence) of pathogenic and/or toxic contaminants. Biosensor-based methods are attractive in the food/water industries, as they can be readily deployed to detect contamination and to ensure food producers take the correct measures to guarantee food safety. To date, the use of biosensors has shown significant promise for food analysis, quality control in agricultural production, and bioprocess and environmental monitoring [8—10].

2.1 Biological Recognition Element

Biosensors are categorized by their biological recognition element and/or their transduction system [11,12]. There are a wide variety of biological recognition elements including enzymes, antibodies, aptamers, ssDNA, RNA, receptors, cells, microorganisms, biological tissue and organelles [13—16]. The biological recognition element predicates the effectiveness of the transduction system and, therefore, the usefulness of the biosensor for detection purposes. Detailed reviews of biological recognition elements have been previously described [4,17,18]. Biosensors are designed to be highly selective as the biological recognition element can be tailored specifically to the analyte of interest whilst the biosensor surface can also be modified to enable correct ligand orientation and to reduce nonspecific binding, increasing the bioreceptor performance in terms of stability and sensitivity [19].

There are many potential variations in signal transduction approaches that can be utilized when developing a biosensor device, which may depend on the performance criteria for the biosensor and its intended use. When antibodies or antibody fragments are applied as the biological recognition element, the device is known as an immunosensor. The most significant obstacles in immunosensor development are related to immobilization, orientation and retention of the specific properties of immunomolecules on transducer surfaces. The optimum density and orientation of antibodies at immunosensor surfaces is a critical parameter. A wide variety of materials can be used as sensing surfaces for immunosensor applications and these include, but are not limited to polymers, metals, glass and plastics. Therefore, antibody immobilization techniques must be tailored for the type of material used at the sensor surface and a suitable method of immobilization derived. It is vital to achieve optimal biomolecule immobilization and, therefore, maximum antibody functionality [20,21].

2.2 Immobilization of Biorecognition Elements in Sensors

Biosensing methods rely on key elements for successful operation and these include a successful biomolecule immobilization strategy, biorecognition in complex matrices, platform design, the presence or absence of labels and the detection strategy employed and all are dependent on the platform used. Various methods of immobilization of biological components at biosensing surfaces have been reported [22]. These include 'adsorption', whereby the biological component is directly adsorbed onto a suitable surface, 'entrapment', where the biological component is trapped within a matrix, 'microencapsulation' in which the biological element is trapped between two membranes, 'covalent attachment' involving direct immobilization via a series of chemical bonds between the surface and the biological component, and 'cross-linking', where the biological component is chemically bonded to the sensor surface either directly or via the use of a bifunctional agent.

2.3 Transduction Systems

The availability of different biosensor transduction systems based on electrochemical and optical phenomena has proven fundamental in the emergence of biosensors for detection and monitoring [23−26]. These transduction methods can be further categorized into subdivisions providing many platforms for the development of a variety of sensitive and specific pathogen and toxin-detecting biosensors.

2.3.1 Electrochemical Biosensors

Electrochemical biosensors currently dominate the biosensing field with the most common approaches including amperometric, potentiometric, conductometric and impedimetric techniques [27−29]. Electrochemical reactions in biosensors function by eliciting a measurable current (amperometric), a measurable charge accumulation or potential (potentiometric), modifying conductive properties of a medium (conductometric) or impedimetric, by measuring resistance and reactance which combine to form impedance. A biosensor using electrochemical transduction typically requires a working electrode, a counter (or auxiliary) electrode and a reference electrode. The reference electrode is maintained at a distance from the site of the biological recognition element and analyte interaction to establish a known and stable potential. The working electrode acts as the transduction component when the interaction occurs whereas the counter electrode measures current and facilitates delivery of electrolytic solution to allow current transfer to the working electrode. The working and counter electrodes should be conductive and chemically stable and are mostly composed of carbon or inert metals like gold and platinum whilst the reference electrode is typically silver or silver chloride [19,30].

2.3.2 Amperometric Transduction

Amperometric techniques achieve signal transduction by exploiting the ability of certain analytes to be oxidized or reduced in a biochemical reaction at the working electrode. The biological recognition element is immobilized at the working electrode and, as analyte binding interactions occur, the current produced reflects the reaction occurring between both. An applied potential serves as the driving force for the electron transfer reaction, and the current produced is a direct measure of the rate of electron transfer. For example, if the working electrode is driven towards a positive potential an oxidation reaction arises and the concentration of the analyte influences the current flow. Conversely, if the working electrode is driven to a negative potential a reduction reaction arises. The current is measured against the reference electrode at a given, known potential [28,30]. However, if the current is measured at known variations of the potential, such transduction approaches are called voltammetry. Not all analytes act as redox partners for electrochemical reactions. Therefore, the use of mediators for direct and indirect transduction is necessitated. However, such systems still maintain excellent sensitivity [19,31]. An amperometric biosensing strip was developed and optimized as an inexpensive, rapid method to detect *Listeria monocytogenes* in food samples [32]. Screen-printed carbon electrode (SPCE) strips were modified with gold nanoparticles (AuNPs) and specific antibodies to *L. monocytogenes* to create a larger electrode reaction surface, assist in electron transfer and increase conductivity, thereby lowering the limit of detection (LOD). The detection limit of this assay was 2 log CFU/mL (or CFU/g) in blueberry samples, and the SPCE strip can detect as little as one bacterial cell.

2.3.2.1 Potentiometric Transduction

Potentiometric techniques measure accruing charge potential at the working electrode against the reference electrode when zero or no significant current flows between them, instead allowing determination of ion activity in a broad range of concentrations [33,34]. Light-addressable potentiometric sensors (LAPS) have garnered recognition as a valuable potentiometric approach [35].

A LAPS is an example of field-effect transducers that implement potentiometric and optical transduction. LAPS are semiconductor systems containing a metal-insulator-semiconductor or electrolyte-insulator-semiconductor that measures an alternating photocurrent when excited by a modulated light source, such as a light-emitting diode. The amplitude of the photocurrent is sensitive to the surface potential, and thus LAPS is able to detect the potential variation caused by an electrochemical event at the sensor surface [36,37]. A LAPS biosensor was used for detecting *E. coli* and permitted detection at as low as 10 cells/mL when the specific primary capture antibody was immobilized on the LAPS flow-through cell with the secondary antibody labelled with urease for sandwich complex formation [38].

2.3.2.2 Conductometric Transduction

Conductometric transduction relies on the direct measurement of conductance variations in electrolytic media containing mobile electric charges. This occurs when an alternating voltage is applied between the working electrode, on which the biological recognition element is immobilized, and a reference electrode [39]. Conductometric biosensors do not make up a significant market share in the field of biosensing; however, they have been exploited as gas biosensors for determining engine oil quality. More recently, their application in food analysis was reported [40].

2.3.2.3 Impedimetric Transduction

Impedance-based transduction has proved to be a promising method for foodborne pathogenic bacteria detection due to its portability, rapidity and sensitivity. Impedimetric biosensors have been designed by immobilizing biorecognition elements at the surface of a solid electrode. Impedance is the opposition to the flow of alternating current in a sensor, and impedimetric transduction measures the impedance change caused by binding of targets to ligands immobilized onto the electrode surface [41].

Electrochemical impedance spectroscopy (EIS) is a method of impedimetric transduction involving the application of a sinusoidal electrochemical perturbation (potential or current) over a wide range of frequencies when measuring a sample. Multifrequency scanning allows measurement of several electrochemical reactions that take place at different rates and measurement of the capacitance of the electrode [42]. The integration of nucleic acid-biorecognition elements with EIS biosensors has emerged as a powerful tool for the detection of bacterial pathogens. Immobilized nucleic acid-probes that specifically hybridize to their complementary sequences in bacteria samples are coupled to the impedance transducer to transform the binding event into an impedance signal. Kara et al. developed nucleic acid-based impedimetric biosensors for rapid and selective detection of *Bacillus anthracis*. Hybridization between probe and target sequences was determined using EIS and delivered a low LOD for *B. anthracis* at 20.15 fmol/mL and illustrates an excellent detection approach for foodborne bacteria [43].

2.3.3 Optical Biosensors

Optical biosensors link biological selectivity with cutting-edge microelectronics and optoelectronics to produce powerful detection and analytical systems. They are particularly appealing in food safety allowing detection of analytes in complex matrices with only minimal sample treatment. Optical biosensors are typically based upon detection and measurement of changes in the surface characteristics of a sensor when an analyte binds the sensing layer directly by adsorption or binds an immobilized biological recognition element

[44]. Optical transduction is achieved most commonly using fluorescence-based detection or label-free detection.

2.3.4 Evanescent Optical-Planar Waveguide Biosensors

Use of an evanescent optical-planar waveguide provides an example of fluorescence-based detection. This method involves either the analyte of interest or the biological recognition element being labelled with fluorescent tags, so that when interaction occurs between the two, the intensity of the fluorescence is measurable and indicates analyte detection. Light propagating through an optical fibre under conditions of total internal reflection (TIR) can generate a thin electromagnetic field referred to as an evanescent wave. This evanescent wave can excite fluorescence in the proximity of the biosensing surface, for example, in fluorescently labelled analytes interacting with biological recognition elements immobilized on the sensor surface. The short range of the evanescent wave enables it to discriminate between unbound and bound fluorescent complexes, hence eliminating normally required washing procedures [45]. Murphy et al. recently reported the detection of freshwater cyanobacterial toxin, microcystin-LR, which causes acute poisoning, using a novel recombinant antibody-based MBio optical-planar waveguide platform. This system incorporated toxin-conjugates printed onto a plastic planar waveguide, which was bonded to a plastic upper component defining the flow of the channel. A single-chain variable fragment against microcystin-LR was biotinylated with Streptavidin-Alexa-647 and applied in a competitive assay format to produce a functional limit of detection (LOD) of 0.19 ng/mL and a detection range of 0.21−5.9 ng/mL [46].

2.3.5 Surface Plasmon Resonance

In contrast, label-free detection using surface plasmon resonance (SPR) permits analytes that are not labelled to be detected in their natural form in 'real time' [45,47,48]. SPR occurs when light undergoes TIR at the gold-coated biosensor surface and the sample solution interface. The angle of reflected light referred to as the SPR angle is sensitive to changes in the refractive index of the analyte solution close to the biosensor surface. The binding event between the analyte and immobilized ligand induces a change in the refractive index of the analyte solution and a measurable shift in SPR angle due to mass change at the biosensor surface. Light energy is then lost at the metal surface, and a measurable decrease in intensity of the reflected light also occurs [6,49,50]. SPR-based systems have been frequently used for the detection and analysis of water and food contaminants (see Fig. 2).

Biosensors offer a means of sensitive and specific detection of very small quantities of contaminants in complex sample matrices, which is vitally important for ensuring safe food and water around the globe. Great strides in the development of quantitative and semiquantitative biosensors have been

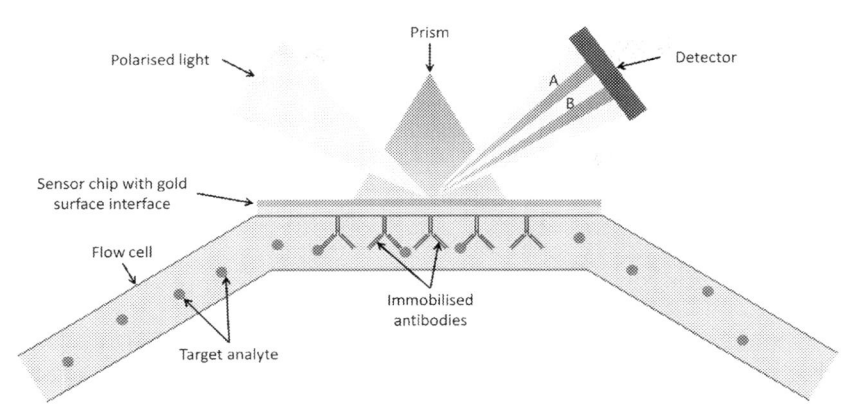

FIGURE 2 Illustration of surface plasmon resonance (SPR). Antibodies are immobilized onto the gold surface of a sensor chip. When polarized light is shone through a prism onto the sensor chip, it reflects at angle A and is measured by a detector. A sample containing the target analyte is passed through the flow cell and across the sensor chip surface. As the target analyte binds to the antibody-coated surface, the refractive index close to the sensor surface changes, resulting in a change in the angle of reflected light (B). This change in SPR angle is proportional to the mass bound to the sensor chip surface.

made in the last two decades and have laid the foundation for a generation of small, cheap, robust, simple, detection systems that will allow users to realize true 'point-of-need' testing and food monitoring. Traditional biosensor types and formats developed for pathogens and toxins have been discussed here. However, advances in nanomaterials, nanoparticles, microfluidics and the generation of excellent often highly engineered, biological recognition elements (recombinant antibodies, recombinant aptamers, cells, cell receptors, etc.) will ensure that biosensors underpin the next generation of detection systems for pathogen and toxin control in food.

3. BIOSENSOR-BASED DETECTION METHODS FOR FOOD CONTAMINANT TOXINS

With the explosive growth in world population, the demand for basic, contaminant-free food sources has never been as urgent. Governments and food safety authorities now recognize the need to regulate and closely monitor foods for human and animal consumption. The analysis of naturally occurring toxins is one of the most important areas in food quality and safety [51]. Food adulterants such as mycotoxins, botulinum neurotoxin and ricin pose a significant threat to both health and economic sectors [52]. The stringent regulation of these toxins in food aims to control and eliminate the risk of human and animal exposure. Therefore, the ability to detect and monitor toxins at legislative limits is a global priority.

Traditional methods for toxin analysis in food include mammalian bioassays, chromatography-based methods or ELISA [53,54]. Each of these brings their own disadvantages in areas such as ethical issues, lack of portability, time consumption and cost-effectiveness. The quintessential method of toxin analysis should be rapid, portable and cost-effective and be able to match the established methods in areas such as sensitivity and selectivity. Biosensor-based methods of analysis are now able to fulfil these requirements and also compete with traditional techniques. A broad range of biosensing platforms have been developed and are reported in the literature, utilizing a variety of transducers, sensing platforms, recognition elements and assay formats (Table 1). It is evident that electrochemical immunosensors have proven the most popular approach, evolving from ELISA-based methods and exploiting the potential of antibody affinity and specificity to achieve LODs well below legislative limits. Additionally, recent advances in DNA and RNA technology have yielded the production of aptamer-based biosensors with LODs similar to their antibody counterparts. Emerging research in this area is now beginning to focus on the utilization of nanotechnology. As this technology progresses, the incorporation of nanomaterials in the development of biosensors has the potential to make these devices more highly sensitive and more applicable for rapid, on-site analyses [55]. Overall, biosensors have provided the foundation for fulfilling the requirements needed for food analysis; however, several issues need to be addressed to ensure these technologies reach the market [56]. Challenges include sample preparation, matrix effects and system integration. Sample preparation and extraction is dependent on the type of sample, the sample volume required for assay and the target analyte concentration necessary for detection and often represents a time-consuming, bottleneck. Pathogen-testing methods, for example, typically use a one- or two-stage broth enrichment step to increase the target pathogen to levels that can be detected during subsequent steps, whilst toxins may need to be extracted from their respective food/water source prior to testing. Alternatively, the biosensor may need to permit multiplex analysis for detection of multiple targets in complex matrices that present strong interferences. The complexity of food matrices presents a major obstacle to developing effective biosensing platforms with the efficient separation of target pathogens or toxins posing difficulties. Furthermore, raw foods contain high levels of naturally occurring, flora, which may interfere with the performance of a wide array of detection technologies. Fully integrated systems that can incorporate sample preparation and analyte detection remain a critical challenge. Microfluidic strategies present a real solution to this problem and include multilayer soft lithography, multiphase microfluidics, electrowetting-on-dielectric, electrokinetics and centrifugal microfluidics. Microfluidic sample preparation steps such as concentration, mixing, pumping and separation can be utilized to make system integration an achievable task necessary for biosensing technology to evolve from the lab to the 'point-of-need' [57,58].

TABLE 1 Examples of Biosensors Used for Detection of Food Contaminant Toxins

Target	Matrix	Biosensor Type	LOD	References
AFB1	Rice	Electrochemical immunosensor	0.06 μg/L	[59]
	Buffer	QCM	0.01 μg/L	[60]
	Buffer	QCM	0.3 ng/mL	[61]
	Bee pollen	Electrochemical immunosensor	0.001 ng/mL	[62]
	Groundnut	EQCM-CV	0.08 ng/mL	[63]
	Buffer	SPR	1 mg/mL	[64]
	Peanuts	Electrochemical aptasensor	0.40 nM	[65]
	Buffer corn powder	Electrochemical immunosensor Electrochemical immunosensor	3.5 pg/mL 13.5 pg/mL	[66]
AFB2	Groundnut	Adsorptive stripping voltammetry	0.1 ng/mL	[67]
	Almond extract	SPR	0.9 ng/mL	[68]
	Peanut oil	Fluorescence-based aptasensor	50 ng/L	[69]
AFM1	Milk	Electrochemical immunosensor	0.039 μg/L	[70]
	Milk	Electrochemical microsensor	0.008 μg/L	[71]
	Milk	Electrochemical immunosensor	0.01 ng/mL	[72]
	Milk	Electrochemical immunosensor	0.001 ng/mL	[73]
	Buffer	Cellular biosensor	5 pg/mL	[74]

Continued

TABLE 1 Examples of Biosensors Used for Detection of Food Contaminant Toxins—cont'd

Target	Matrix	Biosensor Type	LOD	References
DON	Cereals	Electrochemical immunosensor	0.063 ng/mL	[75]
	Buffer	REP	6.25 ng/mL	[76]
	Maize	Quantum dot immunoassay	5 mg/L	[77]
	Buffer	Electrochemiluminescence	1 pg/mL	[78]
	Buffer	Multiplex electrochemical	0.03 μg/mL	[79]
FB1	Corn	Electrochemical immunosensor	5 ng/mL	[80]
OTA	Wine	Electrochemical immunosensor	0.05 μg/L	[81]
	White wine	Electrochemical immunosensor	0.01 ng/mL	[82]
	Wheat	Electrochemical aptasensor	0.07 ng/mL	[83]
	Red wine	Electrochemical immunosensor	0.008 ng/mL	[84]
	Buffer	Electrochemical apatsensor	0.05 ng/mL	[85]
	Buffer	Optical aptasensor	0.39 μgL	[86]
ZEN	Corn silage	Electrochemical immunosensor	0.77 ng/mL	[87]
	Sorghum	SPR	7.8 ng/mL	[88]
	Buffer	Multiplex electrochemical	0.05 μg/mL	[79]
BoNT/A	Buffer	Electrochemical aptasensor	40 pg/mL	[89]
	Buffer and skimmed milk	Electrochemical	8.6 pg/mL	[90]
	Buffer, skimmed milk and apple juice	FRET	1 fg/mL	[91]

BoNT/E	Avian blood	FRET	30 pg/mL	[92]
	Buffer	Electrochemical immunosensor	5 pg/mL	[93]
Ricin	Buffer Milk	XMAP	0.01 ng/mL 0.03 ng/mL	[94]
	Buffer	SPR	30 ng/mL	[95]

AFB1, Aflatoxin B1; *AFB2*, aflatoxin B2; *AFM1*, aflatoxin M1; *BoNT/A*, botulinum neurotoxin A; *BoNT/E*, botulinum neurotoxin E; *DON*, deoxynivalenol; *EQCM-CV*, electrochemical quartz crystal microbalance-cyclic voltammetry; *FB1*, fumonisin B1; *FRET*, fluorescence resonance energy transfer; *LOD*, limit of detection; *OTA*, ochratoxin A; *QCM*, quartz crystal microbalance; *REP*, real-time electrochemical profiling; *SPR*, surface plasmon resonance; *xMAP*, multianalyte profiling; *ZEN*, zearalenone.

4. BIOSENSOR-BASED DETECTION METHODS FOR MARINE TOXINS

Many biosensor assays for hapten molecules such as marine toxins are based on a competitive immunoassay format, utilizing the highly specific interaction between an antibody and its target antigen. Plate-based ELISA is amenable for high-throughput analysis and has excellent performance characteristics. However, it is not truly rapid and requires numerous reagent addition and wash steps, incubations, trained personnel and dedicated equipment for result measurement. Other systems for toxin detection use the same competitive format principle, but have improved read-out or incorporate high-density spotting in a microarray format. SPR biosensors allow for automated analysis as well as regeneration and reusability of assay surfaces and provide the possibility for multiplexed analysis. Luminex microsphere assays use flow cytometry principles alongside the competitive format to allow for high sensitivity, multiplexed detection of toxins. However, both of these systems require dedicated, expensive equipment and trained personnel, limiting their use at the 'point-of-need'. The MBio optical-planar waveguide platform used by McNamee et al. allows for higher sensitivity than lab-based analytical methods, a 15-min turnaround time and multiplexed detection of amnesic, diarrhetic and paralytic shellfish poisoning marine toxins [96]. Lateral flow immunoassays (LFIAs), such as the MBio optical-planar waveguide platform, are cheap to manufacture, easy to use, rapid, have good performance characteristics and, with the incorporation of simple reader systems, are easy to interpret. These benefits, coupled with excellent shelf-life and temperature stability, suggest LFIAs have great potential to be used as dock-side screening assays of newly caught shellfish for the presence of marine toxins. Due to the qualitative output of some of these assays, further confirmatory lab-based analysis may be required. However, such platforms can and will be incorporated into sensors. A number of assays for marine toxin detection are listed in Table 2.

5. BIOSENSOR-BASED DETECTION METHODS FOR MICROORGANISMS AND PATHOGENS

Microbial and pathogenic contamination has severe health and economic implications in a plethora of different environments. These are primarily noted in the food and water industry and within healthcare environments. Biosensors for the detection of contaminating pathogens offer fast and efficient assessment of the qualitative standard of specific produce or environments. Recent biosensors developed for the detection of Salmonella, *L. monocytogenes*, *E. coli*, Campylobacter, *C. perfringens, S. aureus* and *T. gondi* are outlined in Table 3. The majority of the methods listed below are based on immunosensors (antibody-based) or DNA-based sensors. A significant number

TABLE 2 Examples of Assays for Marine Toxins

Target	Biorecognition Element	Biosensor Type	LOD	References
OA/DTX	Antibody	Competitive ELISA	0.41 ng/mL	[97]
	Antibody	Competitive ELISA, SPCE	1.99 ngL	[98]
	Antibody	Multiplex luminex microsphere	16 µg/kg	[99]
	Antibody	Multiplex microarray	1 µg/L	[100]
	Antibody	Multiplex planar waveguide microarray	0.44 ng/mL	[96]
	Antibody	LFIA	50 ng/mL	[101]
	Antibody	SPR	0.24 µg/L	[102]
	Antibody	Chemiluminescent competitive ELISA	0.03 ng/mL	[103]
	Antibody	Competitive ELISA, SPCE	0.15 µg/L	[104]
	Antibody	Multiplex SPR	0.34 ng/mL	[97]
	Antibody	Competitive ELISA	0.02 ng/mL	[105]
	Antibody	Competitive ELISA	0.45 ng/mL	[106]
	Enzyme	Enzymatic inhibition assay	44 µg/mL	[107]
	Aptamer	Electrochemical	70 pg/mL	[108]
	Enzyme	FIA-based enzymatic inhibition assay	70 pg/mL	[109]
	Cell	Cell viability assay	47 µg/kg	[110]
	Enzyme	Enzymatic inhibition assay	0.25 ng/mL	[111]
	Enzyme	Enzymatic inhibition assay	34.8 µg/kg	[112]

Continued

TABLE 2 Examples of Assays for Marine Toxins—cont'd

Target	Biorecognition Element	Biosensor Type	LOD	References
DA	Antibody	Competitive ELISA	3 ng/mL	[113]
	Antibody	Multiplex luminex microsphere	0.1 ng/mL	[99]
	Antibody	Multiplex microarray	0.5 µg/L	[100]
	Antibody	Multiplex SPR	1.66 ng/mL	[97]
	Antibody	LFIA	20 ng/mL	[114]
	Antibody	Competitive ELISA	10 ng/mL	[115]
	Antibody	LFIA	2 ppm	[116]
	Antibody	Multiplex planar waveguide microarray	0.37 ng/mL	[96]
	Antibody	Competitive ELISA	0.01 mg/kg	[117]
	Antibody	CE-EIA	0.02 ng/mL	[118]
STX and analogues	Antibody	Multiplex luminex microsphere	0.5 ng/mL	[99]
	Antibody	Multiplex microarray	0.4 µg/L	[100]
	Antibody	Multiplex SPR	0.82 ng/mL	[97]
	Antibody	Multiplex planar waveguide microarray	0.05 ng/mL	[96]
	Antibody	Competitive ELISA	50 µg/kg	[117]
	Antibody	Bio-barcode assay	<0.74 µg/mL	[119]

	Antibody	Multiplex luminex microsphere	2.2 ng/mL	[120]
	Antibody	Open-sandwich immunoassay	0.05 ng/mL	[121]
	Sodium receptor	Receptor-based assay	45 µg/kg	[122]
	Sodium receptor	Receptor-based assay	1.2 nM	[123]
AZA	Antibody	Luminex microsphere	0.8 ng/mL	[124]

AZA, Azaspiracid; *CE-EIA*, capillary electrophoresis-based enzyme immunoassay; *DA*, domoic acid; *DTX*, dinophysistoxin; *FIA*, flow-injection analysis; *LFIA*, lateral flow immunoassay; *LOD*, limit of detection; *OA*, okadaic acid; *SPCE*, screen-printed carbon electrode; *SPR*, surface plasmon resonance.

TABLE 3 Examples of Biosensors Assays for Microbial Pathogens

Target	Biorecognition Element	Biosensor Type	LOD	References
Salmonella	Polyclonal antibody	Quartz crystal acoustic wave device	10^2 cells/mL	[125]
	Polyclonal antibody	SPR	10^3 cells/mL	[126]
	IgG-protein G complexes	FRET	10^3 cells/mL	[127]
	Salmonella SSeC gene	Fluorescence-based SWNT	10^5 cells/mL	[128]
	Aptamer	Electrochemical	3 CFU/mL	[129]
Listeria monocytogenes	Aptamer	Optical	10^3 CFU/mL	[130]
	ssDNA	Electrochemical DNA	2.9×10^{-13} mol/L	[131]
	Bacteriophage endolysin	Electrochemical	1.1×10^4 CFU/mL	[132]
	DNA probe	Chemiluminescent	6.3×10^{-2} pmol/L	[133]
Escherichia coli	T4 bacteriophage	SPR	10^3 CFU/mL	[134]
	Antibody	Electrochemical	10^2 CFU/mL	[135]
	Monoclonal antibody	Capillary flow microfluidic biosensor	10^6 CFU/mL	[136]
	Monoclonal antibody	Optical	50 CFU/mL	[137]
	Peptide conjugate	Electrochemical	10^3 CFU/mL	[138]
Campylobacter	ssDNA	SPR	2.5 and 5.0 nM	[139]
	Biotin-conjugated antibodies	Fluorescence	30–50 CFU/mL	[140]
	Flagellin gene	Electrochemical	90 pM	[141]
	Conjugated DNA probe	OLED-based DNA biochip	0.37 ng/μL	[142]

Clostridium perfringens	B lymphocyte Ped-2E9 cells	Cell based	10–40 ng	[143]
	ssDNA	SPR	0.02 nM	[144]
	Monoclonal antibody	Electronic detection with SWNT	2 nM	[145]
	DNA via rolling circle amplification	Electrochemiluminescence	10^{-15} M	[146]
	DNA aptamer	Electrochemical	10^{-12} M	[147]
Streptococcus aureus	Lytic phage	SPR	10^4 CFU/mL	[148]
	Carbon nanotubes and aptamers	Electrochemical	10^7 CFU/mL	[149]
	ssDNA aptamers	Electrochemical	10^5 CFU/mL	[150]
	*mec*A gene by PCR	FRET	1 nM	[151]
	Lytic phage	Magnetoelastic	3.0 log CFU/mL	[152]
Toxoplasma gondii	anti-*T. gondii* immunoglobulin	Electrochemical	1:5500 dilution	[153]
	Aptamers	Quantum dots-labelled dual aptasensor	0.1 IU	[154]
	T. gondii-specific DNA	Magnetic molecular beacon probe via quantum dots	2.7×10^9 mol/L	[155]
	T. gondii-specific antibodies	Electrochemical	0.0375–1.2 AU/mL	[156]
	DNA oligonucleotides	Fluorescence	8.3×10^{-9} M	[157]
Gram-negative bacterial endotoxins	Mab-coated monolayer	Microcantilever array	50 µg/mL	[158]

FRET, Fluorescence resonance energy transfer; *IgG*, immunoglobulin G; *Mab*, monoclonal antibody; *OLED*, organic light-emitting diode; *SPR*, surface plasmon resonance; *ssDNA*, single-stranded DNA; *SWNT*, single-wall nanotubes.

of amplification systems based on nanoparticles or magnetic particles have been successfully reported to increase assay sensitivity.

6. CHALLENGES ASSOCIATED WITH BIOSENSORS

Biosensors are powerful tools for the analysis of food contaminants, and their development in this area has seen rapid advances in recent years. There are, however, many challenges and obstacles to overcome in the design, optimization and application of biosensors, particularly for the analysis of biological contaminants.

In terms of assay development, there are generally two approaches: label-free assays and labelled assays. Each of these methods has advantages and limitations. Generally, label-free methods provide simpler sample preparation and can allow real-time measurement, as described for SPR-based methods, but these assays tend to have issues with matrix effects and nonspecific binding. Labelled assays, on the other hand, are usually more complicated, involving multiple steps, but often increase specificity and improve sensitivity. There is no ideal method and the choice often depends on the target analyte and matrix involved [58].

As previously mentioned, a major issue for any assay is matrix effects. Food samples often present complex matrices and require extraction or sample cleanup prior to analysis. This is not ideal when a rapid method is desired, and sometimes the harsh conditions of the extraction procedure can have adverse effects on the assay [54]. Most biosensors perform optimally with buffers or spiked samples but translating preparation of real-world samples imposes a challenging bottleneck for biosensor development. Matrix effects in biological samples present an important problem for many biosensor devices and need to be addressed with each biosensor/analyte/matrix scenario.

Biosensors offer the ability to miniaturize assays to the micro- and nano-scale. This miniaturization can offer many advantages in areas such as sample volume and assay time. However, assays will behave differently at these levels due to the unique physical properties affecting areas such as liquid movement and binding events. It is therefore important to maintain the integrity of the assay during the miniaturization and preserve, or even improve areas such as the assay's sensitivity and selectivity whilst competing with issues such as flux and signal-to-noise ratio [159]. Moreover, multiplexing is viewed as key for food analysis, as samples are often contaminated with multiple adulterants. However, implementation of multiplex analysis on these miniaturized systems has proven to be complex.

Finally, system integration remains the greatest challenge for the transfer of biosensors from research laboratories into the field. Although a great deal of progress has been made in each individual aspect of biosensor development, the integration of these aspects into an automated, user-friendly, 'stand-alone' platform for field analysis remains difficult [51]. Microfluidic platforms offer an answer to these challenges and, if successful, it will significantly increase the likelihood of translating research grade biosensors from the lab to the field.

7. FUTURE DIRECTIONS

A highly important step in toxin detection is their initial extraction from contaminated material. This represents a major challenge to the development of completely autonomous toxin-monitoring systems, as currently most extraction protocols are laboratory-based and require 'hands-on' input by trained personnel. Automated lab-based extraction systems are available [160,161]. For example, there is increasing use of immunoaffinity columns for toxin extraction, and automation of such systems can allow for streamlined workflow, safer toxin handling and improved sensitivity [162,163]. Pioneering work is being carried out by the Monterey Bay Aquarium Research Institute and collaborators at the National Oceanic and Atmospheric Administration who have developed the Environmental Sample Processor, which can autonomously collect, extract and measure toxins and can determine levels of harmful algal bloom species in seawater samples [113,164]. It is the authors' opinion that this area of integrated toxin extraction in automated monitoring systems will receive significant attention in the future.

Continuing on the trend of fully automated analysis systems for toxin detection, it is likely that the area of lab-on-a-chip technology will gain greater prominence. These systems can allow for miniaturized, total analysis assays to measure toxin levels at the 'point-of-need' without the necessity for specialized instrumentation, trained personnel or extended turnaround times. Some systems allow for sensitive detection of toxins coupled with complex matrix cleanup, encompassing the requirement for toxin-extraction with toxin measurement [165,166].

Mobile phone-based diagnostics are another exciting approach that may be successfully applied for the quantification of environmental toxin contaminants. Modern day smartphones boast significant computational power, are equipped with high-quality resolution cameras and have open source operating systems. Therefore, they have excellent potential for the following: interpreting results of 'point-of-need' tests, for example, with LFIA testing, thus removing subjective user-interpretation; actively controlling the electronic and/or mechanical components of more advanced 'point-of-need' assays, such as microfluidic assays, and negating the need for computer-based controllers. The ubiquity of mobile phone networks in developed and developing countries means that diagnostic assays could easily be performed in remote locations and results transferred to localized data centres. Following on from this, inbuilt Global Positioning Systems in smartphones could allow for 'real-time' epidemiological tracking of toxins.

Pathogens and toxins present a significant risk to human and animal health whilst also negatively impacting on the safety of water, food and feed commodities worldwide. Detection and monitoring of these contaminants is of vital importance to avoid acute and chronic illnesses that can lead to mortality, as well as ensuring food security. In recent years, biosensor development has emerged as a field of intense research and underpins the advancement of detection methods that comply with legislative and quality standards for numerous pathogens and toxins. Biosensors are potentially ideal as they enable sensitive and specific detection of very small quantities of contaminants in complex sample matrices.

ACKNOWLEDGEMENTS

The authors of this work are supported by the MARIABOX project (European Union Seventh Framework Programme), the Dublin City University Daniel O'Hare Research Scholarship Scheme, the Irish Research Council and Science Foundation Ireland (14/1A/2646).

The authors have declared no conflicts of interest.

REFERENCES

[1] United States Environmental Protection Agency, Office of Water Report 820-F-12-058, Environmental Science Center, 701 Mapes Road, Fort Meade, MD 20755-5350, 2012.

[2] F. Yeni, S. Acar, O.G. Polat, Y. Soyer, H. Alpas, Food Control 40 (2014) 359−367.

[3] Y. Lei, W. Chen, A. Mulchandani, Anal. Chim. Acta 568 (1−2) (2006).

[4] B. Byrne, E. Stack, N. Gilmartin, R.J. O'Kennedy, Sensors 9 (6) (2009) 4407−4445.

[5] P.J. Conroy, S. Hearty, P. Leonard, R.J. O'Kennedy, Sem. Cell. Dev. Biol. 20 (1) (2009) 10−26.

[6] D. Ozkan-Ariksoysal (Ed.), Biosensors and Their Application in Healthcare, 2013. Unitec House, 2 Albert Place, London N3 1QB, UK.

[7] A.P.F. Turner, Chem. Soc. Rev. 42 (8) (2013) 3175−3648.

[8] I.E. Tothill, Comput. Electron. Agric. 30 (1) (2001) 205−218.

[9] B.D. Malhotra, R. Singhal, A. Chaubey, S.K. Sharma, A. Kumar, Curr. Appl. Phys. 5 (2) (2005) 92−97.

[10] J. Kirsch, C. Siltanen, Q. Zhou, A. Revzin, A. Simonian, Chem. Soc. Rev. 42 (2013) 8733−8768.

[11] D.R. Thevenot, K. Toth, R.A. Durst, G.S. Wilson, Biosens. Bioelectron. 16 (1−2) (2001) 121−131.

[12] B. Van Dorst, J. Metha, K. Bekaert, E. Rouah-Martin, W. De Coen, P. Dubruel, R. Blust, J. Robbens, Biosens. Bioelectron. 26 (4) (2010) 1178−1194.

[13] D.R. Heldman (Ed.), Encyclopedia of Agricultural, Food, and Biological Engineering, 2003. Madison Avenue, New York, New York 10016.

[14] E. Hamidi-Asl, I. Palchetti, E. Hasheminejad, M. Mascini, Talanta 115 (2013) 74−83.

[15] S. Sagadevan, M. Periasamy, Rev. Adv. Mater. Sci. 36 (2014) 62−69.

[16] W.W. Zhao, J.J. Xu, H.Y. Chen, Chem. Rev. 114 (15) (2014) 7421−7441.

[17] R.E. Armstrong, M.D. Drapeau, C.A. Loeb, J.J. Valdes (Eds.), Bio-inspired Innovation and National Security, 2010. Washington, DC.

[18] Y. Liu, Z. Matharu, M.C. Howland, A. Revzin, A.L. Simonian, Anal. Bioanal. Chem. 404 (4) (2012) 1181–1196.

[19] D. Grieshaber, R. Mackenzie, J. Voros, E. Reimhult, Sensors (Basel) 8 (3) (2008) 1400–1458.

[20] A. Makaraviciute, A. Ramanaviciene, Biosens. Bioelectron. 50 (2013) 460–471.

[21] S. Sharma, H. Byrne, R.J. O'Kennedy, Essays Biochem. 60 (2) (2016) (in press).

[22] T.M.S. Chang (Ed.), Biomedical Applications of Immobilized Enzymes and Protein, Vol. I, 2013, 227 West 17th Street, New York, New York 10011.

[23] R.S. Sethi, Biosens. Bioelectron. 9 (4) (1994) 243–264.

[24] M. Gerard, A. Chaubey, B.D. Malhotra, Biosens. Bioelectron. 17 (5) (2002) 345–359.

[25] P. Leonard, S. Hearty, J. Brennan, L. Dunne, J. Quinn, T. Chakraborty, R.J. O' Kennedy, Enzyme Microb. Technol. 32 (2003) 3–13.

[26] R. Singh, M.D. Mukherjee, G. Sumana, R.K. Gupta, S. Sood, B.D. Malhotra, Sensor Actuat. B-Chem. 197 (2014) 385–404.

[27] D.R. Thevenot, K. Toth, R.A. Durst, G.S. Wilson, Pure Appl. Chem. 71 (12) (1999) 2333–2348.

[28] N.J. Ronkainen, H.B. Halsall, W.R. Heineman, Chem. Soc. Rev. 39 (2010) 1747–1763.

[29] B.J. Privett, J.H. Shin, M.H. Schoenfisch, Anal. Chem. 82 (12) (2010) 4723–4741.

[30] A.C. Mongra, A. Kaur, R.K. Bansal, Int. J. Eng. Res. Appl. 2 (2) (2012) 743–749.

[31] S.V. Dzyadevych, V.N. Arkhypova, A.P. Soldatkina, A.V. El'Skaya, C. Martelet, N. Jaffrezic-Renault, IRBM 29 (2–3) (2008) 171–180.

[32] D. Davies, X. Guo, L. Musavi, C.S. Lin, S.H. Chen, V.C.H. Wu, Ind. Biotechnol. 9 (1) (2013) 13–36.

[33] R. Koncki, Anal. Chim. Acta 599 (1) (2007) 7–15.

[34] S. Li, Y. Ge, S.A. Piletsky, J. Lunec (Eds.), Molecularly Imprinted Sensors: Overview and Applications, Elsevier, The Boulevard, Langford Lane, Kidlington, Oxford, OX5, 1GB, UK, 2012.

[35] A. Bratov, N. Abramova, A. Ipatov, Anal. Chim. Acta 678 (2) (2010) 149–159.

[36] P.A. Serra (Ed.), Biosensors for Health, Environment and Biosecurity, InTech, Unit 405, Office Block, Hotel Equatorial Shanghai No.65, Yan An Road (West), Shanghai, 200040, China, 2011.

[37] N. Hu, T. Wang, J. Cao, K. Sua, J. Zhou, J. Wu, P. Wang, Sensor Actuat. B-Chem. 185 (2013) 238–244.

[38] C. Ercole, M.D. Gallo, M. Pantalone, S. Santucci, L. Mosiello, C. Laconi, A.A. Lepidi, Sensor Actuat. B-Chem. 83 (2002) 48–52.

[39] V. Somerset (Ed.), Environmental Biosensors, InTech, University Campus STeP Ri Slavka Krautzeka 83/A 51000 Rijeka, Croatia, 2011.

[40] A.K. Bhunia, M.S. Kim, C.R. Taitt (Eds.), High Throughput Screening for Food Safety Assessment, Woodhead Publishing, 80 High St, Sawston, Cambridge CB22 3HJ, United Kingdom, 2015.

[41] Y. Wang, Z. Ye, Y. Ying, Sensors 12 (3) (2012) 3449–3471.

[42] J.G. Guan, Y.Q. Miao, Q.J. Zhang, J. Biosci. Bioeng. 97 (4) (2004) 219–226.

[43] P. Kara, B. Meric, M. Ozsoz, Electroanal. 20 (24) (2008) 2629–2634.

[44] K. Narsaiah, S.N. Jha, R. Bhardwaj, R. Sharma, R. Kumar, J. Food Sci. Tech. 49 (4) (2012) 383–406.

[45] F. Long, A. Zhu, H. Shi, Sensors 13 (10) (2013) 13928–13948.

[46] C. Murphy, E. Stack, S. Krivelo, D.A. Mcpartlin, B. Byrne, C. Greef, M.J. Lochhead, G. Husar, S. Devlin, C.T. Elliot, R.J. O'Kennedy, Biosens. Bioelectron. 64 (2015) 708–714.

[47] H. Koistinen, U.H. Stenman (Eds.), Novel Approaches in Immunoassays, 2014. Unitec House, 2 Albert Place, London N3 1QB, UK.

[48] S.G. Patching, Biochim. Biophys. Acta 838 (1 Pt A) (2014) 43−55.

[49] X. Fan, I.M. White, S.I. Shopova, H. Zhu, J.D. Suter, Y. Sun, Anal. Chim. Acta 620 (1−2) (2008) 8−26.

[50] P. Leonard, S. Hearty, R.J. O'Kennedy, Methods Mol. Biol. 681 (2011) 403−418.

[51] V. Scognamiglio, F. Arduini, G. Palleschi, G. Rea, Trend Anal. Chem. 62 (2014) 1−10.

[52] H. Sharma, R. Mutharasan, Sensor Actuat. B-Chem. 183 (2014) 535−549.

[53] V.L. Pereira, J.O. Fernandes, S.C. Cunha, Trends Food Sci. Technol. 36 (2) (2014) 96−136.

[54] X. Wang, X. Lu, J. Chen, Trend Environ. Anal. Chem. 2 (2014) 25−32.

[55] I.E. Tothill, World Mycotoxin J. 4 (4) (2011) 361−374.

[56] E.B. Bahadir, M.K. Sezgintürk, Anal. Biochem. 478 (2015) 107−120.

[57] B. Ge, J. Meng, J. Lab. Autom. 14 (4) (2009) 235−241.

[58] M.L.Y. Sin, K.E. Mach, P.K. Wong, J.C. Liao, Expert Rev. Mol. Diagn. 14 (2) (2014) 225−244.

[59] Y. Tan, X. Chu, G. Shen, R. Yu, Anal. Biochem. 387 (1) (2009) 82−86.

[60] X. Jin, X. Jin, X. Liu, L. Chen, J. Jiang, G. Shen, R. Yu, Anal. Chim. Acta 645 (1−2) (2009) 92−97.

[61] L. Wang, X. Gan, Bioprocess. Biosyst. Eng. 32 (1) (2009) 109−116.

[62] L. Zaijun, W. Zhongyun, S. Xiulan, F. Yinjun, C. Peipei, Talanta 80 (5) (2010) 1632−1637.

[63] R. Chauhan, P.R. Solanki, J. Singh, I. Mukherjee, T. Basu, B.D. Malhotra, Food Cont. 52 (2015) 60−70.

[64] J.H. Park, Y. Kim, I. Kim, S. Ko, Food Cont. 36 (1) (2015) 183−190.

[65] G. Castillo, K. Spinella, A. Poturnayová, M. Šnejdárková, L. Mosiello, T. Hianik, Food Cont. 52 (2015) 9−18.

[66] X. Zhang, C. Li, W. Wang, J. Xue, Y. Huang, X. Yang, B. Tan, X. Zhou, C. Shao, S. Ding, J. Qiu, Food Chem. 192 (2016) 97−202.

[67] R. Hajian, A.A. Ensafi, Food Chem. 115 (3) (2009) 1034−1037.

[68] S.R. Edupuganti, O.P. Edupuganti, R.J. O'Kennedy, World Mycotoxin J. 6 (2013) 273−280.

[69] X. Ma, W. Wang, X. Chen, Y. Xia, N. Duan, S. Wu, Z. Wang, Food Cont. 47 (2015) 545−551.

[70] C. Parker, I.E. Tothill, Biosens. Bioelectron. 24 (2009) 2452−2457.

[71] C. Parker, Y. Lanyon, M. Manning, D.W.M. Arrigan, I.E. Tothill, Anal. Chem. 81 (2009) 5291−5298.

[72] N. Paniel, A. Radoi, J. Marty, Sensors 10 (10) (2010) 9439−9448.

[73] G. Bacher, S. Pal, L. Kanungo, S. Bhand, Sensor Actuat. B-Chem. 168 (2012) 223−230.

[74] E. Larou, I. Yiakoumettis, G. Kaltsas, A. Petropoulos, P. Skandamis, S. Kintzios, Food Cont. 29 (1) (2013) 208−212.

[75] D. Romanazzo, F. Ricci, G. Volpe, C.T. Elliott, S. Vesco, K. Kroeger, D. Moscone, J. Stroka, H. Van Egmond, M. Vehniäinen, G. Palleschi, Biosens. Bioelectron. 25 (12) (2010) 2615−2621.

[76] Z. Olcer, E. Esen, T. Muhammad, A. Ersoy, S. Budak, Y. Uludag, Biosens. Bioelectron. 62 (2014) 163−169.

[77] E.S. Speranskaya, N.V. Beloglazova, P. Lenain, S. De Saeger, Z. Wang, S. Zhang, Z. Hens, D. Knopp, R. Niessner, D.V. Potapkin, I.Y. Goryacheva, Biosens. Bioelectron. 53 (2014) 225−231.

[78]   X. Lv, Y. Li, T. Yan, X. Pang, W. Cao, B. Du, D. Wu, Q. Wei, Biosens. Bioelectron. 70 (2015) 28–33.

[79]   W. Gu, P. Zhu, D. Jiang, X. He, Y. Li, J. Ji, L. Zhang, Y. Sun, X. Sun, Biosens. Bioelectron. 70 (2015) 447–454.

[80]   M.K. Abdul Kadir, I.E. Tothill, Toxins 2 (4) (2010) 382–389.

[81]   M. Heurich, M.K. Abduk Kadir, I.E. Tothill, Sensor Actuat. B-Chem. 156 (2011) 162–168.

[82]   L. Zamfir, J. Geana, S. Bourigua, L. Rotariu, C. Bala, A. Errachid, N. Jaffrezic-Renault, Sensor Actuat. B-Chem. 159 (1) (2011) 178–184.

[83]   L. Bonel, J.C. Vidal, P. Duato, J.R. Castillo, Biosens. Bioelectron. 26 (7) (2011) 3254–3259.

[84]   R.R. Perrotta, F.J. Arévalo, N.R. Vettorazzi, M.A. Zón, H. Fernández, Sensor Actuat. B-Chem. 162 (1) (2012) 327–333.

[85]   A. Rhouati, A. Hayat, D.B. Hernandez, Z. Meraihi, R. Munoz, J. Marty, Sensor Actuat. B-Chem. 176 (2013) 1160–1166.

[86]   L. Liu, X. Zhou, H. Shi, Biosens. Bioelectron. 72 (2015) 300–305.

[87]   N.V. Panini, F.A. Bertolino, E. Salinas, G.A. Messina, J. Raba, Biochem. Eng. J. 51 (1–2) (2010) 7–13.

[88]   S.R. Edupuganti, O.P. Edupuganti, R.J. O'Kennedy, World Mycotoxin J. 34 (2) (2013) 668–674.

[89]   F. Wei, C. HO, Anal. Bioanal. Chem. 393 (8) (2009) 1943–1948.

[90]   C. Chan, J. Guo, C. Sun, M. Tsang, F. Tian, J. Hao, S. Chen, M. Yang, Sensor Actuat. B-Chem. 220 (2015) 131–137.

[91]   J. Shi, J. Guo, G. Bai, C. Chan, X. Liu, W. Ye, J. Hao, S. Chen, M. Yang, Biosens. Bioelectron. 65 (2015) 238–244.

[92]   T.M. Piazza, D.S. Blehert, F.M. Dunning, B.M. Berlowski-Zier, F.N. Zeytin, M.D. Samuel, W.C. Tucker, Appl. Environ. Microbiol. 77 (2011) 7815–7822.

[93]   J. Narayanan, M.K. Sharma, S. Ponmariappan, M. Sarita, Shaik, S. Upadhyay, Biosens. Bioelectron. 69 (2015) 249–256.

[94]   M.A. Simonova, T.I. Valyakina, E.E. Petrova, R.L. Komaleva, N.S. Shoshina, L.V. Samokhvalova, O.E. Lakhtina, I.V. Osipov, G.N. Philipenko, E.K. Singov, E.V. Grishin, Anal. Chem. 84 (15) (2012) 6326–6333.

[95]   T. Nagatsuka, H. Uzawa, K. Sato, S. Kondo, M. Izumi, K. Yokoyama, I. Ohsawa, Y. Seto, P. Neri, H. Mori, Y. Nishida, M. Saito, E. Tamiya, ACS Appl. Mater. Inter. 5 (10) (2013) 4173–4180.

[96]   S.E. McNamee, C.T. Elliot, B. Greer, M. Lochhead, K. Campbell, Environ. Sci. Technol. 48 (22) (2014) 13340–13349.

[97]   S.E. McNamee, C.T. Elliott, P. Delahaut, K. Campbell, Environ. Sci. Pollut. R. 20 (10) (2012) 6794–6807.

[98]   A. Hayat, L. Barthelmebs, A. Sassolas, J. Marty, Talanta 85 (1) (2011) 513–518.

[99]   M. Fraga, N. Vilariño, M.C. Louzao, P. Rodríguez, K. Campbell, C.T. Elliott, L.M. Botana, Anal. Chem. 85 (16) (2013) 7794–7802.

[100] A. Szkola, K. Campbell, C.T. Elliott, R. Niessner, M. Seidel, Anal. Chim. Acta 787 (2013) 211–218.

[101] S. Lu, C. Lin, Y. Li, Y. Zhou, X. Meng, S. Yu, Z. Li, L. Li, H. Ren, Z. Liu, Anal. Biochem. 422 (2) (2012) 59–65.

[102] B. Prieto-Simón, H. Miyachi, I. Karube, H. Saiki, Biosens. Bioelectron. 25 (6) (2010) 1395–1401.

[103] M.M. Vdovenko, C. Hung, I.Y. Sakharov, F. Yu, Talanta 116 (2013) 343−346.

[104] R.D. Dominguez, A. Hayat, A. Sassolas, G.A. Alonso, R. Munoz, J. Marty, Talanta 99 (2012) 232−237.

[105] C. Desmet, L.J. Blum, C.A. Marquette, Anal. Chem. 84 (23) (2012) 10267−10276.

[106] S. Lu, Y. Zhou, Y. Li, C. Lin, X. Meng, D. Yan, Z. Li, S. Yu, Z. Liu, H. Ren, Environ. Sci. Pollut. R. 19 (7) (2012) 2619−2626.

[107] H.G.F. Smienk, D. Calvo, P. Razquin, E. Domãnguez, L. Mata, Toxins 4 (5) (2012) 339−352.

[108] S. Eissa, A. Ng, M. Siaj, A.C. Tavares, M. Zourob, Anal. Chem. 85 (24) (2013) 11794−11801.

[109] G. Volpe, E. Cotroneo, D. Moscone, L. Croci, L. Cozzi, G. Ciccaglioni, G. Palleschi, Anal. Biochem. 385 (1) (2009) 50−56.

[110] E. Cañete, M. Campàs, P. De La Iglesia, J. Diogène, Toxicol. In Vitro 24 (2) (2010) 611−619.

[111] A. Sassolas, G. Catanante, A. Hayat, J. Marty, Anal. Chim. Acta 702 (2) (2011) 262−268.

[112] T. Ikehara, S. Imamura, A. Yoshino, T. Yasumoto, Toxins 2 (1) (2010) 195−204.

[113] G.J. Doucette, C.M. Mikulski, K.L. Jones, K.L. King, D.I. Greenfield, R. Marin III, S. Jensen, B. Roman, C.T. Elliott, C.A. Scholin, Harmful Algae 8 (6) (2009) 880−888.

[114] L.L. Gao, J.P. Cheng, Y.Y. Liu, Q. Wang, W.H. Wang, Huan Jing Ke Xue (J. Environ. Sci. China) 32 (8) (2011) 2492−2496.

[115] R. Liu, D.F. Xu, Y.F. Dong, Y. Liang, Wei ShengYan Jiu (J. Hyg. Res.) 38 (5) (2009) 622−624.

[116] W. Jawaid, J. Meneely, K. Campbell, M. Hooper, K. Melville, S. Holmes, J. Rice, C.T. Elliott, Talanta 116 (2013) 663−669.

[117] E. Garet, Ã. González-Fernández, J. Lago, J.M. Vieites, A.G. Cabado, J. Agric. Food Chem. 58 (3) (2010) 1410−1415.

[118] X. Zhang, Z. Zhang, Toxicon 59 (6) (2012) 626−632.

[119] Y. Tang, H. Wang, J. Xiang, Y. Chen, W. He, N. Deng, H. Yang, Anal. Chem. Acta 657 (2) (2010) 210−214.

[120] M. Fraga, N. Vilariño, M.C. Louzao, K. Campbell, C.T. Elliott, K. Kawatsu, M.R. Vieytes, L.M. Botana, Anal. Chem. 84 (10) (2011) 4350−4356.

[121] Y. Hara, J. Dong, H. Ueda, Anal. Chem. Acta 793 (2013) 107−113.

[122] F.M. Van Dolah, S.E. Fire, T.A. Leighfield, C.M. Mikulski, G.J. Doucette, J. AOAC Int. 95 (3) (2012) 795−812.

[123] F.M. Van Dolah, T.A. Leighfield, G.J. Doucette, J. AOAC Int. 92 (6) (2009) 1705−1713.

[124] L.P. Rodriguez, N. Vilariño, M.C. Louzao, T.J. Dickerson, K.C. Nicolaou, M.O. Frederick, L.M. Botana, Anal. Chem. 447 (2014) 58−63.

[125] S.T. Pathirana, J. Barbaree, B.A. Chin, M.G. Hartell, W.C. Neely, V. Vodyanoy, Biosens. Bioelectron. 15 (3−4) (2000) 135−141.

[126] G.C.A.M. Bokken, R.J. Corbee, F. Van Knapen, A.A. Bergwerff, FEMS Microbiol. Lett. 222 (1) (2003) 75−82.

[127] S. Ko, S.A. Grant, Biosens. Bioelectron. 21 (7) (2006) 1283−1290.

[128] Y. Ning, Z.J. Li, Y.F. Duan, Z.H. Peng, L. Deng, J. Nanomater. Mol. Nanotechnol. 2 (5) (2013) 1−10.

[129] X. Ma, Y. Jiang, F. Jia, Y. Yu, J. Chen, Z. Wang, J. Microbiol. Met 98 (2014) 94−98.

[130] S. Ohk, O. Koo, T. Sen, C. Yamamoto, A. Bhunia, J. Appl. Microbiol. 109 (3) (2010) 808−817.

[131] W. Sun, X. Qi, Y. Zhang, H. Yang, H. Gao, Y. Chen, Z. Sun, Electrochim. Acta 85 (2012) 145−151.

[132] M. Tolba, M.U. Ahmed, C. Tlili, F. Eichenseher, M.J. Loessner, M. Zourob, Analyst 37 (24) (2012) 5749−5756.

[133] F. Liu, C. Zhang, Sensor Actuat. B-Chem. 209 (2015) 399−406.

[134] N. Tawil, E. Sacher, R. Mandeville, M. Meunier, Biosens. Bioelectron. 37 (1) (2012) 24−29.

[135] A.D. Chowdhury, A. De, C.R. Chaudhuri, K. Bandyopadhyay, P. Sen, Sensor Actuat. B-Chem. 171 (2012) 916−923.

[136] S. Jin, M. Dai, B. Ye, S.R. Nugen, Microsyst. Technol. 19 (12) (2013) 2011−2015.

[137] L. Jinhua, L. Tao, M. Rizeng, N. Dandan, L. Yang, X. Ligang, J. Food Saf. Qual. 5 (4) (2014) 1142−1146.

[138] Y. Li, R. Afrasiabi, F. Fathi, N. Wang, C. Xiang, R. Love, Z. She, H. Kraatz, Biosens. Bioelectron. 58 (2014) 193−199.

[139] T.J. Gnanaprakasa, O.A. Oyarzabal, E.V. Olsen, V.A. Pedrosa, A.L. Simonian, Sensor Actuat. B-Chem. 156 (1) (2011) 304−311.

[140] H. Wang, Y. Li, M. Slavik, Int. J. Poult. Sci. 13 (11) (2014) 611−618.

[141] M.C. Morant-Miñana, J. Elizalde, Biosens. Bioelectron. 70 (2015) 491−497.

[142] M. Manzano, F. Cecchini, M. Fontanot, L. Iacumin, G. Comi, P. Melpignano, Biosens. Bioelectron. 66 (2015) 271−276.

[143] P. Banerjee, A.K. Bhunia, Biosens. Bioelectron. 26 (1) (2010) 99−106.

[144] J. Wang, Y. Luo, B. Zhang, M. Chen, J. Huang, K. Zhang, W. Gao, W. Fu, T. Jiang, P. Liao, J. Transl. Med. 9 (1) (2011) 85−94.

[145] A. Palaniappan, W. Goh, D. Fam, G. Rajaseger, C. Chan, B. Hanson, S. Moochhala, S. Mhaisalkar, B. Liedberg, Biosens. Bioelectron. 43 (2013) 143−147.

[146] D. Jiang, F. Liu, C. Liu, L. Liu, Y. Liu, X. Pu, Anal. Methods 6 (5) (2014) 1558−1562.

[147] D. Jiang, F. Liu, L. Zhang, L. Liu, C. Liu, X. Pu, RSC Adv. 4 (100) (2014) 57064−57070.

[148] S. Balasubramanian, I.B. Sorokulova, V.J. Vodyanoy, A.L. Simonian, Biosens. Bioelectron. 22 (6) (2007) 948−955.

[149] G.A. Zelada-Guillén, J.L. Sebastián-Avila, P. Blondeau, J. Riu, F.X. Rius, Biosens. Bioelectron. 31 (1) (2012) 226−232.

[150] F. Jia, N. Duan, S. Wu, X. Ma, Y. Xia, Z. Wang, X. Wei, Microchim. Acta 181 (9−10) (2014) 967−974.

[151] J. Shi, C. Chan, Y. Pang, W. Ye, F. Tian, J. Lyu, Y. Zhang, M. Yang, Biosens. Bioelectron. 67 (2015) 595−600.

[152] N. Hiremath, R. Guntupalli, V. Vodyanoy, B.A. Chin, M. Park, Sensor Actuat. B-Chem. 210 (2015) 129−136.

[153] H. Wang, C. Lei, J. Li, Z. Wu, G. Shen, R. Yu, Biosens. Bioelectron. 19 (7) (2004) 701−709.

[154] Y. Luo, X. Liu, T. Jiang, P. Liao, W. Fu, Anal. Chem. 85 (17) (2013) 8354−8360.

[155] S. Xu, C. Zhang, L. He, T. Wang, L. Ni, M. Sun, H. Miao, J. Zhang, Z. Dai, B. Wang, J. Nanomater. 2013 (2013) 62−68.

[156] S. Jiang, E. Hua, M. Liang, B. Liu, G. Xie, Colloids Surf. B Biointerfaces 101 (2013) 481−486.

[157] L. He, L. Ni, X. Zhang, C. Zhang, R. Li, S. Xu, Int. J. Biochem. Res. Rev. 6 (3) (2015) 130−139.

[158] K. Nieradka, K. Kapczyńska, J. Rybka, T. Lipiński, P. Grabiec, M. Skowicki, T. Gotszalk, Sensor Actuat. B-Chem. 198 (2014) 114−124.

[159] A.B. Dahlin, Sensors 12 (3) (2012) 3018−3036.

[160] A. Bacaloni, C. Cavaliere, A. Faberi, E. Pastorini, R. Samperi, A. Laganà, J. Agric. Food Chem. 53 (14) (2005) 5518−5525.

[161] O.G. Cabrices, F.F. Foster, E.A. Pfannkoch, Glob. Anal. Solutions (2013) 10/2−10/9.

[162] R.A. Devlin, K. Campbell, K. Kawatsu, C.T. Elliot, Harmful Algae 10 (2011) 542−548.

[163] T.R. Hofhine, E.K. Krantz, P. Doolittle, C. Barta, Agro Food Ind. Hi-Tech 26 (1) (2015) 12−15.

[164] D.I. Greenfield, R. Marin, G.J. Doucette, C. Mikulski, K. Jones, S. Jensen, B. Roman, N. Alvarado, J. Feldman, C. Scholin, Limnol. Oceanogr: Methods 6 (12) (2008) 667−679.

[165] R.R.G. Soares, P. Novo, A.M. Azevedo, P. Fernandes, M. Aires-Barros, V. Chu, J.P. Conde, Lab Chip 14 (21) (2014) 4284−4294.

[166] M. Kim, S. Choi, Biosens. Bioelectron. 66 (2015) 136−140.

Chapter 5

Biosensor to Ensure Food Security and Environmental Control

G.A. Evtugyn
Chemistry Institute of Kazan Federal University, Kazan, Russian Federation
E-mail: Gennady.Evtugyn@kpfu.ru

Chapter Outline

1. INTRODUCTION

There is an urgent need in the simple and reliable instruments for the assessment of food quality and potential hazards related to the food processing and contamination. Although modern technologies in agriculture sufficiently decrease direct risks related to the toxic species in foodstuffs, the variety of chemicals applied in agriculture as well as underestimated contamination sources call for the further efforts in the above area.

Biosensors are portable analytical instruments that integrate biological recognition element (enzyme, antibody, DNA, whole cells) with appropriate transducer converting biochemical recognition in the electric signal [1]. To

Comprehensive Analytical Chemistry, Vol. 74. http://dx.doi.org/10.1016/bs.coac.2016.03.017

some extent, the choice of a biosensor for the solution of particular analytical problem is based on the biochemical function of an analyte molecule. In this sense, biosensor mimics real biochemical paths that are moved from a lining being on the biosensor interface.

Application of biosensors has found predominant attention in medicine for the determination of metabolites and biomarkers of various diseases including direct diagnostics of pathogenic microorganisms and viruses [2−4]. Biosensors market, and mainly that of diagnostics devices, was estimated in US$8.5 billion in 2012 and up to US$16.8 billion by 2018 [5]. Regarding other areas of biosensor application, the interest to environmental problems has been initiated in 1980s by adaptation of the alarm systems previously designed for chemical warfare detection [6]. Organophosphate and carbamate pesticides exert toxic effect similar to that of nerve gases and are detected by the quantification of the inhibition of cholinesterase enzyme [7]. Less specific toxic effect of heavy metals can be recorded by various oxidoreductases and hydrolases [8]. Although the maximum of the interest to the above analytes in environmental monitoring is now in the past due to their reduced intake in the environment, the interest to the biosensor-based detection of individual chemicals remains in the scope of research efforts because of the chemical terrorism and industrial accidence threatening.

Previously in the chapters of the monograph, the possibilities for the biosensor-based determination of pesticides, heavy metals, toxins and other pathogens have been carefully considered. In this chapter, biosensors-based approaches have been extended to detection of some native food components (spices, flavour additives, antioxidants and preservatives) as well as to the general estimation of potential risks related to food contamination (toxicity biosensors). In the latter case, main attention is paid to the estimation of the total impact of toxicants expressed in the quantities of a standard contaminant or directly referred to the biological tests applied for the same purpose.

In most cases, the chapter covers biosensors developed within last 10 years. Main attention is paid to the devices that meet the definition given previously, ie, those with biological component directly attached to the transducer surface and providing information on the sample content in a real-time scale. Among signal detection system, electrochemical methods have obvious advantage of simple design and variety of measurement techniques available for laboratory and field applications. A rare exclusion, bioluminescence described in the toxicity testing, is mostly based on the techniques, which are rather far from traditional biosensor design. In this part, compact devices applicable for field measurements are mostly considered.

2. BIOSENSORS FOR FOOD ADDITIVES AND QUALITY CONTROL

Food additives are intentionally added to food for maintaining its appearance, flavour and nutrition value and for prevention of the microbial spoilage. In addition to food preparing and storage, additives can be required also in packaging and transportation of food. Some additives like glutamate or sweeteners can be used in special food for consumers with specific dietary needs, or in accordance with local or cultural requirements including those related to religious and ethnic limitations (the laws of kashrut or halal food). In accordance with classification, the following types of the food additives are specified: texturizing agents, colourants, flavouring agents and miscellaneous additives. Some of them, eg, alcohol, vinegar, oils and spices have been used for several thousand years [9]. Meanwhile the majority of food additives were introduced for less than 100 years in the period of accelerated urbanization related to industrial revolution. Processed food and new cooking technologies resulted in the food-related poisoning that occurred more and more often. The use of many additives (2500 different chemicals since 1950 in the United States) diminished the requirements to raw materials used for prepared food and masked the deficiency in nutrients and vitamins, or microbial spoilage often happen in prolonged storage period.

Modern application of food additives is a subject of strict legislation [10] including the analytical methods for their determination in appropriate matrices, which are mostly based on chromatography and to some extent on immunoassay techniques. Nevertheless, portable sensing devices including biosensors are denuded for preliminary testing food and raw food materials to exclude the income of toxic species and be sure in high quality of the products. The analysis of food additives is also directed on the detection of chemicals, which were recently excluded from the list of additives allowed for use because of new information of their toxicity and human health effect. The summary of biosensors developed for food additives analysis is presented in Table 1.

It should be mentioned that most of the biosensors were tested with standard substrate solutions and only few of them were validated on the analysis of spiked samples of fruits, juices or pharmaceutical preparations. Modification of the transducer is mostly directed on decrease of the working potential observed in the presence of mediators of electron transfer placed on the transducer (Co phthalocyanine, Meldola's blue, tetrathiafulvalene) or added to the solution (ferricyanide ions). In addition to mediators, the electrodes can be modified with membrane films limiting access of interferences (Clark-type oxygen electrodes) or mechanically preventing damage of the biolayer. The use of electropolymerized materials simplifies the assembling of the surface layer and can combine the stages of mediator deposition and enzyme immobilization. However, some of the polymerization products like poly(phenylene diamine) do not participate in the electron transduction and decrease both permeability of the surface layer and currents recorded.

In addition to biosensors, some species can be also detected by electrochemical techniques with the electrodes modified with mediators and electrocatalysts. This is especially true for antioxidants that are rather easily oxidized on such electrodes [33]. Meanwhile the use of biochemical elements improves either selectivity of the response or sensitivity of the assay.

TABLE 1 Biosensors for Food Additives Testing (2005–2015)

Analyte	Biorecognition Element	Signal Detection Conditions	Concentrations Tested	References
Sweeteners				
Aspartame	Alcohol oxidase–carboxylesterase in BSA matrix	Screen-printed carbon electrode modified with Co phthalocyanine	LOD 0.2 µM, conc. range 5–600 µM (FIA regime)	[11]
Aspartame	Alcohol oxidase–carboxylesterase in gelatin film	Graphite epoxy composite electrode	Conc. range 2.5–400 µM, win, beer and diet cola testing	[12]
Aspartame	Alcohol oxidase–α-chymotrypsin, glutaraldehyde binding	Pt electrode (H_2O_2 detection)	LODs 0.2, 0.4 and 1.0 µM in batch, flow-through and FIA regimes, beverages testing	[13]
Sorbitol	D-sorbitol dehydrogenase	Carbon paste electrode with NAD^+ and enzyme covered with poly(o-phenylene diamine)	LOD 40 µM, conc. range up to 0.8 mM	[14]
Sorbitol	D-sorbitol dehydrogenase	Glassy carbon (GCE) covered with carbon nanotubes (CNTs) dispersed in hyaluronic acid	LOD 16 µM, conc. range up to 1.2 mM	[15]
Sorbitol	D-sorbitol dehydrogenase on CNTs with Nafion/chitosan outer layer	GCE modified with CNTs dispersion	LODs 119 (Nafion) and 15 (chitosan), conc. range up to 2 mM	[16]
Acesulfame-K, cyclamate and saccharin	Langmuir–Blodgett egg phospholipid membrane	Ion current signal after exposure to sweetener	Conc. range 1–15 (acesulfame-K), 10–160 (cyclamate) and 0.4–7 (saccharin) µM, spiked samples testing	[17]

Flavour additives

Glutamate	Glutamate oxidase in poly(carbamoyl-sulphonate) hydrogel	Thick-film Pt electrode (H_2O_2 detection)	LOD 1.01 µM, conc. range 0.1−5 mM	[18]
Glutamate	Glutamate oxidase in BSA matrix with glutaraldehyde or poly(ethylene glycol)diglycidyl-ether cross-linking	Pt or Pt/Ir wire electrodes covered with Nafion/poly(phenylene diamine) coating	LOD 1 µM, conc. range up to 100 µM	[19]
Glutamate	Glutamate oxidase in BSA matrix cross-linked with glutaraldehyde	Pt microelectrode array on silicon base covered with Nafion−polypyrrole layers	Conc. range 5−300 µM	[20]
Glutamate	Glutamate oxidase cross-linked with glutaraldehyde	Au electrode covered with polypyrrole nanoparticles and polyaniline	LOD 0.1 nM, conc. range 0.02−400 µM	[21]
Glutamate	Glutamate oxidase cross-linked with glutaraldehyde	Carbon paste or screen-printed carbon electrode with tetrathiafulvalene-tetracyanoquinodimethane	LOD 0.05 mM, conc. range 0.15−3 mM (logistics curve), spiked juice, ketchup and paste samples	[22]
Glutamate	Glutamate oxidase co-immobilized with NADP$^+$ in siloxane polymer	Carbon paste electrode covered with poly(Methylene green)	LOD 5 µM, conc. range 50 µM−10 mM	[23]
Glutamate	Glutamate dehydrogenase and diaphorase in chitosan matrix	CNTs and melted N-eicosane paste electrode, measurements in the presence of NAD$^+$ and ferricyanide ions added to the solution	LOD 5.4 µM, conc. range 10 µM−3.5 mM, spiked samples of soya sauce, liquid spice, chicken bouillon, garlic instant soup and herb butter spice	[24]
Glutamate	Glutamate dehydrogenase in CNTs/chitosan matrix	Screen-printed carbon electrode covered with Meldola's blue	LOD 3 µM, conc. range 7.5−105 µM, spiked samples of beer, foetal bovine serum	[25]

Continued

TABLE 1 Biosensors for Food Additives Testing (2005–2015)—cont'd

Analyte	Biorecognition Element	Signal Detection Conditions	Concentrations Tested	References
Capsaicin	Horseradish peroxidase co-immobilized with ferrocene in poly(2-hydroxyethyl methacrylate)	Screen-printed carbon electrode	LOD 1.94 µM, conc. range 2.5–99.0 µM, spiked chilli samples	[26]
Vitamins				
Thiamine	Whole cells of *Saccharomyces cerevisiae* in gelatin on Teflon membrane	Clark-type oxygen sensor (respiration detection)	LOD 0.005 µM, conc. range 0.005–0.1 µM, pharmaceutical preparations assay	[27]
Vitamin B12	Whole cells of *Tetrasphaera duodecadis* physically adsorbed on cellulose filter	Clark-type oxygen sensor (respiration detection)	Conc. range $1 \times 10^{-7} – 1 \times 10^{-5}$ M, commercial vitamin B12 tablets	[28]
Ascorbic acid	Ascorbate oxidase entrapped in poly(3,4-ethylene dioxythiophene) film	Pt disk electrode	LOD 0.464 µM, conc. range 2.0 µM–15 mM, orange and grapefruit juices, cabbage, tomato, cucumber, turnip	[29]
Ascorbic acid	Ascorbate oxidase cross-linked in egg shell membrane	Au electrode	LOD 10 µM, conc. range 10 µM–0.4 mM, orange, apple, lemon and grapefruit juices	[30]
Ascorbic acid	Cucumber tissues	Clark-type oxygen sensor	LOD 1.4 mM, conc. range 1.4–22.7 mM, orange, lemon, sweet potato, tomato, grapes, chilli, apple, banana	[31]
Ascorbic acid	Ascorbate oxidase bound to poly-maleimidostyrene in polystyrene membrane	Aminated GCE and Au electrode as oxygen sensors	LOD 2 µM, conc. range 5 µM–0.4 mM	[32]

The examples presented in Table 1 do not exhaust the variety of biosensors applied for antioxidant detection. Thus, the use of uricase makes it possible to detect uric acid [34]. Glutathione peroxidase catalyses oxidation of reduced form of glutathione [35]. However, such biosensors are mainly intended for the analysis of biological fluids for diagnostics purposes.

A number of biosensors have been designed to the detection of *preservatives* used in processed food. Thus, benzoic acid can be detected by its inhibitory effect on tyrosinase. Thus, amperometric sensor with enzyme immobilized in polyaniline—polyacrylonitrile matrix with catechol as substrate made it possible to detect down to 0.2 μM of benzoic acid [36]. In similar conditions with enzyme entrapped in polyaniline matrix, the limit of detection (LOD) value of 0.3 μM was reported for sodium benzoate [37]. Glassy carbon electrode (GCE) modified with $CaCO_3$ nanocrystals with adsorbed tyrosinase was successfully applied for the determination of 0.56—92 μM of benzoic acid with catechol as substrate. The results of benzoate determination in spiked samples of Coca-Cola, Pepsi-Cola, Sprite and yoghurt were validated by the HPLC analysis [38]. The use of extract from fungi *Agaricus bisporus* with Clark-type oxygen sensor made it possible to detect tyrosinase activity with L-tyrosine as substrate and determine from 6 to 14 mg of sodium benzoate in model solution [39]. The biosensor was applied to test the samples of refreshment guarana.

Sulphite is used as preservative in many products like processed meats, wine and beer, soft drinks, fruit juices, dried fruits and shellfish [40]. Sulphites can be directly determined in the food samples with biosensors based on the sulphite oxidase, isolated or presented in the biological tissue homogenate. The immobilization in gelatin film by cross-linking with glutaraldehyde makes it possible to detect 0.2—2.8 mM of sulphite by the current related to reduction of dissolved oxygen. The reaction scheme involves interaction of sulphite anion with oxygen resulting in formation of sulphate and hydrogen peroxide. The biosensors results were compared with conventional analytical tools for sulphite determined in vinegar, creeker, sesame and ready soup [41]. In another sulphite biosensor based on the signal of hydrogen peroxide oxidation, the enzyme was mixed with the carbon paste and placed in flow channel of the thin cell as working electrode [42]. The current was corrected on the possible interference contribution estimated with additional working electrode with no enzyme. Signal linearly depended on sulphite concentration in the range from 0.1 to 1.00 mM. The flow-through chip was applied for the detection of sulphite in beer. Direct immobilization of *Malva vulgaris* homogenate by cross-linking in gelatin on Teflon membrane of Clark-type oxygen sensor provided the signal on sulphite concentration from 0.2 to 1.8 mM [43]. The results of the sulphite determination in the pickle water, biscuit, beer, soup and vinegar were in agreement with the enzymatic determination with the spectrophotometric signal measurement. Alternatively, sulphite can be determined by their inhibitory effect exerted on tyrosinase [44].

2.1 Taste Sensing

The detection of sweeteners and flavour additives can be considered as a particular case of taste sensing. The use of such biomimetic systems for food quality and safety characterization is reviewed in Refs [45,46]. Generally, such systems include a number of sensors with cross-selectivity of the response towards various species that determine the taste (electronic tongue) or smell (electronic nose). The signals of the sensors are processed with the software based on nonlinear statistical treatment (pattern recognition). The system is calibrated on standard samples with appropriate taste scale or on individual chemicals used as taste carriers (glucose for sweetness, sodium chloride as saltiness, etc.). The taste systems are commonly used to solve two kinds of problems, ie, qualitative classification of the samples based on a priori expressed rules (freshness, maturation step, areas of origin) and multivariate calibration of the content (determination of sugars, bitterness agents, etc.). In the first case, the similarity of unknown samples to the standard(s) is established with the following estimation of the degree of similarity. The detection of fruit/vegetable ripeness [47] or wine maturity [48] is an example of such an approach. In the second case, the content of appropriate compound important for food quality assessment is determined. Many of electronic tongue systems are based on conventional electrochemical sensors, eg, ion-selective electrodes [49] or voltammetric techniques [50]. They are beyond the scope of the chapter. Meanwhile, some of the taste sensors utilize biosensor approach.

Thus, one of the first taste sensors developed by Toko [51], included modifiers mimicking lipophilic membranes of the human taste receptors. The membranes of individual electrodes contained phosphoric acid di-*n*-hexadecyl ester and tetra-dodecylammonium bromide, and a plasticizer, dioctyl phenylphosphonate. The detection of electric charge of membrane surface made it possible to estimate sweetness of the sample expressed in sucrose concentration [52].

Natural and artificial sweeteners were distinguished using microelectrode array combined with intact taste epithelium [53]. Electrophysiological measurements performed with sweetness stimuli made it possible to detect separately sucrose, glucose, saccharin and cyclamate. The classification of the signals was performed on the base of temporal resolution of responses. Similar investigations were performed for establishment of dose—response relationship with salt receptor cells from rat placed on microelectrode array [54].

Horseradish peroxidase and tyrosinase were used in the assembly of the amperometric biosensors based on screen-printed electrodes developed for the detection of phenols referred to the bitterness and pungency of virgin olive oils [55]. The content of taste components was independently determined by HPLC and the bitterness score of oil samples by trained sensory panel in accordance with EU regulations. The taste intensity of the olive oil was mainly attributed to the content of the following phenols: hydroxytyrosol, tyrosol, and dialdehyde forms of decarboxymethyl oleuropein aglycone; of decarboxymethyl ligstroside aglycone, of oleuropein aglycone; and of ligstroside aglycone. High correlation between the total content of phenols and signals of biosensors was established with principal component analysis and partial least square regression. Peroxidase biosensor showed significant correlation with pungency specified by sensory score. Meanwhile the correlation between the biosensor signals and sensory panel scores was found to be weaker than that with chemical analysis data. The improvement of extraction procedure and extension of the number of taste components is recommended to improve the taste prediction.

A number of biosensors are based on the direct use of human receptors of taste. Thus, colorectal carcinoma cells expressing α-gustducin and sweet taste receptor T1R1/T1R3 were used for modification of screen-printed carbon electrode [56]. Impedance spectra detected at fixed frequency of electric polarization made it possible to separately specify four taste carriers (sucrose, hydrochloric acid, sodium chloride and magnesium sulphate). Negative control with similar sensor bearing mammalian COLO-205 cells did not show ability of taste detection. Later on, the same detection system was applied for the separate detection of sweetness and bitterness using human enteroendocrine NCI-H716 cells expressing G protein-coupled receptors and sweet receptors together with human enteroendocrine STC-1 cells expressing G protein-coupled receptors and bitter receptors [57].

3. TOXICITY BIOSENSORS

Contrary to detection of particular chemical species, toxicity biosensors are intended to estimate potential hazards related to specific biochemical paths or poisoning of the whole organism. In some cases, appropriate biosensor devices are called as 'alarm systems' because they should control severe contamination of the foodstuffs or environment that might be dangerous for living beings immediately after intake (acute toxicity).

3.1 Cholinesterase-Based Biosensors

Cholinesterase-based biosensors are one of the first examples of such an approach developed for preventing the intoxication with species suppressing the nerve pulse transmission [58]. This enzyme catalyses the hydrolysis of acetylcholine, a natural neurotransmitter participating in the nerve impulse transduction (1).

$$(CH_3)_3N^+CH_2CH_2OCCH_3 + H_2O \longrightarrow (CH_3)_3N^+CH_2CH_2OH + \text{Acylated cholinesterase}$$

Acetylcholine   O   Choline   ⌐ H_2O

⌐ CH_3COOH (1)

Acetylcholinesterase

Depending on the substrate specificity, several cholinesterase types are classified among them acetylcholinesterase and butyrylcholinesterase are most known. The term 'cholinesterase' is mainly used for raw preparations containing several types of the enzyme or in the case when all the enzyme types are meant.

Although organophosphates and carbamates are main analytes detected by the cholinesterase-based sensors, surfactants [59], heavy metal salts [60,61], fluorides [62], glycoalkaloids [63], aflatoxin [64] and urea [65] also affect the enzyme activity. Such interactions interfere with pesticide determination but make it possible to consider the cholinesterase inhibition as a measure of total contamination of the samples tested.

First alarm systems based on cholinesterase inhibition were adapted from the military devices developed for the detection of chemical warfare. Thus, Cholinesterase Antagonist Monitor, CAM-1, has been developed for 3 min detection of toxic and subtoxic levels of cholinesterase inhibitors present in water [66]. Cholinesterase was immobilized in a polyurethane foam placed between two net electrodes. Acetylthiocholine iodide was passed through the system, and the electrode potential was measured at a constant current polarization. Fast increase of the potential indicated the enzyme inhibition. A minicomputer was used to automate the detection and to signal an alarm when necessary. Next prototype (CAM-2) included the pumps for aqueous extraction of anticholinesterase agents from the air. Automatic replacement of the inhibited enzyme reactor was also provided.

The possibility of adequate assessment of the pollution based on acetylcholinesterase inhibition was later discredited because many highly toxic species, eg, polyaromatic hydrocarbons or dioxins, did not inhibit cholinesterase. Even petroleum components at rather high concentrations did not affect the enzyme activity. For this reason, the cholinesterase-based sensors were combined with some other biochemical receptors or conventional hydrochemical parameters, eg, chemical (COD) and biochemical oxygen demand (BOD) to extend the number of the contaminants detected. Thus, the use of urease [67], tyrosinase and peroxidase [68,69] in combination with acetylcholinesterase and/or butyrylcholinesterase made it possible to increase relative sensitivity of the signal towards heavy metals and phenolics, respectively. To distinguish the contribution of various groups of toxicants, chemometric data treatment was successfully applied together with some toxicological parameters, eg, acute toxicity determined on *Vibrio fischeri* and other biological test organisms. Luminescent toxicity testing was used for comparison with the characteristics of the cholinesterase biosensors developed for the detection of the parathion photodegradation products [70]. Acute toxicity of industrial wastewaters for *Paramecium caudatum* was compared with the inhibition degree determined with butyrylcholinesterase potentiometric sensor [71]. The use of artificial neural nets or linear discriminant analysis made it possible to classify the sources of wastewaters in accordance with their toxic components. For this purpose, BOD and COD values of industrial wastewaters were used together with inhibition degree as input signals. Nevertheless, such approaches do not guarantee reliable prediction of toxicity because of great variety of the factors affecting the response of cholinesterase sensors. Irreversible character of inhibition exerted by most toxic species also prevents continuous operation of such alarm systems.

3.2 Other Enzymatic Sensors

Total amounts of inhibitors can be measured using some other enzyme systems sensitive to a variety of factors affecting their activity. In biosensor format, two examples should be mentioned, ie, horseradish peroxidase and tyrosinase.

Horseradish peroxidase catalyses oxidation of various organic substrates AH_2 by hydrogen peroxide and organic hydroperoxides (2).

$$AH_2 + H_2O_2 \xrightarrow{Peroxidase} A + 2H_2O \tag{2}$$

Heavy metals [72,73], phenolics [74−76], cyanides [77,78] and thiourea derivatives [79] inhibit peroxidasea activity. Mechanism of inhibition includes complexation of Fe in the haem group of the enzyme active site, interaction with the thiol groups followed by the distortion of the protein structure, co-oxidation of the phenolic compounds together with the substrate used for the peroxidase activity measurement. Horseradish peroxidase is widely used as a label in the immunochemical assays, the DNA hybridization detection and as a second enzyme in the oxidoreductase-based biosensors. For this reason, the conditions for the peroxidase activity measurements and potential inhibition elimination are well known. Meanwhile kinetic consideration of the inhibition especially that observed in the mixtures of oxidizable phenols calls for further consideration and application in total toxicity estimation. Thus, the influence of wastewaters constituents on the peroxidase oxidation of phenols was referred to the sulphur containing intermediates and changes in the ionic strength of the diluted wastewaters [80].

Tyrosinase catalyses hydroxylation and oxidation of various phenols as shown in scheme (3) as example.

$$\tag{3}$$

Carbamate [81], organophosphate [82] insecticides, atrazine [83,84] and chlortoluron [85] herbicides, sulphite [86] and fluoride [87] inhibit tyrosinase activity. Together with possibilities of the total estimation of phenols in foodstuffs [88], this offers good opportunities for the application of tyrosinase sensors to monitor toxicity of various samples.

A concept of bioelectronics tongue with biochemical, hydrochemical and toxicological parameters has been developed and applied for wastewaters classification [89]. For this purpose, two types of electrochemical sensor arrays were designed. In the first array, butyryl- and acetylcholinesterase and oxidureductases (horseradish and soya bean peroxidases, cellobiose dehydrogenase and tyrosinase) were immobilized on individual electrodes of screen-printed eight-electrode array. The signal of individual biosensors was measured in the mixture of thiocholine and catechol as specific enzyme substrates. Another array involved glucose dehydrogenase, alone and co-immobilized with catalase, haemoglobin and lactate oxidase immobilized in the polyurethane film on the Pt electrodes. In addition, the cathodic and anodic currents were recorded with carbon and Pt bare electrodes. The relationship between the enzyme responses and general pollution parameters including BOD, COD, total organic carbon (TOC), Microtox test and algae growth inhibition was established using principal component analysis and partial least square regression. The use of the biosensors response made it possible to predict the toxicity level of real samples of wastewaters of a Swedish chemi-thermo-mechanical pulp. Moreover, the use of hydrochemical parameters (COD, BOD and TOC), which determination is time- and labour-consuming, can be substituted with voltammetric measurements with carbon electrode.

3.3 Bioluminescent Microbial Biosensors

Among other enzyme systems, bioluminescence enzyme has found attention due to ability to generate light in the ATP assisted oxidation of a specific substrate. Scheme (4) illustrates mechanism of firefly bioluminescence based on luciferin oxidation performed in the presence of ATP and luciferase.

$$\text{Luciferin} + \text{ATP} + O_2 \xrightarrow[\text{Mg}^{2+}]{\text{Firefly luciferase}} \text{Oxyluciferin} + \text{AMP} + \text{PP}_i + CO_2$$

$$(4)$$

Native luciferase is instable and cannot be used in biosensor assembly. Instead, genetically modified microorganisms with firefly luciferase (*luc*) gene have been produced and used in the toxicity tests [90]. The principal idea consists of monitoring the reporter gene product activity which decreases with various stress factors, including intoxication. The ATP synthesis plays significant role in the luminescence intensity. Any disruption of the electron transport inhibits the ATP production and hence reduces the light emission.

In a similar manner, the luciferase from marine bacterium *V. fischeri* catalyses oxidation of long-chain fatty aldehydes (5). The catalytic cycle needs NAD(P)H/flavin mononucleotide (FMN) system to recover $FMNH_2$ in the presence of flavin reductase. The reaction requires a set from two to five reporter genes, the protein produced is heat labile and does not work in the absence of oxygen, ie, in anaerobic conditions.

$$
\begin{array}{c}
\text{Luciferase} \\
\text{FMNH}_2 + \text{O}_2 + \text{RCHO} \xrightarrow{\hspace{3cm}} \text{FMN} + \text{H}_2\text{O} + \text{RCOOH} + h\nu \\
\text{Flavin reductase} \\
\text{NAD(P)H} + \text{H}^+ + \text{FMN} \xrightarrow{\hspace{3cm}} \text{NAD(P)}^+ + \text{FMNH}_2
\end{array}
\tag{5}
$$

Bioluminescence inhibition assay is applicable for single toxicants and their mixtures present in almost all kinds of samples, eg, surface and groundwater, municipal effluents, extracts from sediments, etc. [91]. In most cases, toxic effect results in the suppression of luminescence. Universal mechanism of action, application to the mixtures of chemicals, short-time exposure are often mentioned as advantages of bioluminescent assays. Unfortunately, false-positive results can occur in case of any decrease in the metabolic activity. For this reason, tests require careful establishment of optimal media conditions for culture growth and measurement protocol.

Conventional testing is performed by incubation of lyophilized microorganisms in the sample tested followed by the addition of the substrates. The same scheme can be realized in portable devices that simplify the assay and decrease the amounts of reagents needed. The summary of portable devices for the cell-based bioluminescent inhibition assay is presented in the review [92]. Among other commercial microbial devices, Microtox (Azure, Bucks, UK) [93] and ToxAlert (Merck, Darmstadt, Germany) are mostly known in toxicity testing. Correlation of the toxicity assessment established with commercial bioluminescence tests utilizing *V. fischeri* was found for more than 1300 individual compounds [94]. Over the years, alternative tests based on more robust *Photobacterium leiognathi* [95] and *Janthinobacterium lividum* [96] have been developed.

The *lux* operon responsible for bioluminescence can be introduced in many microorganisms. Coupled to specific promoters, it allows following the expression in presence of specific analytes or environmental factors and screening both the environmental conditions and specific contaminants. The activity of toxic species results mainly in increasing light emission observed during a certain period after incubation of the sample in the cell suspension. The review of bioluminescence assays related to genetically modified bacteria is presented in Ref. [97].

Immobilization of luciferase or whole cells of appropriate microorganisms normally improves their performance for toxicity testing due to stabilization of the signal, possibilities for co-immobilization of auxiliary reagents and enzyme systems, multiple use and reduced cost of measurement. The immobilization protocols do not differ from those developed for other enzymatic biosensors. Most attention is paid to accessibility of the analytes to enzymes and providing appropriate microenvironment favourable for light emitting. In simple devices, enzymes are implemented with the reagent in the gelatine or starch gel to be dissolved in the cuvette of bioluminometer [98]. In a similar manner, bacteria can be adsorbed on solid supports mainly in the presence of stabilizers like agarose or polyelectrolytes.

Later, some examples of optical biosensors are briefly reviewed with particular emphasis to integrated devices that meet the definition of biosensor and able to conduct toxicity assessment in real-time scale.

Photosensitive PVA-SbQ hydrophilic polymer makes it possible to prepare a membrane with luminescent microorganisms to be attached to the optical fibre connected to luminometer [99]. The analytical performance of the biosensor equipped with flow-through microcell was tested using hydrogen peroxide as oxidative stress model.

Bioluminescent *Escherichia coli* strains with plasmids of *luc* gene were immobilized on the walls of the channels of microfluidic chip by physical adsorption from agarose suspension [100]. A perforated five-by-five microwell array was filled with luciferin/ATP solution and with a model mutagen (mitomycin C) and then the intensity of luminescence was recorded. Solid-state photodetectors were used for measurements of light emission from *E. coli* strain with *lux*-CDABE genes and *E. coli lac* promoter activated by isopropyl-β-D-thiogalactopyranoside as a model of toxicant [101]. Modulation of the optical signal with a special electrically driven shutters increased sensitivity of the detection system.

Real-time biomonitor of wastewaters pollution has been developed from microfluidic chambers harbouring microwells with genetically engineered *E. coli* and aligned single photon avalanche diode detectors [102]. The chambers are connected to four feeding tubes and four tubes for solutions discharge. Three inducible bacterial reporter strains incorporating fusions between the lux reporter genes and promoters activated by DNA damage, oxidative stress and heavy metals. The system was tested using arsenic (6 mg/L), nalidixic acid (20 mg/L), and paraquat (50 mg/L) and the response period ranged from 0.5 to 2 h.

Vibrio fischeri flow-through chip has been designed for continuous detection of water pollution [103]. The chip included two counter-flow micro-mixers, a T-junction droplet generator and six spiral microchannels to mix together the sample, buffer solution and cell suspension and form a droplet within the air flow. The signal was recorded with photomultiplayer tube located at the top of observation chamber (20 μL). The experiments were performed with various model toxicants, ie, Zn(II) and Cu(II) salts, cetyl-pyridinium chloride, potassium dichromate, and 3,5-dichlorophenol taken alone and in various mixtures dissolved in deionized water and artificial drinking water. The response was found to be about 20 min and the detectable concentrations were in 2–20 mg/L range.

Online continuous monitoring of the water contamination has been proposed with the porous alumina chip with recombinant reporter bacteria [104].

High porosity of the material (40 vol.%) provided the support for microorganisms and their access to nutrients diffusing through the chip. Gene promoters applied made it possible to detect DNA damaging factors, oxidative stress, 2,4-dinitrotoluene and some by-products. Besides *lux*, genes of mutated green fluorescent protein (*gfp* and *yqjF*) with green and red fluorescence emission were introduced to distinguish the response. The chip was tested on model solution of nalidixic acid and hydroquinone. In freeze-dried storage conditions, the biosensor retained its activity for at least 12 weeks.

The *lux* and *gfp* genes are often used in the microbial sensors with specific reporters to detect individual contaminants and the products of their degradation. The *lux* gene is commonly faster than *gfp* but less stable. The *gfp* gene can be used only for long-term exposure but normal GFP protein is very stable and cannot be used as dynamic reporter in bacteria [105].

Magnetospirillum gryphiswaldense strain MSR-1 was bioengineered with a gene of red-emitting click beetle luciferase [106]. Magnetic properties of this bacterium were applied for its positioning in the detection area of measurement device called MAGNETOX chip. Light emission measurements were performed in the microfluidic chip in four chambers made in the middle layer and connected to two detection areas equipped with constant magnets and the Sony ICX285 monochrome CCD sensor. The portable device was tested on dimethyl sulphoxide and representative of bile acids (taurochenodeoxycholic acid).

3.4 Respiratory Activity Inhibition

In such microbial biosensors, the oxygen consumption is recorded with appropriate transducer as a measure of metabolic activity and amount of organic species involved in microbial oxidation. The respiratory activity can be applied for the detection of organic contamination of natural and wastewaters as an analogue of BOD, a parameter determined in preliminary testing of industrial wastewaters and municipal sewage prior to their biological treatment. Standard closed bottom test described for BOD estimation includes incubation of the sample diluted with water containing native microorganisms taken from the environment and determination of the oxygen concentration decreased due to microbial oxidation of organic matter. This standardized method is quite reliable but time-consuming (the incubation covers several days). Microbial BOD sensors can utilize either specific strands of microorganisms or their consortia, eg, sewage sludge or soil extracts [107]. The changes in the oxygen concentration can be determined directly by the Clark-type oxygen sensor or with mediated sensors mainly utilizing ferricyanide ions as artificial electron acceptor. The reaction can be presented by scheme (6).

$$\text{Organic matter} + \text{Fe(CN)}_3{}^{3-} \xrightarrow{\textit{Microorganisms}} \text{Oxidation products} + \text{Fe(CN)}_6{}^{4-}$$

$$(6)$$

The use of mediators avoids some technical problems typical for closed bottle test, ie, unexpected contact of the mixture with atmospheric oxygen or dependencies of the degradation rate on the dissolved oxygen concentration. To accelerate the electron exchange, lipophilic mediator, menadione, able to penetrate biological membranes was added to hydrophilic ferricyanide present in the sample [108].

A high BOD value indicates danger level of pollution that can cause severe consequences for hydrobionts and ecological systems, eg, anaerobic regime and eutrophication of the basins. The use of microorganisms assumes strict control of their growth conditions to take into account possible changes in the population affecting the oxygen consumption. Many toxic species, eg, heavy metals, can suppress microbial reactions and hence display decreased BOD levels against conventional methods. This might be important because inhibition of respiratory activity can cause potential damage of sewage sludge and hence decrease the quality of wastewaters purification. There are many reports on the possible influence of water pollution on the BOD sensors. Numerous works consider the aspects of inhibition related to particular toxicants or novel methods for respiration estimation.

Escherichia coli is often used in toxicity biosensors based on respiratory activity. Thus, chronoamperometric detection of ferricyanide signal with microelectrode array was used for comparison of the toxic effect exerted by 3,5-dichlorophenolcyanides, As(III), Hg(II), Cu(II), Pb(II) and Ni(II) salts in 60 min incubation. The IC_{50} (concentration exerting 50% changes in the recorded parameter) varied from 4 to 40 mg/L [109].

In the colorimetric sensor, the electron exchange between *E. coli* and ferricyanide ion was monitored by formation of Prussian blue after Fe^{3+} ions addition. Inhibitors suppressing respiration do not produce $[Fe(CN)_6]^{4-}$ ions and hence prevent the dye formation [110]. The colour shift can be followed by naked eye. The detection limits of 12.5 ppm were obtained for 3,5-dichlorophenol, As(III) and Cr(VI) species. The use of simple spectrophotometer instead of naked eye decreased the LOD values by about one order of magnitude.

A number of microorganisms including *E. coli*, *Enterobacter cloacae*, *Alcaligenes faecalis*, *Pseudomonas fluorescens*, *Pseudomonas putida* and *Trichosporon cutaneum* were tested in the detection of phenol, 3,5-dichlorophenol and nitrophenols [111]. The electrochemical response was estimated by the inhibition of the current related to the mediated respiratory activity and recorded in the presence of ferricyanide ions. The IC_{50} values measured for the 60 min incubation showed similarity of the response obtained among the representatives of one family. *Pseudomonas fluorescens* was found to be most sensitive for the phenol derivatives tested.

Multivariate analysis was performed with the microbial BOD biosensors utilizing different microorganisms, some of which were semispecific to cellulose and milk (*Aeromonas hydrophila*, *P. putida*, *E. coli*, *Bacillus subtilis*,

Paenibacillus sp. and *Microbacterium phyllosphaerae*) in order to increase estimation of the full BOD value and specify the source of pollution [112]. The results showed correlation between the prediction of BOD_7 value obtained by the array of microbial biosensors and real results obtained by conventional technique.

3.5 Microbial Fuel Cells

Microbial fuel cells have found increasing interest during the past decade in the toxicity testing [113]. The microbial fuel cell consists of anode where bacteria grow and cathode separated from anode with ion-exchange membrane. Wastewaters or pollutant solutions are directed in anodic compartment. Anode and cathode are connected with external circuit with a potentiostat or external resistance. The microbial oxidation of organic matter generates current recorded by external device. The exposure of the microorganisms in the pollutant solution decreases the electric current recorded. In many cases, bacteria are immobilized on the electrodes by formation of biofilm similarly to that utilized in wastewaters treatment plants. Electron mediators are added to cathodic compartment to accelerate the electron flow in the fuel cell.

Two-compartment cell with activated sludge and artificial wastewater was applied for the estimation of toxic effect of selected species [114]. Coulometric measurements were performed prior to and after toxicant addition and relative inhibition was calculated as a measure of toxicity. Diazinon, an anticholinesterase pesticide, decreased the coulometric yield to 61%, Hg to 28%, Pb to 46%, polychlorobiphenyls to 38% (at 1 mg/L of each toxic substance). The recovery of microorganisms after the contact with toxic species was not full and reached 86% of initial value for most toxic organophosphate.

Biofilms obtained on the polarized carbon felt electrodes contacted with real wastewaters were used for the detection of some biocides, ie, sulphonamides, chloramine B and heavy metal cations [115]. Inhibition of microbial activity was estimated by changes in the potential or current output recorded in the presence of ferricyanide ions as mediators. As was shown, immobilization of bacterial cells suppressed the toxic influence of all the model toxicants against planktonic cells present in wastewaters. Similar microbial fuel cells were tested on standard toxicant solutions: nickel [116] and copper [117] salts, sodium dodecyl sulphate [118], formaldehyde [119] and β-lactam antibiotics [120].

4. GENOTOXICITY BIOSENSORS

As was mentioned in the previous sections, molecular level of recognition rarely provides reliable detection of toxic species due to variety of biochemical paths involving hazards and various mechanism of their toxic action. Besides anticholinesterase agents, the detection of DNA damage is another important exclusion from this general rule. This is related to unique function of DNA in encoding and transfer of genetic information as well as importance of the DNA damage phenomena in carcinogenesis [121]. Although many factors affect the statistics of oncological diseases, most cancer results from the interaction of genetics and the environment. In accordance with some estimates, less than 5% of cancer cases are explained only by genetics. External factors including chemical contamination is responsible for the majority of the remained cases [122]. They include many chemicals existing in the workplace, air, drinking water, foodstuffs and drugs, especially those produced in anthropogenic activity, ie, those belonging to xenobiotics. The contaminants specifically interact with DNA and initiate different types of the DNA damage, eg, oxidation, alkylation and demethylation of nucleic bases, hydrolysis of a main chain, formation of bulky adducts, removal of single nucleic bases (formation of abasic sites), etc. [123].

Unrepaired DNA damage blocks its replication and causes point mutations that affect the DNA functioning. Thus, the detection of DNA affinity interactions with xenobiotics as well as oxidative DNA damage caused by radical species formed in the living beings is significant for the evaluation of genotoxicity in the early stages of development. Such genotoxicity sensors can be applied as early warning devices for preliminary screening of potentially dangerous contamination of the environment and foodstuffs and for the assessment of potential hazards related to new chemical compounds introduced to industrial manufacture.

It should be clearly stated that such DNA sensors do not take into account many biochemical consequences of contaminations, eg, lipid and lipoprotein oxidation and certain enzymes inhibition, as well as estimation of the efficiency of natural DNA repair mechanisms. For this reason, the genotoxicity biosensors cannot substitute long-term tests on living cells and test organisms common in traditional toxicology and environmental monitoring [92]. The advantage of DNA biosensors approach is mainly referred to the screening applications when the conclusion on potential mutagenic/cancerogenic activity is required for the following comprehensive investigation of certain chemicals. In these cases, biosensors provide faster information on the risks mainly demanded in food safety control or extreme contamination detection.

4.1 DNA Damage of Reactive Oxygen/Nitrogen Species

The formation of radical species in the reactions of transient metal cations with hydrogen peroxide is often used as a model reaction mimicking the initial stages of oxidative DNA damage. The Fenton's reagent consisting of Fe(II) and H_2O_2 is mainly used in laboratory experiments [124]. The reaction is performed mainly in basic media in the presence of EDTA preventing losses of iron ions in hydrolysis. Ascorbic acid is also added to catalytically recover

Fe(II) ions level and increase the efficiency of the production of hydroxyl radical OH and superoxide anion HO_2 (7) [125].

$$Fe^{2+} + H_2O_2 \rightarrow Fe^{3+} + OH^- + OH$$
$$OH + H_2O_2 \rightarrow H_2O + HO_2 \qquad (7)$$
$$HO_2 + Fe^{3+} \rightarrow Fe^{2+} + O_2 + H^+$$

Besides Fe(II), Cu(II) and their complexes with organic ligands [126], Cr(VI)/glutathione/H_2O_2 system [127] and V_2O_5 dispersion [128] participate in the generation of hydroxyl radicals. Superoxide radicals O_2^{-} are generated in enzymatic reactions including xanthine oxidase [129] (8).

$$(8)$$

Hydrogen peroxide utilized in such reactions, is produced in vivo in the reactions catalysed with oxidoreductases. Besides, it is transferred from exogenous sources. For this reason, an increase in the level of hydrogen peroxide is considered as a sign of severe environment or metabolism distortion. In genotoxicity sensors, H_2O_2 can be generated in oxidation of glucose by immobilized glucose oxidase [130,131].

Reactive nitrogen species (RNS) include nitrogen oxides NO_x ($x = 1, 2$) and the products formed in their reaction with radical species including superoxide anion [132]. RNS exert mutagenic effect on native DNA by nitration, nitrosation and deamination of nucleic bases. The reactive oxygen species (ROS)/RNS initiate the DNA cleavage [133] and transformation of supercoiled DNA in a single-stranded form [134]. Although RNS are not generated in model reactions, they are formed as intermediates in electrochemical reduction of nitrates, both organic and inorganic, and can participate in the DNA damage processes followed by biosensors in aqueous media. The formation of NO_2^{-} has been described for the reaction of nitrite anions with horseradish peroxidase in the presence of hydrogen peroxide [135]. Its interaction with DNA immobilized in polyelectrolyte complexes was monitored by the signal of Ru complex.

4.2 Electrochemical Detection of the DNA Damage

Changes in the DNA structure caused by genotoxic factors are mainly detected by placing the DNA sample on the electrode followed by recording the signal related to intrinsic DNA activity or affinity of specific interactions with electrochemically active indicators. The compactness of the surface layer can be monitored by electrochemical impedance spectroscopy (EIS). The reaction of DNA probe with genotoxic species can be also performed prior to implementation in the biosensor assembly although such a protocols does not correspond to biosensor format.

In all the cases main attention is paid to the reliability of the response, ie, to the selection of the conditions modelling natural DNA environment and in vivo conditions of DNA—analyte interactions. Thus, the distortion of the DNA conformation during the immobilization limits the access of genotoxic species and can result in underestimation of potential risks. Metabolic activation of potentially toxic species can be achieved by combination of DNA sensors with enzymes involved in xenobiotics oxidation like cytochrome P_{450} or myoglobin [136].

Direct electrochemistry of DNA is mainly associated with guanine which is oxidized in conventional conditions of the DNA sensor operation at high anodic potential [137]. Appropriate signals are mainly recorded on carbonaceous materials and amalgamated silver electrode using differential pulse voltammetry (DPV) or differential stripping chronopotentiometry. The influence of DNA damaging factors on guanine signal is attributed to higher accessibility of the guanine residues caused by the DNA unwinding, the DNA chain cleavage and other processes affecting flexibility of the DNA helix. For the same reason, the use of the single-stranded (ss-) DNA increases the signal of guanine oxidation against that of double-stranded (ds-) DNA. The DNA damage is also identified by appearance and growth of 8-oxoguanine, a stress biomarker formed in many radical reactions of guanine residues. For adenine, diimine derivative was identified using superoxide anion generated by xanthine oxidase reaction [138].

Normally the guanine signal increases with the intensity of external factor. Correlation between the changes of the guanine oxidation current and the mutagenic effect observed for wastewaters and soil extracts was found for a number of microbial tests (*mutatox*, Toxalert and *umu* test) [139,140].

Among other nucleotides, adenine was electrochemically oxidized to monitor DNA-specific processes [141]. Normally, changes in the appropriate peaks recorded by DPV are proportional to the content of nucleotides but in some cases, the comparison of the changes makes it possible to identify damaging factors. Thus it was shown that the Cr(III) damage is mainly resulted in increase of adenine oxidation against analogous guanine reaction [142].

The guanine oxidation can be accelerated by some organic complexes of transient metals with electrochemically active planar ligands, eg, bipyridine (bpy), tetraazatriphenylene (tatp) and phenanthroline (phen) [143—145]. Scheme (9) represents the reaction for the Ru complex as an example. Carbon nanotubes exert similar effect due to partial unwinding of the DNA helix [146].

$$\begin{aligned} &\text{Ru(bpy)}_3^{2+} \overset{-e}{\rightleftarrows} \text{Ru(bpy)}_3^{3+} \\ &\text{Ru(bpy)}_3^{3+} + \text{Guanine(DNA)} \rightarrow \text{Ru(bpy)}_3^{2+} + \text{8-oxoguanine(DNA)} \end{aligned} \qquad (9)$$

Some of the complexes mentioned can participate in the generation of OH radicals similarly to Fenton's reagent [147]. This increases the sensitivity of the DNA damage detection. The efficiency of mediation depends also on the accumulation of the mediators in the DNA helix which is higher for Co(II) and Ru(II) in comparison with Cu(II) complexes.

Besides guanine oxidation, some mediators exert their own redox activity that is changing due to the intercalation of ds-DNA molecules. Such electrochemically active intercalators decrease their signal due to intercalation and partially retain activity after the DNA damage. Thus, Methylene blue often used as indicator for hybridization detection was successfully applied for quantification of the DNA damage caused by the Fenton's reagent [148] or by endocrine disrupting compounds [149]. In addition to intercalator capacity, Methylene blue changes the signal due to contaminant influence on hybridization efficiency. Thus, Cd(II) ions suppressed the hybridization of 20-mer DNA probe with complementary DNA sequence and changed the signal of Co(phen)$_3$ and Methylene blue reduction depending on the incubation period and presence of oxygen or hydrogen peroxide in the working solution [143]. In a similar manner, methylation products were detected in the products of polymerase chain reaction by the signal of ferrocenylnaphthalene diimide affected by hybridization of the DNA probe with modified sequences [150].

The performance of the DNA sensor assembled by layer-by-layer deposition of polyelectrolytes on the electrode covered with electropolymerized phenothiazine dyes (Methylene blue and Methylene green) has been compared by EIS after the incubation in doxorubicin and Fenton's reagent [151]. Oxidative DNA damage increased the resistance of the outer interface of the modifier with the maximal effect for the polymeric Methylene green and direct contact of the DNA and polyphenothiazine layer. Meanwhile the selectivity of the response was found higher for the poly(Methylene blue)-based biosensor. The difference in the specific changes of the EIS parameters made it possible to distinguish the response related to the DNA damage and changes in the redox status of polyphenothiazines.

9,10-Anthraquinone-2,6-disulphonic acid intercalates the DNA helix by its flat anthraquinone fragment. Meanwhile it retains its redox activity increasing with accumulation of the indicator in the DNA layer dispersed in chitosan matrix at the electrode surface. As a result, Fenton's reagent decreases the voltammetric signal due to lowering the accumulation degree [152].

Neutral red was suggested to use for the detection of oxidative DNA damage. The interaction of ds-DNA with Fenton's reagent resulted in significant increase of the charge transfer resistance and capacity measured by EIS [153]. The indicator was covalently attached to GCE by carbodiimide binding. The signal was amplified by silver nanoparticles participating in the electron exchange and generation of ROS.

The characteristics of the DNA sensors for the detection of the DNA damage by various chemicals are summarized in Table 2 for the period from

TABLE 2 Potential Contaminants Exerting DNA Damage Detected With Electrochemical Biosensors

Analyte	DNA Immobilization Conditions	Signal Detection Conditions	Concentrations Tested	References
Metronidazole	Electrostatic adsorption on GCE	Direct reduction of metronizadole	LOD 20 nM, conc. range up to 60 μM	[154]
Ciprofloxacin	Physical adsorption on GCE	DPV peak of guanine oxidation	LOD 9 μM, conc. range 40–80 μM	[155]
Styrene oxide	Physical adsorption on tin dioxide nanoparticles	Guanine oxidation by photogenerated $[Ru(bpy)_2]^{3+}$ complex, detection of the Ru complex photocurrent signal (phenothiazine derivative of $Ru(bpy)_2$)	Incubation of 2% styrene oxide	[156]
Styrene oxide	Polyelectrolyte complex of DNA, PSS and PDDA	Methylene blue reduction current	Incubation of 0.015% styrene oxide	[157]
Arsenic(III) and tin(II) compounds	Physical adsorption of DNA and DNA–CNTs composite on screen-printed carbon electrodes	$[Co(phen)_3]^{3+}$ and guanine oxidation DPV signals	10 μM–1 mM testing solutions	[158]
Cd(II)	Thiolated 20-mer oligonucleotide on Au nanoparticles deposited on GCE	$[Co(phen)_3]^{3+}$ and Methylene blue DPV signals after hybridization of DNA probe	0.1 mM testing solution	[143]
Cr(VI)	Supercoiled DNA adsorbed on mercury electrode	Direct current DNA signal	Conc. range 50–250 μM	[159]
Cr(VI), pesticides	Thiolated DNA probes attached to Au electrode	Voltammetric and QCM signal recorded after assembling of ds-hybridized DNA helix and avidine–ferrocene conjugate binding	0.1 mM solutions of atrazine, 2,4-D, glufosinate, paraoxon-ethyl, diflubezuran and carbofuran	[160]

2-Aminofluorene	Polyelectrolyte complex of DNA, PAH and N-acetyltransferase	SWV signal of $[Ru(bpy)_3]^{2+}$ after metabolic activation	1.5 mM testing solution	[161]
Benzo(a)anthracene and phenantrene	Electrostatic adsorption on screen-printed carbon electrode	DSCP signal of guanine	20 and 50 ng/mL testing solutions	[162]
Benzo(a)pyrene photodegradation products	Thiolated DNA probe on Au screen-printed electrode	Hybridization efficiency detection using streptavidine–alkaline phosphatase conjugate and naphthol detection and DSCP measurement of guanine oxidation peak	40 nM testing solution	[163]
Benzo(a)pyrene	Layer-by-layer deposition of FNA and haemin on graphene–Nafion covered GCE	DPV peak of haemin oxidation in the presence of H_2O_2 providing metabolic activation of analyte	Conc. range $(2.0–22.0) \times 10^{-8}$ M, LOD 1.12×10^{-8} M	[164]
7,12-Dimethyl-benz[a]anthracene	Interaction with ds-DNA in solution	SWV signal of adenine on pencil graphite electrode after electrochemical preoxidation of the analyte	LOD 0.194 nM, conc. range 2–10 nM	[165]
Acrylonitrile	Adsorption in chitosan film	Direct current signal of 9,10-anthraquinone-2,6-disulphonate as intercalator, synergetic effect of DNA oxidative damage caused by Fenton's reagent	0.5% added to Fenton's reagent	[152]
Quinazoline derivative	Adsorption on screen-printed carbon electrode alone or together with CNTs	$[Co(phen)_3]^{3+}$, $[Ru(bpy)_3]^{2+}$ and Methylene blue DPV signals and EIS parameters	10–200 µg/mL	[166]
N-nitrosopyrrolidine, styrene	ds-DNA in polyelectrolyte complex with rat liver microsomes	SWV signal of $[Ru(bpy)_3]^{2+}$ after metabolic activation	1 mM testing solution	[167]

Continued

TABLE 2 Potential Contaminants Exerting DNA Damage Detected With Electrochemical Biosensors—cont'd

Analyte	DNA Immobilization Conditions	Signal Detection Conditions	Concentrations Tested	References
2-Nitrofluorenone and 2,7-dinitrofluorenone	Physical adsorption on screen-printed carbon electrode	SWV signal of guanine and adenine oxidation	5 and 10 μM testing solutions	[168]
Nitrofurazone	Physical adsorption of ds-DNA	$[Co(phen)_3]^{3+}$ guanine and adenine oxidation DPV signals	Conc. range 2.5×10^{-6}–3.75×10^{-5} M, LOD 8.0×10^{-7} M	[169]
Surfactants	Electrostatic adsorption on screen-printed carbon electrode	SWV signal of guanine oxidation	Conc. range tested 0.02–5 mM	[170]
Chlorobenzenes	Thiolated DNA probe on Au electrode	Methylene blue signal after hybridization with complementary sequence	Conc. range 100 pM–100 nM, LOD 30 pM hexachlorobenzene	[171]
Chloroaromatics (dioxin-like chemicals)	ss- and ds-DNA on Au	Methylene blue signal after hybridization with complementary sequence	LODs from 0.3 tp 3.0 pg/mL, conc. range 1–1000 pg/mL	[172]

CNTs, carbon nanotubes; *DPV,* differential pulse voltammetry; *EIS,* electrochemical impedance spectroscopy; *GCE,* glassy carbon electrode; *SWV,* square wave voltammetry; *DSCP,* differential stripping chronopotentiometry; *QCM,* quartz crystal microbalance.

2006 to 2015. It should be noted that the biosensors described were mainly tested on certain samples of toxicants so that the concentration range and LOD values are often not reported.

All the examples of genotoxicity biosensors summarized in Table 2 can be subdivided into two groups. Nitroaromatic compounds, nitrofurazone, metronidazole and quinazoline derivatives produce radical species and RNS like intermediates after electrochemical conversion and then exert DNA damaging effect detected by a common way. Other analytes are electrochemically inactive and form adducts or oxidation products with ds-DNA deposited on the electrode, which are detected preferably by electrochemically active intercalators. In some cases, DNA molecules are co-immobilized with metabolic activators that amplify signal due to formation of more active species similarly to the processes of lethal synthesis in human beings. Adducts and/or active intermediates were determined with LC-MS technique. It should be noted that in most cases of simultaneous 'activation' of mutagenic species their concentrations tested with biosensors are very high and cannot be compared with traces commonly observed in real samples. Such DNA sensors are reported mainly for styrene detection.

The DNA repair mechanisms are rarely used for the damage detection. The application of exonuclease III is an exception [173]. In this method, the DNA damage caused by methyl sulphate was attributed to N-methylation of guanine residues followed by spontaneous release of the damaged base. Os tetroxide formed adducts with ss-DNA fragment that were cathodically detected on amalgamated silver electrode. In a similar manner, displacement protocol was suggested for DNA damage detection on the base of ferrocene-labelled signalling DNA probe [174]. It interacted with partially complementary ss-DNA sequence. In the absence of genotoxic species, the following reaction with another DNA probe and treatment with exonuclease resulted in the release of the ferrocene label probe and its capture with the complementary DNA probe attached to Au electrode. The peak of ferrocene oxidation decreased after the DNA damage due to inhibition of exonuclease-assisted digestion step.

Low selectivity of the DNA damage detection does not allow identification of the hazards. Some information can be obtained from the specific paths of metabolic activation and comparison of the signals attributed to different damage effects. Thus, comparison of adenine and guanine oxidation peaks and appropriate oxidation products showed the difference related to certain analytes. However, the number of examples is too limited for suggestion on reliability of such an approach.

5. CONCLUSION

Biosensors offer unique possibilities for the estimation of the food quality and potential hazards related to contaminants and food additives due to flexibility of assembly and variety of the biological receptors involved in biorecognition process. Although the analysis of individual chemicals remains main topic of biosensor application, their use for the general assessment of foodstuffs and raw food materials offers very important opportunities for the improvement of the food quality and diminishing risks of poisoning. It seems important that such an integral food assessment is based on mimicking the biological reactions of a human being with no respect of the chemical nature of damaging factors. This means biosensors are potentially able to respond on chemical species unrecognized by chemical analysis or unexpected to be found in particular food.

Considering the prospects of the previously mentioned approaches to biosensors application, the following trends can be mentioned. First, it is rather obvious that individual enzymes cannot provide reliable detection of hazards except high contamination levels expected in extreme situations (chemical terrorism or industrial accidents). Monitoring of common residues present in food can be performed with microbial biosensors based on gene expression that is rather easily adapted for particular analytes. Although the number of such examples is limited to few chemicals, the possibilities of gene expression in biosensor format will grow.

This is particularly true for such an emerging area as cancer prevention and diagnostics. It can be expected that DNA sensors will be further combined with biochemical mechanisms of DNA repair and metabolic activation of potential carcinogens to increase the reliability of the response.

Multisensor approaches based on a number of biosensors manufactured on the same platform and utilizing a family of biosensing elements with similar recognition principles find growing area of application starting from identification of chemical pollutants and improved resolution of the signals to structurally similar analytes. The use of microfluidic devices and chip technologies offer excellent opportunities to automation of the manufacture and operation and decrease the cost of a single measurement. The estimation of taste, flavour and other organoleptic properties is one of the promising directions of the progress on the above area.

Finishing the consideration of biosensor technologies in food quality assessment, it should be noted that all of the progress is tightly dependent on the progress in biotechnology providing genetically modified microorganisms and enzymes to meet the requirements of the assessment becoming stricter and stricter. The experience accumulated on the manufacture of biosensors and their application for food quality control allows optimistic vision of the prospects in the reviewed topics of application.

ACKNOWLEDGEMENTS

Financial support of Russian Foundation for Basic Research (RFBR), grant No 14-03-00409 is gratefully acknowledged.

REFERENCES

[1]   D.R. Thévenot, K. Toth, R.A. Durst, G.S. Wilson, Biosens. Bioelectron. 16 (2001) 121−131.

[2]   J. Kirsch, C. Siltanen, Q. Zhou, A. Revzin, A. Simonian, Chem. Soc. Rev. 42 (2013) 8733−8768.

[3]   M. Mascini, S. Tombelli, Biomarkers 13 (2008) 637−657.

[4]   M.S. Belluzo, M.É. Ribone, C.M. Lagier, Sensors 8 (2008) 1366−1399.

[5]   V. Scognamiglio, F. Arduini, G. Palleschi, G. Rea, TrAC Anal. Chem. 62 (2014) 1−10.

[6]   M. Pohanka, P. Dobes, L. Drtinova, K. Kuča, Electroanalysis 21 (2009) 1177−1182.

[7]   F. Arduini, A. Amine, D. Moscone, G. Palleschi, Microchim. Acta 170 (2010) 193−214.

[8]   N. Verma, M. Singh, Biometals 18 (2005) 121−129.

[9]   W. Helferich, C.R. Winter (Eds.), Food Toxicity, CRC Press, Boca Raton, 2001.

[10] General Standard for Food Additives. Codex Stan 192−1995, Codex Alimentarus − WHO, 2015.

[11] M.-C. Radulescu, B. Bucur. M.-P. Bucur, G.L. Radu, Sensors 14 (2014) 1028−1038.

[12] Ü.A. Kirgöz, D. Odaci, S. Timur, A. Merkoçi, S. Alegret, N. Beşün, A. Telefoncu, Anal. Chim. Acta 570 (2006) 165−169.

[13] D. Compagnone, D. O'Sullivan, G.G. Guilbault, Analyst 122 (1997) 487−490.

[14] S.B. Saidman, M.J. Lobo-Castañón, A.J. Miranda-Ordieres, P. Tuñón-Blanco, Anal. Chim. Acta 424 (2000) 45−50.

[15] J. Filip, J. Šefčovičová, P. Tomčík, P. Gemeiner, J. Tkac, Talanta 84 (2011) 335−361.

[16] J. Šefčovičová, J. Filip, P. Tomčík, P. Gemeiner, M. Bučko, P. Magdolen, J. Tkac, Microchim. Acta 175 (2011) 21−30.

[17] D.P. Nikolelis, S. Pantoulias, Electroanalysis 12 (2000) 786−790.

[18] A.W.K. Kwong, B. Gründig, J. Hu, R. Renneberg, Biotechnol. Lett. 22 (2000) 267−272.

[19] D. Sirca, A. Vardeu, M. Pinna, M. Diana, P. Enrico, Biosens. Bioelectron. 61 (2014) 526−531.

[20] V.M. Tolosa, K.M. Wassum, N.T. Maidment, H.G. Monbouquette, Biosens. Bioelectron. 42 (2013) 256−260.

[21] B. Batra, S. Kumari, C.S. Pundir, Enz. Microb. Technol. 57 (2014) 69−77.

[22] R. Pauliukaite, G. Zhylyak, D. Citterio, U.E. Spichiger-Keller, Anal. Bioanal. Chem. 386 (386) (2006) 220−227.

[23] S.P. Gomes, J.D. Žalová, A.N. Araújo, C.M.C.M. Couto, M.C.B.S.M. Montenegro, J. Anal. Chem. 68 (2013) 794−800.

[24] R. Monošík, M. Streďanský, E. Šturdík, Food Anal. Methods 6 (2013) 521−527.

[25] G. Hughes, R.M. Pemberton, P.R. Fielden, J.P. Hart, Sens. Actuators B 216 (2015) 614−621.

[26] R. Mohammad, M. Ahmad, L.Y. Heng, Sensors 13 (2013) 10014−10026.

[27] E. Akyilmaz, İ. Yaşa, E. Dinçkaya, Anal. Biochem. 354 (2006) 78−84.

[28] M. Ovalle, E. Arroyo, M. Stoytcheva, R. Zlatev, L. Enriquez, A. Olivasa, Anal. Methods 7 (2015) 8185−8189.

[29] Y. Wen, J. Xu, M. Liu, D. Li, L. Lu, R. Yue, H. He, J. Electroanal. Chem. 674 (2012) 71−82.

[30] N. Chauhan, T. Dahiya, Priyanka, C.S. Pundir, J. Mol. Catal. B 67 (2010) 66−71.

[31] K. Rekha, B.N. Murthy, Food Agricultur. Immun. 21 (2) (2010) 103−111.

[32] X. Wang, H. Watanabe, S Uchiyama, Talanta 74 (2008) 1681−1685.

[33] A. Karadag, B. Ozcelik, S. Saner, Food Anal. Methods 2 (2009) 41−60.

[34] M.E. Ghica, C.M.A. Brett, Talanta 130 (2014) 198−206.

[35] H. Xu, J. Xiao, B. Liu, S. Griveau, F. Bedioui, Biosens. Bioelectron. 66 (2015) 438−444.

[36] D. Shan, Q. Shi, D. Zhu, H. Xue, Talanta 72 (2007) 1767−1772.

[37] S. Li, Y. Tan, P. Wang, J. Kan, Sens. Actuators B 144 (2010) 18−22.

[38] D. Shan, Q. Li, H. Xue, S. Cosnier, Sens. Actuators B 134 (2008) 1016−1021.

[39] V.P.S. dos Santos, L.M.C. Silva, A.M. Salgado, K.S. Pereira, Chem. Eng. Trans. 32 (2013) 1831−1836.

[40] A. Isaac, J. Davis, C. Livingstone, A.J. Wain, R.G. Compton, TrAC Trends Anal. Chem. 25 (2006) 589−598.

[41] E. Dinçkaya, M.K. Sezgintürk, E. Akyılmaz, F.N. Ertaş, Food Chem. 101 (2007) 1540−1544.

[42] M. Zhao, D.B. Hibbert, J.J. Gooding, Anal. Chim. Acta 556 (2006) 195−200.

[43] M.K. Sezgintürk, E. Dinçkaya, Talanta 65 (2005) 998−1002.

[44] M. Teke, M.K. Sezgintrk, E. Dinkaya, Artific. Cells Blood Subst. Biotechnol. 37 (2009) 138−142.

[45] M. Ghasemi-Varnamkhasti, S.S. Mohtasebi, M. Siadat, J. Food Eng. 100 (2010) 377−387.

[46] D. Ha, Q. Sun, K. Su, H. Wan, H. Li, N. Xu, F. Sun, L. Zhuang, N. Hu, P. Wang, Sens. Actuators B 207 (2015) 1136−1146.

[47] K. Beullens, P. Mészáros, S. Vermeir, D. Kirsanov, A. Legin, S. Buysens, N. Cap, B.M. Nicolaï, J. Lammertyn, Sens. Actuators B 131 (2008) 10−17.

[48] J. Zeravik, A. Hlavacek, K. Lacina, P. Skládal, Electroanalysis 21 (2009) 2509−2520.

[49] A. Rudnitskaya, S.M. Rocha, A. Legin, V. Pereira, J.C. Marques, Anal. Chim. Acta 662 (2010) 82−89.

[50] X. Cetó, J.M. Gutiérrez, M. Gutiérrez, F. Céspedes, J. Capdevila, S. Mínguez, C. Jiménez-Jorquera, M. del Valle, Anal. Chim. Acta 732 (2012) 172−179.

[51] K. Toko, Meas. Sci. Technol. 9 (1998) 1919−1936.

[52] M. Habara, H. Ikezaki, K. Toko, Biosens. Bioelectron. 19 (2004) 1559−1563.

[53] F. Zhang, Q. Zhang, D. Zhang, Y. Lu, Q. Liun, P. Wang, Biosens. Bioelectron. 54 (2014) 385−392.

[54] Q. Liu, F. Zhang, D. Zhang, N. Hu, H. Wang, K.J. Hsia, P. Wang, Biosens. Bioelectron. 40 (2013) 115−120.

[55] J.H.C. Busch, K. Hhrncirik, E. Bulukin, C. Boucon, M. Mascini, J. Agric. Food Chem. 54 (2006) 4371−4377.

[56] T.-H. Wang, G.-H. Hui, S.-P. Deng, Biosens. Bioelectron. 26 (2010) 929−934.

[57] G.-H. Hui, S.-S. Mi, S.-P. Deng, Biosens. Bioelectron. 35 (2012) 429−438.

[58] A. Amine, F. Arduini, D. Moscone, G. Palleschi, Biosens. Bioelectron. 76 (2016) 180−194.

[59] G.A. Evtugyn, H.C. Budnikov, E.B. Nikolskaya, Analyst 121 (1996) 1911−1915.

[60] O.O. Soldatkin, O.S. Pavluchenko, O.L. Kukla, I.S. Kucherenko, V.M. Peshkova, V.M. Arkhypova, S.V. Dzyadevych, A.P. Soldatkin, A.V. El'skaya, Biopolymer Cell 25 (2009) 204−209.

[61] O. Domínguez-Renedo, M. Asunción Alonso-Lomillo, M.J. Arcos-Martínez, Crit. Rev. Environ. Sci. Technol. 43 (2013) 1042−1073.

[62] M. Ovalle, M. Stoytcheva, R. Zlatev, B. Valdez, Z. Velkova, Electrochim. Acta 53 (2008) 6344−6350.

[63] M.A. Espinoza, G. Istamboulie, A. Chira, T. Noguer, M. Stoytcheva, J.-L. Marty, Anal. Biochem. 457 (2014) 85−90.

[64] K.V. Stepurska, O.O. Soldatkin, V.M. Arkhypova, A.P. Soldatkin, F. Lagarde, N. Jaffrezic-Renault, S.V. Dzyadevych, Talanta 144 (2015) 1079−1084.

[65] H. Tsai, R. Doong, Biosens. Bioelectron. 20 (2005) 1796−1804.

[66] L.H. Goodson, W.B. Jacobs, Methods Enzymol. 44 (1976) 647−658.

[67] N.F. Starodub, J.M. Shirshov, A.L. Kukla, N.I. Karyuk, A.V. Prokhorovich, R. Merker, Electrochem. Soc. Proc. 97 (1997) 799−808.

[68] E. Tønning, S. Sapelnikova, J. Christensen, C. Carlsson, M. Winther-Nielsen, E. Dock, R. Solna, P. Skladal, L. Nørgaard, T. Ruzgas, J. Emnéus, Biosens. Bioelectron. 21 (2005) 608−617.

[69] R. Solná, S. Sapelnikova, P. Skladal, M. Winther-Nielsen, C. Carlsson, J. Emnéus, T. Ruzgas, Talanta 65 (2005) 349−357.

[70] S.V. Dzyadevych, J.-M. Chovelon, Mater. Sci. Eng. C 21 (2002) 55−60.

[71] G.A. Evtugyn, E.A. Rizaeva, E.E. Stoikova, V.Z. Latipova, H.C. Budnikov, Electroanalysis 9 (1997) 1124−1128.

[72] S. Han, M. Zhu, Z. Yuan, X. Li, Biosens. Bioelectron. 16 (2001) 9−16.

[73] T.N. Shekhovtsova, S.V. Muginova, Anal. Bioanal. Chem. 381 (2005) 1328−1335.

[74] V.H.-T. Šukalović, M. Vuletić, Ž. Vučinić, S. Veljović-Jovanović, J. Plant Res. 121 (2008) 115−123.

[75] I.V. Naumchik, E.I. Karasyova, D.I. Metelitza, I.P. Edimecheva, V.L. Sorokin, O.I. Shadyro, Biochemistry 70 (2005) 322−329.

[76] D.I. Metelitza, E.I. Karasyova, E.E. Grintsevich, R.N.F. Thorneley, J. Inorg. Biochem. 98 (2004) 1−9.

[77] A. Attar, L. Cubillana-Aguilera, I. Naranjo-Rodríguez, J.L. Hidalgo-Hidalgo de Cisneros, J.M. Palacios-Santander, A. Amine, Bioelectrochemistry 101 (2015) 84−91.

[78] M. Ghanavati, R.R. Azad, S.A. Mousavi, Sens. Actuators B 190 (2014) 858−864.

[79] O. Adeyoju, E.I. Iwuoha, M.R. Smyth, Anal. Chim. Acta 305 (1995) 57−64.

[80] M. Wagner, J.A. Nicell, J. Chem. Technol. Biotechnol. 77 (2002) 419−428.

[81] T.M.B.F. Oliveira, M.F. Barroso, S. Morais, M. Araújo, C. Freire, P. de Lima-Neto, A.N. Correi, M.B.P.P. Oliveira, C. Delerue-Matosa, Bioelectrochemistry 98 (2014) 20−29.

[82] T. Liu, M. Xu, H. Yin, S. Ai, X. Qu, S. Zong, Microchim. Acta 175 (2011) 129−135.

[83] C. Tortolini, P. Bollella, R. Antiochia, G. Favero, F. Mazzei, Sens. Actuators B (2015). http://dx.doi.org/10.1016/j.snb.2015.10.095.

[84] Z. Yu, G. Zhao, M. Liu, Y. Lei, M. Li, Environ. Sci. Technol. 44 (2010) 7878−7883.

[85] M. Haddaoui, N. Raouafi, Sens. Actuators B 219 (2015) 171−178.

[86] J. Kochana, M. Strzałka, J. Kozak, Instrument. Sci. Technol. 42 (2014) 532−546.

[87] E. Asav, E. Yorganci, E. Akyilmaz, Talanta 78 (2009) 553−556.

[88] M.R. Montereali, L.D. Seta, W. Vastarella, R. Pilloton, J. Mol. Catal. B 64 (2010) 189−194.

[89] I. Czolkos, E. Dock, E. Tønning, J. Christensen, M. Winther-Nielsen, C. Carlsson, R. Mojzíková, P. Skládal, U. Wollenberger, L. Nørgaard, T. Ruzgas, J. Emnéus, Biosens. Bioelectron. 75 (2016) 375−382.

[90] H.J. Shin, Appl. Microbiol. Biotechnol. 89 (2011) 867−877.

[91] S. Parvez, C. Venkataraman, S. Mukherji, Environ. Intern. 32 (2006) 265−268.

[92] V. Kokkali, W. van Delft, TrAC Trends Anal. Chem. 61 (2014) 133−155.

[93] A.A. Bulich, D.L. Isenberg, ISA Trans. 20 (1981) 29−33.

[94] V.L.K. Jennings, M.H. Rayner-Brandes, D.J. Bird, Water Res. 35 (2001) 3448−3456.

[95] S. Ulitzur, T. Lahav, N. Ulitzur, Environ. Toxicol. 17 (2002) 291−296.

[96] J.C. Cho, K.J. Park, H.S. Ihm, J.E. Park, S.Y. Kim, I. Kang, K.H. Lee, D. Jahng, D.H. Lee, S.J. Kim, Biosens. Bioelectron. 20 (2004) 338−344.

[97] S. Girotti, E.N. Ferri, M.G. Fumo, E. Maiolini, Anal. Chim. Acta 608 (2008) 2−29.

[98] A. Bezrukikh, E. Esimbekova, E. Nemtseva, V. Kratasyuk, O. Shimomura, Anal. Bioanal. Chem. 406 (2014) 5743−5747.

[99] S.K. Yoo, J.H. Lee, S.-S. Yun, M.B. Gu, J.H. Lee, Biosens. Bioelectron. 22 (2007) 1586−1592.

[100] H. Tani, K. Maehana, T. Kamidate, Anal. Chem. 76 (2004) 6693−6697.

[101] N.M. Elman, H. Ben-Yoava, M. Sternheim, R. Rosen, S. Krylov, Y. Shacham-Diamand, Biosens. Bioelectron. 23 (2008) 1631−1636.

[102] T. Elad, S. Belkin, Bioengineered Bugs 3 (2012) 124−128.

[103] X. Zhao, T. Dong, Int. J. Environ. Res. Public Health 10 (2013) 6748−6763.

[104] S. Yagur-Kroll, E. Schreuder, C.J. Ingham, R. Heideman, R. Rosen, S. Belkin, Biosens. Bioelectron. 64 (2015) 625−632.

[105] Y.-F. Li, F.-Y. Li, C.-L. Ho, V.H.-C. Liao, Environ. Pollut. 152 (2012) 123−129.

[106] A. Roda, L. Cevenini, S. Borg, E. Michelini, M.M. Calabretta, D. Schüler, Lab. Chip 13 (2013) 4881−4889.

[107] S. Jouanneau, L. Recoules, M.J. Durand, A. Boukabache, V. Picot, Y. Primault, A. Lakel, M. Sengelin, B. Barillon, G. Thouand, Water Res. 49 (2014) 62−82.

[108] H. Nakamura, K. Suzuki, H. Ishikuro, S. Kinoshita, R. Koizumi, S. Okuma, M. Gotoh, I. Karube, Talanta 72 (2007) 210−216.

[109] C. Liu, T. Sun, X. Xu, S. Dong, Anal. Chim. Acta 641 (2009) 59−63.

[110] J. Zhai, D. Yong, J. Lia, S. Dong, Analyst 138 (2013) 702−707.

[111] C. Liu, D. Yong, D. Yu, S. Dong, Talanta 84 (2011) 766−770.

[112] M. Raud, T. Kikas, Water Res. 47 (2013) 2555−2563.

[113] X.C. Abrevaya, N.J. Sacco, M.C. Bonetto, A. Hilding-Ohlsson, E. Cortón, Biosens. Bioeelectron 63 (2015) 591−601.

[114] M. Kim, M.S. Hyun, G.M. Gadd, H.J. Kim, J. Environ. Monit. 9 (2007) 1323−1328.

[115] S. Patil, F. Harnisch, U. Schröder, ChemPhysChem 11 (2010) 2834−2837.

[116] N.E. Stein, H.V.M. Hamelers, C.N.J. Buisman, Sens. Actuators B 163 (2012) 1−7.

[117] Y. Jiang, P. Liang, C. Zhang, Y. Bian, X. Yang, X. Huang, P.R. Girguis, Bioresour. Technol. 190 (2015) 367−372.

[118] N.E. Stein, H.V.M. Hamelers, C.N.J. Buisman, Sens. Actuators B 171−172 (2012) 816−821.

[119] X. Wang, N. Gao, Q. Zhou, Biosens. Bioelectron. 43 (2013) 264−267.

[120] G. Schneider, M. Czeller, V. Rostás, T. Kovács, Enz. Microb. Technol. 73−74 (2015) 59−64.

[121] H.-G. Neumann, J. Cancer Res. Clin. Oncol. 112 (1986) 100−106.

[122] C.C. Abnett, Cancer Invest. 25 (2007) 189−196.

[123] M. Fojta, Electrochemistry of nucleic acids and proteins − towards electrochemical sensors for genomics and proteomics, in: E. Palecek, F. Scheller, J. Wang (Eds.), Perspectives in Bioanalysis, 2005, pp. 386−422.

[124] E.S. Henle, S. Linn, J. Biol. Chem. 272 (1995) 19095−19098.

[125] M.F. Barroso, N. de-los-Santos-Álvarez, C. Delerue-Matos, M.B.P.P. Oliveira, Biosens. Bioelectron. 30 (2011) 1−12.

[126] J. Kang, S. Dong, X. Lu, B. Su, H. Wu, K. Sun, Bioelectrochemistry 69 (2006) 58−64.

[127] A.A. Ensafi, M. Amini, B. Rezaei, Sens. Actuators B 177 (2013) 862−870.

[128] W. Zhang, T. Yang, W. Li, G. Li, K. Jiao, Biosens. Bioelectron. 25 (2010) 2370−2374.

[129]   P. Kuppusamy, J.L. Zweier, J. Biol. Chem. 264 (1989) 9880−9884.

[130]   M. Liang, S. Jia, S. Zhu, L.-H. Guo, Environ. Sci. Technol. 42 (2008) 635−639.

[131]   Y. Zu, H. Liu, Y. Zhang, N. Hu, Electrochim. Acta 54 (2009) 2706−2712.

[132]   H. Wiseman, B. Halliwell, Biochem. J. 313 (1996) 17−29.

[133]   A.P. Breen, J.A. Murphy, Free Radic. Biol. Med. 18 (1995) 1033−1077.

[134]   P. Johnson, L.I. Grossman, Biochemistry 16 (1977) 4217−4225.

[135]   Y. Zhang, H. Liu, N. Hu, Bioelectrochemistry 86 (2012) 67−71.

[136]   M. So, E.G. Hvastkovs, J.B. Schenkman, J.F. Rusling, Biosens. Bioelectron. 23 (2007) 492−498.

[137]   R. Fadrna, B. Yosypchuk, M. Fojta, T. Navratil, L. Novotny, Anal. Lett. 37 (2004) 399−413.

[138]   M.F. Barroso, N. de-los-Santos-Álvarez, M.J. Lobo-Castañón, A.J. Miranda-Ordieres, C. Delerue-Matos, M.B.P.P. Oliveira, P. Tuñón-Blanco, J. Electroanal. Chem. 659 (2011) 43−49.

[139]   A.M. Tencaliec, S. Laschi, V. Magearu, M. Mascini, Talanta 69 (2006) 365−369.

[140]   I. Palchetti, M. Mascini, Analyst 133 (2008) 846−854.

[141]   B. Meric, K. Kerman, D. Ozkan, P. Kara, M. Ozsoz, Electroanalysis 14 (2002) 1245−1250.

[142]   S.C.B. Oliveira, A.M. Oliveira-Brett, Anal. Bioanal. Chem. 398 (2010) 1633−1641.

[143]   J.Q. Zhang, P. Dai, Z. Yang, Microchim. Acta 173 (2011) 347−352.

[144]   J.F. Rusling, Biosens. Bioelectron. 20 (2004) 1022−1028.

[145]   S. Xu, C. Lu, J. Shao, Q. Li, H. Li, W. Li, J. Electroanal. Chem. 661 (2011) 287−293.

[146]   R.N. Goyal, S. Bishnoi, Biosens. Bioelectron. 26 (2010) 463−469.

[147]   Z.-S. Yang, Y.-L. Wang, Y.-Z. Zhang, Electrochem. Commun. 6 (2004) 158−163.

[148]   A.M. Nowicka, A. Kowalczyk, F. Scholz, Z. Stojek, Electroanalysis 23 (2011) 55−62.

[149]   X. Lin, Y. Ni, S. Kokot, Anal. Chim. Acta 867 (2015) 29−37.

[150]   S. Sato, K. Hokazono, T. Irie, T. Ueki, M. Waki, T. Nojima, H. Kondo, S. Takenaka, Anal. Chim. Acta 578 (2006) 82−87.

[151]   G.A. Evtugyn, V.B. Stepanova, A.V. Porfireva, A.I. Zamaleeva, R.F. Fakhrullin, J. Nanosci. Nanotechnol. 14 (2014) 6738−6747.

[152]   Y. Liu, N. Hu, Electroanalysis 20 (2008) 2671−2676.

[153]   Y. Kuzin, A. Porfireva, V Stepanova, V. Evtugyn, I. Stoikov, G. Evtugyn, T. Hianik, Electroanalysis (2015). http://dx.doi.org/10.1002/elan.201500312.

[154]   X. Jiang, X. Lin, Bioelectrochemistry 68 (2006) 206−212.

[155]   H. Nawaz, S. Rauf, K. Akhtar, A.M. Khalid, Anal. Biochem. 354 (2006) 28−34.

[156]   M. Liang, L.-H. Guo, Environ. Sci. Technol. 41 (2007) 658−664.

[157]   Y. Zhang, N. Hu, Electrochem. Commun. 9 (2007) 35−41.

[158]   A. Ferancová, M. Adamovski, P. Gründler, J. Zima, J. Barek, J. Mattusch, R. Wennrich, J. Labuda, Bioelectrochemistry 71 (2007) 33−37.

[159]   J. Vacek, T. Mozg, K. Cahová, H. Pivoňková, M. Fojta, Electroanalysis 19 (2007) 2093−2102.

[160]   A.M. Nowicka, A. Kowalczyk, Z. Stojek, M. Hepel, Biophys. Chem. 146 (2010) 42−53.

[161]   M. So, E.G. Hvastkovs, B. Bajrami, J.B. Schenkman, J.F. Rusling, Anal. Chem. 80 (2008) 1192−1200.

[162]   M. del Carlo, M. di Marcello, M. Perugini, V. Ponzielli, M. Sergi, M. Mascini, D. Compagnone, Microchim. Acta 163 (2008) 163−169.

[163]   M. del Carlo, M. di Marcello, M. Giuliani, M. Sergi, A. Pepe, D. Compagnone, Biosens. Bioelectron. 31 (2012) 270−276.

[164]   Y. Ni, P. Wang, H. Song, X. Lin, S. Kokot, Anal. Chim. Acta 821 (2014) 34−40.

[165] Y. Yardım, E. Keskin, A. Levent, M. Özsöz, Z. Şentürk, Talanta 80 (2010) 1347–1355.

[166] J. Labuda, R. Ovádeková, J. Galandová, Microchim. Acta 164 (2009) 371–377.

[167] S. Krishnan, B. Bajrami, V. Mani, S. Pan, J.F. Rusling, Electroanalysis 21 (2009) 1005–1013.

[168] V. Vyskočil, J. Labuda, J. Barek, Anal. Bioanal. Chem. 397 (2010) 233–241.

[169] Y. Ni, P. Wang, S. Kokot, Biosens. Bioelectron. 38 (2012) 245–251.

[170] F. Cugia, A. Salis, A. Barse, M. Monduzzi, M. Mascini, Chem. Sensors 1 (2011) 3.

[171] L. Wu, X. Lu, J. Jin, H. Zhang, J. Chen, Biosens. Bioelectron. 26 (2011) 4040–4045.

[172] L. Wu, X. Lu, X. Wang, Y. Song, J. Chen, Anal. Methods 7 (2015) 3347–3352.

[173] L. Havran, J. Vacek, K. Cahová, M. Fojta, Anal. Bioanal. Chem. 391 (2008) 1751–1758.

[174] W. Wei, Q. Ni, Y. Pu, L. Yin, S. Liu, J. Electroanal. Chem. 714-715 (2014) 25–29.

New Generations of Synthetic Receptors and Functional Materials for Food Biosensors

Chapter 6

Aptamers as Synthetic Receptors for Food Quality and Safety Control

R. Miranda-Castro, N. de-los-Santos-Álvarez and M.J. Lobo-Castañón*

Universidad de Oviedo, Oviedo, Spain
*Corresponding author: E-mail: mjlc@uniovi.es

Chapter Outline

1. INTRODUCTION

Food security is one of the challenges faced by the world whose population is expected to be close to 10 billion by 2050 [1]. Providing a sustainable, safe and healthy food supply requires innovative solutions through research, including the development of fast and precise analytical methods to ensure food quality and safety, thus supporting regulatory compliance and meeting the consumers' expectations and needs.

Within this global context, many scientists have endeavoured to develop chemical sensors to monitor the presence in the food supply chain of chemical and microbial contaminants as well as unauthorized additives, illegal adulterants or even natural components such as allergens that can be life-threatening to highly sensitive people. A chemical sensor can be ideally considered as an artificial sensing organ able to provide real-time information related to the molecular status of the device's environment, for example, to instantly detect and report the quality of our food. This is accomplished by integrating a natural or artificial receptor, which is designed to bind with high affinity and selectivity to a specific target, with a transducer capable of converting the chemical energy involved in the binding event between receptor and target into an electronic signal. The research in this field is partially driven by the search for proper molecular recognition elements able to specifically capture a given analyte, triggering sensitive transduction [2]. Molecular recognition plays a major role in biological systems, which by evolutionary process have refined natural receptors of high affinity and specificity, such as enzymes and antibodies, extensively used in the construction of chemical sensors (biosensors) [3]. Despite the advantages of these natural systems, issues such as high cost and low stability have spurred the development of synthetic receptors that mimic natural ones. Recognition by natural receptors relies on key molecular interactions conferring specificity to the receptor—target association, including specific noncovalent bonding and molecular shape complementarities. This principle has been successfully mimicked to obtain synthetic receptors using both artificial polymeric systems such as molecular imprinted polymers and naturally occurring polymers like nucleic acid aptamers.

Besides their crucial roles in biological processes, providing the molecular basis for all life through their unique ability to store and propagate information, nucleic acids can artificially undergo Darwinian evolution to acquire new functions such as folding and specific ligand binding. Aptamers are nonnatural single-stranded oligonucleotides (DNA or RNA) that mimicking antibodies can bind with high affinity and selectivity to almost any analyte [4]. They undergo an adaptive conformational change in the presence of the target, resulting in an aptamer—target complex through discriminatory molecular interactions similar to those involved in natural recognition events. Aptamers have been generated against a wide variety of targets relevant in food analysis such as proteins, drugs, toxic molecules or even whole cells by means of a process of directed chemical evolution known as SELEX (systematic evolution of ligand by exponential enrichment) [5,6]. Once selected, they are readily produced by chemical synthesis that provides advantages in terms of cost-effectiveness and quality control. In addition, they often show similar affinity to those of existing bioreceptors while exhibiting higher stability and easier combination with labels and other modifiers that provide flexibility for adaptation to a great collection of analytical platforms. All these features make aptamers highly versatile receptors for the development of a variety of analytical assays not only chemical sensors. While RNA aptamers dominated the scene in the early years when research was mainly focused on therapeutic and diagnostic applications, the increased stability against nucleases and consequently its easier handling, tipped the scale in favour of DNA aptamers. This fact along with the shortening of the SELEX procedure when using DNA libraries (see next section) justifies the current dominance of DNA aptamers in bioanalysis. Analytical applications of aptamers are currently evolving at an enormous pace, with an exponential growth of research and development in food analysis during the last 10 years. Although as stated previously the generally accepted definition of chemical sensor requires a direct interface between the recognition element and the transducer, other assay formats in which recognition and transduction are not coupled are frequently referred as sensors in the literature. While not true sensors, many of these analytical assays show the potential for further development as chemical sensors and because of this they will be included for the purpose of the present discussion.

Within this perspective, this chapter aims at summarizing the major advances in the development of sensors and analytical assays based on aptamers to address control of food quality and safety, and ensure food authenticity and traceability. The first part is devoted to present the SELEX process and the most prominent improvements since their inception. Break-throughs in SELEX are frequently linked to clinical applications and model targets. The recurrent use of model targets is translated and multiplied in the development of aptamer-based analytical methods. Consequently, not all advancement in SELEX has been already applied to aptamers for food analysis, and only those that seem to be of general application are revised here. Further the discussion is grouped around the various design strategies using aptamers as recognition element with special attention to those assays that demonstrate their usefulness in food analysis applications. Finally, challenges that need to be considered to facilitate these new devices are transferred to the commercial market and future directions to overcome them are discussed.

2. SELECTION OF APTAMERS

The goal of SELEX is the isolation of functional oligonucleotides that bind a specific target with high affinity and selectivity from a random single-stranded three-dimensional structured library of about $10^{14}-10^{15}$ individual DNA or

RNA members. Classical or conventional SELEX relies on iterative cycles of binding, partitioning, elution-amplification and conditioning steps, each one accomplished using an ever-increasing variety of techniques. This provides a great versatility to succeed in, theoretically, any evolution experiment but, at the same time, the lack of a universal protocol for any target makes it highly dependent on the ability and knowledge of the specialized staff. Increasing stringent conditions are applied to each cycle to ensure adequate enrichment in high affinity binding sequences. This strategy along with the efficiency in partitioning, that is, the separation between free and bound sequences, determines the rate of enrichment in sequences of high affinity, usually the only screening criteria guiding the selection, and so the number of iterative rounds before cloning and sequencing to identify the winning sequences. As a result, it takes weeks or even months to complete the overall process.

The last quarter of a century has witnessed significant technological advances in analytical methods and improvements in chemical synthesis that have transformed the way of performing the SELEX process. More than 25 variants have been described by modifying one or several steps simultaneously [7]. Fig. 1 shows a scheme of the process, where improvements in each step are indicated as the name of the SELEX variant. Those implying more than

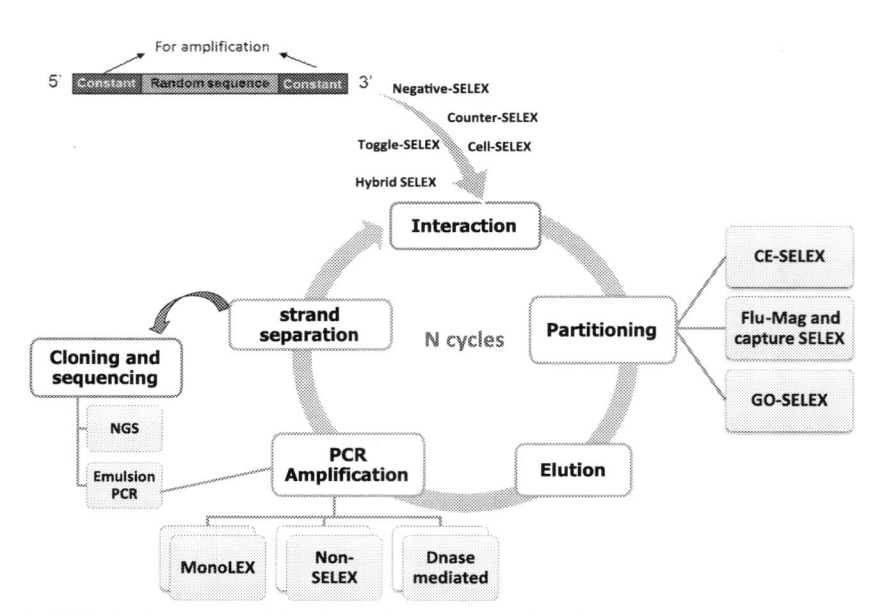

FIGURE 1 Conventional SELEX procedure. It starts with the interaction between the designed library and the target, then bound sequences are separated (partitioning) and eluted from the complex and finally amplified and conditioned (strand separation) for the next cycle. After a certain number of cycles the strands are cloned and sequenced. In each step the most significant advances are indicated. *NGS*, next-generation sequencing; *PCR*, polymerase chain reaction.

one step are omitted in the figure for clarity. Some variants, and historically the earliest, were clearly oriented or even only applicable for the selection of aptamers in the therapeutic field, so they are beyond the scope of this chapter and are not indicated in the figure.

The length of the aptamer is a major issue for market applications because solid-phase synthesis of oligonucleotides becomes inefficient and expensive with increasing size. Common libraries comprise 60–90 nucleotides including the randomized central region (typically 40 nucleotides in length) and two flanking constant regions. Truncation experiments are needed to get the minimum functional sequence. In most cases, such costly and laborious post-SELEX experiments are not carried out and aptamers lacking the fixed regions, originally only intended to facilitate the polymerase chain reaction (PCR) amplification step, are directly employed without testing the affinity. However, it has been early realized but mostly ignored until a decade, that the presence of the constant sequences can greatly impact the screening and be a source of selection bias. On one hand, the fixed regions (primers) can become part of the binding motif or stabilize the active structure accounting for up to 50% of the selected sequences [7], so truncation is difficult or even impossible. On the other hand, they can compromise (bias) the selection of potential aptamers [8] by (1) adopting secondary structures that sterically block the specific interaction between the target and the randomized region, (2) reducing the number of possible folding to those compatible with the stable structure of the primer regions or (3) by self-hybridizing with the randomized fragment limiting the diversity of the library. To avoid these deleterious effects, several strategies to achieve primer-less binding steps were developed [9]. Full removal of primers was achieved on DNA libraries by cleaving them using the adequate endonucleases and re-ligating again after partitioning during each selection round. Recently, complete hybridization of the primers during the binding step was proposed as time-saving, less loss-prone and elegant way of precluding interferences from flanking regions. This strategy screens out all aptamers involving both random and constant regions but it is surpassed by the gain in functional aptamers within the variable region [8,10].

2.1 Improving the Selectivity of the Binding Step

Right after SELEX development, specificity becomes an issue because the high probability of evolving an aptamer against components whose presence is unavoidable in the selection buffer such as target impurities or the solid support and spacers when the target is immobilized. A binding step in the absence of target under identical conditions and removal of bound sequences (negative-SELEX) was the simplest solution proposed. This procedure can no longer be considered as a SELEX variant because it has been readily incorporated in most protocols as a previous step before certain selected rounds. To further increase the discrimination power of aptamers, a conceptually similar stage was added. In this case, a structurally related compound (an isoform, an enantiomer, etc.) is challenged against the library and the unbound sequences used in posterior rounds of selection. This strategy called counter-SELEX or subtractive SELEX is responsible for the exquisite selectivity of aptamers that can recognize their cognate ligand with $>10^4$ higher affinity than related compounds, including enantiomers. On the contrary, in certain applications a low level of cross-reactivity is desirable, so alternating rounds of selection with homologous molecules are carried out (toggle-SELEX). This variant was applied to the food field for aminoglycoside antibiotics [11] and might be extended to other group of compounds. Allergens of animal origin such as crustaceous or fish are associated to proteins with a variety of homology causing that many allergic patients are sensitive to several species. The evolution of aptamers must consider the multispecies issue and toggle-SELEX could be the strategy of choice to ensure the detection in several species simultaneously.

In spite of these strategies, it was apparent that high purity of target molecules is a prerequisite to avoid generating nonspecific aptamers. The difficult purification of membrane-embedded peptides hindered the development of aptamers against cell surface proteins. Aptamers against food pathogens were developed using purified lysates of cell overexpressing surface proteins [12] or purified/commercial toxins [13,14]. Aptamers recognizing complex targets such as living pathogenic organisms had been achieved several years before but the actual target is unknown unless additional experiments are performed. In food analysis, this is not strictly necessary, so currently, cell-SELEX is the standard method for evolution of aptamers against specific pathogen strains. It has the advantage of challenging oligos with molecules in their native state but appropriate counter and negative selections are critical to eliminate cross-reactivity with related species. This SELEX variant incorporates the use of flow cytometry as a technique to assay the binding affinity, but mass spectrometry is used to identify the target. Alternatively, in some cases, the recombinant purified target is available for usage in the final rounds of selection to remove aptamers targeting other markers (hybrid-SELEX). This strategy has not been yet applied in the context of food analysis.

Driving the selection towards a specific epitope is an attractive feature of SELEX procedure that can be also achieved in the case of proteins incubating the library in the first rounds with a short synthetic peptide, corresponding to a fragment of the targeted protein with predicted immunogenicity, and then switching to the whole protein. Selection is then based on specificity rather than affinity but the latter also usually increased. This way aptamers recognizing the native and denatured states of the proteins can be generated. Omitting the challenge towards the entire protein is also possible but may yield aptamers with slightly lower affinity for the native protein. This was the case for gliadin, a constituent of gluten commonly determined in food samples to ensure compliance to gluten-free labelling. That protein is insoluble in aqueous solution so a small soluble immunodominant epitope was chosen as a target for SELEX [15].

2.2 Improving the Conditioning Step

The amplification of the bound sequences by PCR yields double-stranded DNA (dsDNA) that has to be purified and separated to obtain the single-stranded form (ssDNA) of interest. The most widely used strategy in the food field is to incorporate a biotin in the antisense, not desired, strand during PCR, followed by entrapment on streptavidin-coated magnetic beads and alkaline separation. This method does not require previous purification of the ampli-cons. Recently, lambda exonuclease digestion has been suggested to provide higher ssDNA recovery than alkaline denaturation on purified amplicons but purification increases time, cost and the risk of aptamer loss up to 30% [16], so the alkaline method remains the more adequate. Alternatively, asymmetric [13] or unequal length PCR followed by denaturing PAGE [17] has been performed to obtain aptamers for food analysis.

2.3 Accelerating the Selection Process

Since 2004, once the aptamer usefulness consolidated through challenging biomedical applications, the research was focused on making easier and speeding up the evolution of new aptamers. Several improvements have been reported and can be classified as follows: increasing the partitioning efficiency, eliminating the amplification between rounds, decreasing the reagent consumption and handled volumes through miniaturization, improving the identification of the desired sequences, achieving parallel selection (multiplexing) and increasing the level of automation.

2.3.1 Improving the Partitioning Step

Immobilization of the target on a solid support facilitates the partitioning step, allowing stringent rinses to remove most of the weakly bound strands. However, it may change the conformation of the target or interfere in the access of the library to the target. In the first years of SELEX, affinity chromatography and nitrocellulose membrane filtration that allows binding in solution were very popular. Both methods share a low resolution in separation and large number of rounds, although they are still used. Recently, marine toxins were immobilized through hydroxyl group on divinyl sulphone sepharose beads to keep most of the rest of the small structure available for binding [17]. In the same way, affinity chromatography over concanavalin A-modified agarose beads has been employed to select aptamers against the allergen [18].

Introduction of magnetic separation has had a tremendous impact not only on manual SELEX but also on bioassay technology and has already been launched to the market due to easiness of handling and low cost. The commercial availability of magnetic microparticles was essential to adopt this method for partitioning in SELEX. It requires small amounts of bound target offering a rapid and efficient separation by means of stringent washing steps and simple aptamer elution. The most important improvement in this regard was the so-called Flu-Mag SELEX [19] because it replaces the radiolabels for small fluorophore molecules attached to one end that are readily incorporated during the PCR step through a labelled forward primer. In this way, easy and safe monitoring of the enrichment is achieved and a few parallel selections can be also carried out simultaneously in adapted microtitre plates. Magnetic bead-based SELEX is undoubtedly the strategy of choice for developing most aptamers in food analysis including large (toxins [13], allergens [15] or even viruses [20]) and small molecules (marine toxins, antibiotics, pharmaceuticals, mycotoxins [21] or PCBs). The reason behind this is probably related to the wide range of functional groups created on the polymer coating (carboxylic acid, tosylactivated, epoxy, amine, nitrilotriacetic acid, streptavidin or even antibodies) that ensures easy bioconjugation to virtually any kind of molecule.

Immobilization of library instead of target on magnetic beads [22] or avidin-coated agarose beads [23] is an alternative for small molecules, in theory less suitable for immobilization, such as aminoglycoside antibiotics [22,24] or acetamiprid pesticide [23]. It requires a special design consisting of a central fixed sequence (docking sequence) flanked by two randomized regions of different size and the primers at both ends. Simplified library with a continuous randomized region and a docking sequence comprising part of one primer has also been proposed [24]. The aptamer is anchored to the magnetic beads through hybridization between the docking sequence and a complementary strand attached to the magnetic beads. Aptamers that bind the target must switch from duplex state on the particle to the complex state in solution through a target-induced conformational change. Hybridization anchoring, however, requires an overnight incubation before each round and extensive rinses (up to 23) to remove weakly bound sequences before incubation with the target [22].

Immobilization-free screening of aptamers to improve partitioning is usually associated to sophisticated instrumentation. Kinetic capillary electrophoresis (CE-SELEX) also known as nonequilibrium capillary electrophoresis of equilibrium mixtures (NECEEM-SELEX) separates species that interact during electrophoresis at so high efficiency that only between two and four cycles are necessary [25]. It minimizes the nonspecific binding to the point that negative steps can be omitted, works with very low volumes (nL) but this also limits the diversity of the initial library, so the chances to find a very good binder decrease. It must also be considered that low ionic strength and high electric field are not the ideal conditions to favour a specific and strong interaction, so a trade-off between CE and SELEX optimum conditions must be found. A good example of the advantages of this technique was demonstrated with the evolution of an aptamer for ricin toxin. By filter membrane binding, no obvious aptamers were recovered after nine cycles, whereas affinity chromatography and CE provided 38.5% and 87.2% of binding after nine and four cycles, respectively [26]. In spite of this remarkable data, only a few aptamers for food targets have been derived through it. In the case of allergens up to eight rounds of selections were necessary to obtain a very modest aptamer (K_D about 400 nM) against Ara h 1 protein, a peanut allergen [27]. It should also be emphasized that molecules smaller than DNA are expected not to change the mobility of the free DNA, and if it occurs, the collection of the bound fraction barely separated from the free fraction is specially challenging, which complicates obtaining anti-small molecules aptamers. In sum, it is interesting to note that the dramatic increase in partitioning efficiency using this method was not accompanied by a similar increase in binding affinity of the selected aptamers. K_D values are not impressively lower than those obtained with traditional SELEX.

Novel nanomaterials with interesting properties can assist the partitioning step without costly instrumentation. Short ssDNA even those with G-quartets structures strongly adsorb on graphene oxide (GO) sheets through nucleobases whereas dsDNA is only weakly adsorbed because the nucleobases are hidden into the helix. In GO-SELEX, sheets of this material serve to remove unbound free DNA from bound fraction by centrifugation. GO-SELEX requires only one round of selection followed by a combined negative-counter selection that overall takes about 11 h. After counter selection, sequences bound to GO are recovered by affinity elution with the target. It is important to keep the pH of the solution near the isoelectric point of the target to avoid electrostatic interactions with the negatively charged GO sheet [28].

2.3.2 Restriction of Polymerase Chain Reaction Amplification Steps

PCR amplification of a degenerated library is not entirely identical to amplification of a unique specific sequence. It favours the conversion of expected amplicons into longer spurious by-products before primers are exhausted, and eventually the library disappears after no more than five additional cycles [29]. This effect is attributed to heteroduplexes, that is, duplexes between two partially complementary strands usually showing end-complementarity because of the abundant terminal constant regions that can be extended by DNA polymerase without primers [10].

Restriction of PCR amplification to a final step is desirable for the aforementioned reason but also because it biases the selection towards sequences with weak secondary structures, which somehow compromises the selection. As a consequence, too long random regions are not recommended in spite of adding extra diversity. Several SELEX variants are designed to circumvent these problems. MonoLEX uses only one cycle of affinity chromatography selection followed by physical cutting of the column into small fragments to recover the bound aptamers that are finally amplified only once [30]. CE-SELEX can be performed without amplification steps between repetitive rounds, so the entire selection can be finished in about a week (3–4 h just for the binding-partitioning steps). A portion of the eluted complexes is directly incubated with a fresh solution of the target and subjected to partitioning skipping the PCR step, so avoiding the artefacts associated [31]. This method known as non-SELEX uses a fluorescently labelled library which allows the evaluation of affinity after each round. In this way, a single instrument is able to perform the entire SELEX except PCR step, achieving K_D in the subnanometre range in only three rounds. Surface plasmon resonance (SPR) that is frequently used for aptamer characterization also offers that possibility but the partitioning efficiency is lower and immobilization of target is compulsory. In spite of all the advantages, the mechanism behind non-SELEX is unclear and only a few examples of this strategy have been reported, one of them for an enterotoxin [32]. Recently failure in non-SELEX but success in CE-SELEX was published [10].

To develop low-cost one-round approaches, in addition to GO-SELEX, DNAse-mediated digestion of unbound sequences was proposed. A crude lysate of proteins was separated by PAGE and electroblotted on a membrane. The library is incubated on a strip containing the protein of interest and on another strip with matrix proteins for negative selection. This protein immobilization method is amenable to most laboratories and in combination with target-protected DNAse digestion and extensive urea rinses yields high affinity aptamers [33].

2.3.3 Miniaturization and Automation

Integration of several steps of the analytical process in a single small platform is an appealing trend in the field. It offers a range of capabilities such as fast high resolution separations and ultrasensitive detection using small quantities of reagents and samples. Microfluidics and microelectromechanical systems (MEMS) technologies, bloomed in the last two decades, have made this dream to the lab reality with the help of physicists and engineers. Currently commercial exploitation is gaining pace progressively, although designing of microfluidic devices is challenging. A decisive advance was the incorporation of ferromagnetic structures into the microchannels to allow magnetic beads selection. Combining the advantages of both technologies a single-round screening of aptamers is possible and constitute the first step to a fully automated and integrated miniaturized SELEX. Integrated microfluidic devices are designed for the entire SELEX process, including an on-chip PCR step. This chip is complemented with another microfluidic system for competitive assay tests of the selected aptamers to reduce the number of sequences subjected to sequencing and SPR affinity characterization [34]. In this way, a single round of SELEX is shortened to about 70 min, so the entire SELEX process can be finished in 6 h with the continuous device.

In addition to high affinity constant (K_D), a desirable feature of aptamers is a low dissociation rate constant (k_{off}). Re-equilibration of target—ligand complexes in increasing buffer volumes is an effective way of eliminating sequences with fast dissociation rates (high k_{off}) but it does not allow a stringent washing unless they are covalently linked. It has been noted that, even at high flow rates, formation of magnetic bead clusters makes some particles inaccessible to rinses. Incorporating a volume dilution challenge step between the binding and magnetic-based microfluidic separation provides a convenient particle dilution that improves the washing efficiency and biases the selection towards slow off-rate aptamers [35].

2.3.4 Multiplex Selection

Parallel selection of aptamers for multiple targets is another way of shortening their evolution and fulfilling the ever-increasing need for novel aptamers. Recently, two already existing methods: magnetic separation and BEAMing (beads, emulsion, amplification and magnetics) were successfully coupled to yield a semiautomated method that enables up to 12 simultaneous selections in 10 days. Incubation is performed in special microtitre plates that allow introduction of a magnetic separator that transfers magnetic beads from well to well in an automated fashion and was validated for lysozyme, a wine fining agent [36].

2.3.5 Improving the Screening of Selected Aptamers

From all the revised SELEX variants, it is apparent that this combinatorial procedure has been refined to the point to be performed in a few days in an automated or at least semiautomated fashion. However, the final steps out of the cycles, namely cloning, sequencing and affinity screening still consume most of the time and cost of aptamer selection because of their labour-intensive character. Two major strategies have been reported to facilitate these steps: single-molecule emulsion PCR and massive sequencing.

After a SELEX only a few highly repeated sequences containing consensus motifs are expected and characterized. However, the most abundant sequences are not often the best binders, and less repeated good binders are usually ignored and lost. Besides, some high affinity aptamers lack adequate selectivity. Evaluation of binding ability of individual sequences before or without sequencing is therefore advantageous. Single-molecule emulsion PCR enables amplification of single sequences minimizing the generation of PCR by-products and direct visualization of positive binding events. Enriched oligo libraries can be amplified inside agarose or water-in-oil droplets, each containing a single aptamer candidate. In the first case, SYBRGreen fluorophore is added to detect those containing DNA amplicons followed by flow cytometry to evaluate the affinity [37]. In the latter case, the enriched library is individually encapsulated in droplets containing a single bead conjugated to the forward primer. Therefore, beads containing millions of copies of a single sequence are generated by PCR. Then, identification of individual aptamers without sequencing is straightforward by flow cytometry because of the one-bead one-sequence approach derived from 'monoclonal' bead construction [38]. In this way, rare but potent aptamers can be isolated. The main drawback of these sequencing-skipping methods is the still large number of rounds of conventional SELEX required.

The dramatic development of bioinformatics in the last years has driven the availability of high-throughput sequencing at a reasonable price. Next-generation sequencing (NGS) technique does not require the convergence of the oligo pool to a few sequences carrying consensus motifs and definitively eliminates cloning because it is capable of sequencing millions of strands without separation. Therefore, sequencing is no longer a final step but it can assist each round of selection and help to deepen the understanding of SELEX enrichment. For example, it verified that best binders are not the most abundant and the strongest binders accumulate in the early rounds. This finding indicates that selection is biased to best PCR performance sequences instead of binding [39]. Taking into account this along with the probable incorporation of mutation by polymerases encourages the use of high-resolution techniques that avoid numerous PCR steps. The analysis of the first rounds may help in early identification of repeated sequences saving time and expenses. *In silico* analysis of massive sequencing allows clustering-related sequences (by secondary structure, conserved motifs and enrichment fold) independently of their copy number to find strong binders of low frequency. The analysis of NGS data can boost the optimization of the winner aptamers. In this regard availability of easy-to-use software like COMPAS is desirable [7].

3. APTAMER-BASED ASSAYS: DESIGN STRATEGIES

The introduction of nucleic acid aptamers furnishes unprecedented opportunities to detect nonnucleic acid targets and broadens the design strategies typically used with conventional bioreceptors. The objective of all aptamer-based assays is to accurately estimate the concentration of the analyte in the sample, with sensitivity as high as possible, from the observation of the binding reaction between the analyte and an aptamer as recognition element, which provides selectivity to the assay. Different configurations can be adopted to obtain the best analytical performance, and these can be divided into three large groups: direct, sandwich and competitive assays. The two first may be considered as 'aptamer excess' methods, in the sense that an excess of the recognition element is used, so that after the recognition event practically all analyte will be in the form of aptamer—target complex that is quantified (or the aptamer-free fraction) and related to the analyte concentration in the sample. Direct assays rely on a single-site format (only one aptamer is employed for the recognition process), whereas sandwich assays are double-site configurations requiring simultaneous binding of two aptamers to the analyte. By contrast, competitive assays are single-site 'limited reagent' methods as they are based on the competition between the analyte and a tracer for a limited amount of aptamer. After recognition, the bound or the free fraction of the tracer is measured and related to the concentration of analyte in the sample. This is the classification criterion used through this section to give an overview of aptamer-based methods useful for assessing the quality and safety of foodstuffs. In many cases assays depend on the use of labelled substances (tracer) for signal generation and amplification. Therefore, when appropriate, the main strategies in each group are categorized according to the type of marker used, which determines the transduction principle. In all cases both homogeneous and heterogeneous assays will be included. The lack of a separation step before measurement, which is the hallmark of homogeneous assays, can be advantageous in terms of speed. However, the inclusion of a step to separate the aptamer—target complex (heterogeneous assay) can improve the analytical performance, reducing nonspecific interactions, particularly in complex matrices.

3.1 Direct Assays

A great number of direct aptassays have been developed to detect different molecules of interest in food analysis. These assays can be divided into two categories, those that do not use labelled reactants and those that use reporter molecules, harnessing conformational changes of aptamer after the recognition event to detect the amount of complex; these two groups will be discussed through this section.

3.1.1 Label-Free Direct Assays

In the simplest assay format an immobilized aptamer is used to specifically capture the target and any physical change caused by the formation of this complex can be used for transducing the recognition event [40]. This strategy demands very sensitive techniques able to monitor minute physical changes, ie, mass, conductivity, refractive index, which occur at the surface of the sensing layer after binding, as assay sensitivity mainly relies on the sensitivity of the transduction mechanism. It is also important to note that selectivity requirements are specially demanding in this kind of assays. Any nonspecific interaction or cross-reactivity of the aptamer will be more evident than in other format assays, leading to high background noise, and in consequence reduced sensitivity or even false positives. The prime advantages of these approaches are that no label or additional reagents are required, and the assay can provide continuous real-time data, useful for continuous monitoring.

Mass change is the most immediate property accompanying the recognition event. It can be monitored by measuring changes in behaviour of some vibrating device, usually made of a piezoelectric material, coupled to the chemically sensitive layer that incorporates the aptamer. As vibration energy may be perceived as sound, most of these devices are termed acoustic-wave sensors. When mass is loaded onto the surface of these sensors, the propagation velocity of the acoustic wave decreases, resulting in a frequency shift, which is a function of the concentration of the analyte. Two types of acoustic wave sensors have been used to transduce the aptamer—target interaction: bulk acoustic wave (BAW) and surface acoustic wave (SAW) devices, depending on acoustic waves propagate through the whole volume of the device or only on its surface.

Quartz crystal microbalance (QCM) is the typical BAW device. It operates at relatively low frequencies (3—20 MHz), and in the case of aptamer-based assays the required operation in liquid media and the viscoelastic characteristics of the recognition layer cause dissipation of an important portion of the vibrational energy, resulting in loss of sensitivity. Therefore only large targets, ie, proteins, have been detected using this approach, and in general the reached sensitivity is not sufficient for the application of aptasensors in real samples. An attractive possibility, which has only been recently explored to detect low molecular weight compounds, is the use of a particular method of operating the QCM known as quartz crystal microbalance with dissipation monitoring (QCM-D). In this method, viscoelastic properties of the recognition layer in addition to its mass are measured by analysing energy dissipation. In this way, reliable and direct detection of a specific target is achieved through the measurement of nanometric changes of the aptamer layer thickness induced by the conformational change of the aptamer upon interaction with the target [41,42]. This strategy has been applied to the detection of model molecules with sensitivities in micrometre range but not to relevant targets in food analysis; obviously, sensitivity depends on the magnitude of the conformational and dynamical change of aptamers when they bind their recognition partners.

Most suitable for liquid-phase sensing are SAW devices, in particular love-wave sensors where the acoustic wave is guided in an extremely thin layer, parallel to the resonator surface. Compared to QCM, the characteristic of much higher resonance frequency and the fact that vibrations are located near the surface, becoming more affected by surface interactions with very low-energy dissipation, determine its higher sensitivity. Nevertheless, a limited number of attempts have been made to apply them to aptassays, mainly for protein detection [43].

Another mass-sensitive detection method, microcantilever-based aptasensing, has recently attracted attention [44—46]. The non-piezoelectric mass sensors using this principle are arrays of micromachined cantilevers functionalized with aptamer monolayers, usually through a self-assembling process. Although there are two modes in which cantilevers can function as chemical sensors (static or bending mode and dynamic or resonator mode), only the static operation principle has been applied so far in the aptamer field.

In this mode, the recognition reaction over one of the surfaces of the cantilever causes a surface stress change leading to a bending in proportion to the analyte concentration, which is optically measured by means of a position-sensitive detector. Two different surfaces are combined in the array: reference and sensing cantilevers and the differential bending between both is the analytical signal. In this way nonspecific interactions can be eliminated, making possible the detection of multiple species in a single step.

SPR spectroscopy is another well-established technique for direct monitoring of the recognition event. SPR phenomenon occurs when a plane-polarized radiation is totally internally reflected at a thin metal film (typically gold) coated onto a dielectric. At a specific incident angle (resonance angle) a maximum fraction of the energy of the incident radiation interacts with free electrons from metal (plasmons) thus causing a minimum in the reflected light intensity. The resonance angle is dependent on a number of factors that are maintained constant in the biosensing experiment with the exception of the refractive index close to the back of the metal film. When aptamer is immobilized on the metal surface, the recognition event produces minute changes in the refractive index or the local value of the dielectric constant on the metal/sample interface, which are detected as a change in the resonance angle and related to the concentration of target in solution. It is important to realize that for the plasmon resonance to occur a dielectric/metal interface is required, which can be achieved using different couplers such as prisms, gratings or fibre optics. The most widely used instrumental setups are based on prism couplers employing the Kretschmann configuration. This is the system frequently used not only for quantification purposes, especially to detect proteins [47−49] with sensitivity in the range of nanometre, but also for detailed characterization of the aptamer−target interaction [24,27,49]. GO electrophoretically electrodeposited on gold-SPR interfaces allows easy immobilization of nonlabelled aptamers and direct detection of targets as lysozyme [48]. The aptamer was not desorbed after binding contrarily to other works based on competition between GO and target. The detection of small molecules has also been demonstrated in spite of its inherent greater difficulty due to the reduced change in refractive index they cause [50]. High-throughput and multiplexed detection can be performed by using SPR imaging, also called SPR microscopy, although this alternative has not yet been applied for targets of interest in food analysis. Prism-based instruments are however rather costly and bulky, which restricts its application to a laboratory environment. More suitable for miniaturization and cost-effective manufacturing are the systems using as coupler fibre optic and grating [51] or those based on localized transmission [52], with applications so far in the clinical diagnostics field.

Aptamers are charged species (polyanions) and the recognition event will change the charge distribution at the selective layer/sample interface, affecting different electrical properties that can be used for direct label-free transduction purposes. Simple instrumentation is one of the distinguishing features of aptasensors based on this principle. Among the electrochemical techniques allowing direct aptamer—target interaction monitoring, electrochemical impedance spectroscopy (EIS) is specially promising. This technique measures the change in the interfacial electron transfer resistance (R_{et}) for a redox probe in solution as a consequence of steric hindrance and/or electrostatic repulsion caused by the recognition event in the selective layer. It is not possible to predict whether R_{et} will increase or decrease with target concentration since it depends on its molecular weight, surface charge and the ionic nature of the redox probe in solution. In general, more sensitive devices are obtained for larger molecules. To date, numerous impedimetric aptasensors have been described for proteins [53—56]. The introduction in the selective layer of nanomaterials such as carbon nanotubes [53,56] allows incorporating higher amounts of aptamer and leads to more sensitive devices. This is especially relevant in the determination of small molecules [57,58].

The change in the electric charge density can be also detected by means of potentiometry. In this technique, the nonuniform charge distribution that appears on the selective layer—sample interface after recognition gives rise to an electrical potential difference that is measured under equilibrium conditions and constitutes the analytical signal. Two of the most suitable formats for miniaturization have proved useful for the design of aptasensors, ie, ion-selective electrodes with internal solid contact and ion-selective field-effect transistors (ISFET). Both designs take advantage of the extraordinary properties of carbon nanotubes derived from their structure. In the first case, single-walled carbon nanotubes (SWCNT) are used as ion-to-electron transducers and substrates for covalent aptamer immobilization. The aptamer conformational change promoted by target induces a charge change of shielded SWCNT and the subsequent change in the potential. This principle was used for the detection of large living bacteria with excellent sensitivity, but it suffers from problems of strong interference from electrolyte background and small potential changes [59,60] that are expected even lower for smaller targets. The relative small size of aptamers, smaller than the Debye length, has been successfully exploited for the design of label-free protein aptasensors on aptamer-modified carbon nanotube ISFET with subnanometre detectability [61]. In this case, binding of the target, which must be charged, gives rise to a change in the charge distribution within proximity to the CNT and causes a sharp decrease in the source—drain current. This response can also be measured as a potential change under constant drain current conditions.

3.1.2 Target Structural Switching Assays

Selectivity, one of the most important features in chemical sensing, is affected by the correct combination of recognition and transduction mechanisms, and the transduction principles mentioned so far are rather unspecific. Although lack of selectivity of transduction is normally compensated by the high selectivity of aptamers, a different strategy to develop direct aptamer-based assays with improved sensitivity and selectivity is to incorporate a reporter system, ie, a label or indicator that provides the analytical signal, in such a way that the aptamer—target interaction modulates its activity. The reporter system can be an electroactive or optically active molecule or an amplifier, mainly nanoparticles that can even act as enzyme-mimicking systems [62], leading to large signal amplification and improved sensitivity. Modulation of the signal occurs as a result of the recognition event, taking advantage of the conformational changes the aptamer normally undergoes during the process, so this group of assays can be termed target structural switching assays.

Aptamers often comprise unpaired loop regions, which are disordered in the free-state and acquire a defined conformation by adaptive folding around the target [63]. This structural change makes possible to transduce the recognition event using reporter systems able to differentiate single-stranded from double-stranded DNA, such as intercalating agents (dyes or electroactive molecules) or gold nanoparticles. A very simple strategy allowing the semi-quantitative detection of a wide range of analytes by naked eyes is based on this principle [64—67]. The combination of AuNPs with ssDNA aptamers, which are physisorbed to the nanoparticle surface through the bases, has a stabilizing effect on the nanoparticles, reducing salt-induced aggregation that would lead to a colour change from pink to blue. After target interaction, the conformational change causes inhibition of aptamer adsorption and thus salt-induced aggregation of AuNPs occurs, which is easily observed by eye or spectrophotometrically. Intrinsic limitations of this strategy are related to the weak interaction between AuNPs and aptamers, which can be unspecifically displaced by sample components leading to false positives in real samples.

To overcome this difficulty, reporters can be chemically conjugated to aptamers. In this way self-reporting strategies can be developed provided that the recognition event induces a structure switch able to alter the activity of reporter in a significant extension. Both optical and redox-active systems have been exploited.

Using fluorophore-labelled aptamers assays have been developed to give target-dependent fluorescence changes through fluorescence resonance energy transfer. These assays are based on the transfer of nonradiative energy from donors to acceptors (quenchers) when they are in close proximity to each other. The recognition event would change the distance between donors and acceptors, either by decreasing or increasing it, which gives rise to quenching (signal-off assays) [68] or increasing (signal-on assays) [14,69—72] the radiative signal from fluorophore, respectively (Fig. 2A). Energy donors include organic fluorescent dyes [14,69—71], semiconductor quantum dots, and upconversion fluorescence nanoparticles [72], while energy acceptors include organic quenchers, gold nanoparticles [71], graphene [14,69,72] and polymeric films [70].

Likewise, aptamers may be linked at one end to a redox-active molecule (typically methylene blue or ferrocene), while the other end is attached to an

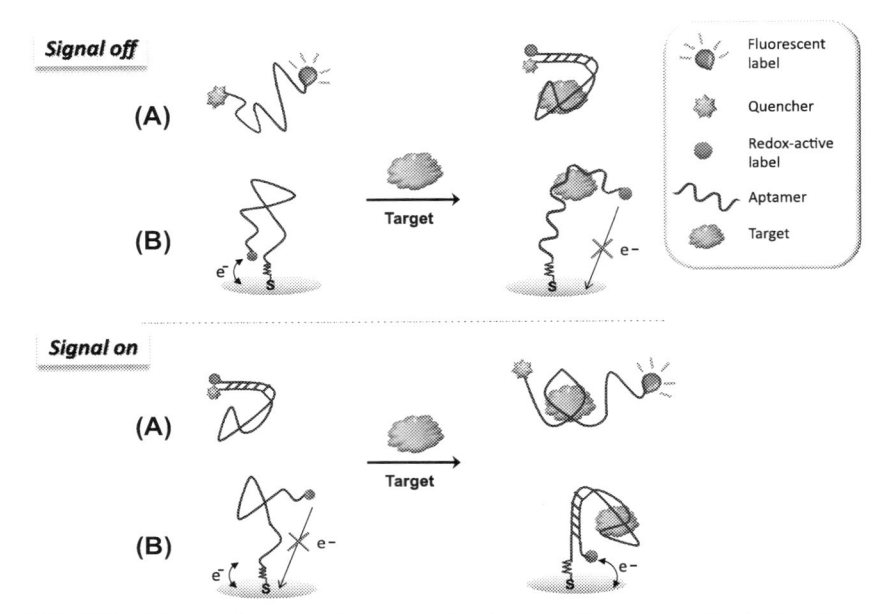

FIGURE 2 Schemes of self-reporting aptamer-based assays. Signal-on and signal-off architectures in combination with (A) optical and (B) electrochemical detection.

electrode surface (Fig. 2B). These aptasensors will lead to an output signal after target recognition provided that it induces a large enough structure switching of the aptamer to alter the distance or dynamics of the redox label with respect to the electrode surface, causing a measurable change in the electrochemical signal of the label [73]. Both, signal-on and signal-off configurations have been developed for the detection of a reduced number of targets, mainly thrombin as representative of large molecules and cocaine for small ones [74]. Similarly, it has been proved that if an enzyme is used as a label of the immobilized aptamer, target binding may induce enzyme inhibition (signal-off assay), which can be electrochemically detected [75].

The success of the aforementioned strategies depends on the unique structure of the aptamer, which should preferably adopt a specific conformation (for example, hairpin, G-quadruplex) and on the extension of its folding—unfolding during biomolecular recognition. Unfortunately, many of the aptamers described so far fail to undergo a significant conformational change upon target binding. For this reason, the rational design of aptamers is a crucial step in the development of such direct assays. Some strategies have been described for re-engineering aptamers, incorporating them into new structures that undergo large scale, binding-induced conformational changes that support self-reporting strategies [76,77], but they cannot be generalized.

An alternative and more easily generalizable strategy is that based on hybridization induced structural transition, also termed target-induced strand displacement. In this case, the conformation of the aptamer is initially fixed by hybridization with a strand completely or partially complementary to the aptamer, which acts as a competitor species. In the absence of target this oligonucleotide binds to the aptamer, and it is displaced upon formation of the aptamer–target complex, which is accompanied by a change in the signal of the reporter or indicator system that can be bound to either the aptamer or the complementary strand. The competitor oligo can be designed in different ways to achieve either signal-off or signal-on detection (Fig. 3). Mycotoxins, especially ochratoxin A (OTA), have been employed as model targets to demonstrate the feasibility of these schemes. By this general approach, a colorimetric aptassay has been developed using two different strands, complementary to anti-OTA aptamers, which are immobilized onto AuNPs. Nanoparticles are linked by hybridization with the aptamer forming AuNP dimers, but in the presence of OTA the AuNP aggregates disassemble, and the colour of the solution changes [78]. Likewise, complementary strand may be linked to a quencher while aptamer is modified with a fluorophore in such a way that hybridization brings fluorophore and quencher in close proximity. In the presence of target the complementary strand is released and an increase in fluorescence is measured [79]. If the complementary strand is designed with a hairpin structure functionalized with both fluorophore and quencher, labelling of the aptamer can be avoided, although the design results in signal-off

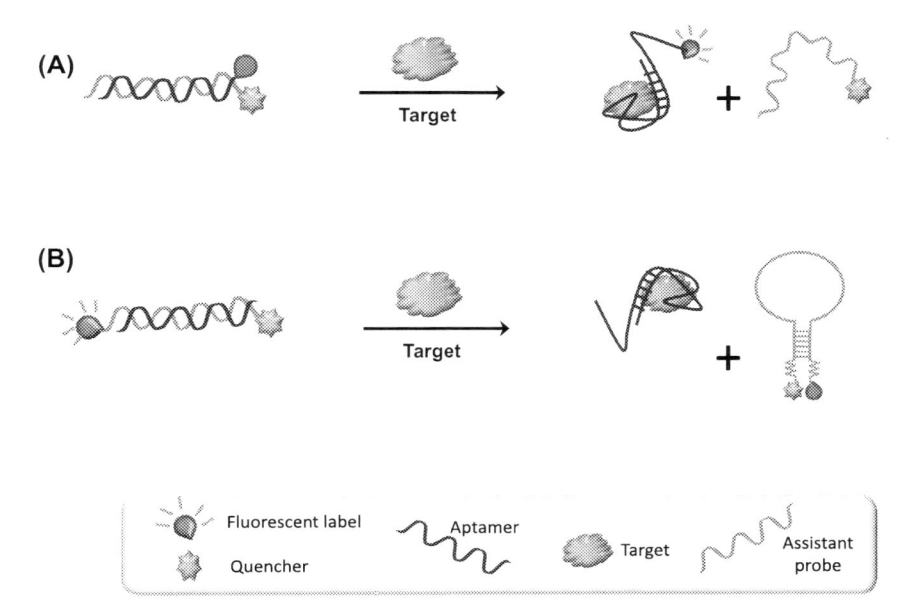

FIGURE 3   Signal-on (A) and signal-off (B) aptamer-based approaches using a target-induced strand displacement format with optical detection.

detection. This scheme has been applied for detecting organophosphorous pesticides [80]. The assay can be transformed into signal-on detection by using spatially sensitive fluorophores, such as pyrenes, attached to both ends of the complementary hairpin. In the presence of target, the competitor strand is displaced from the duplex with the aptamer, forming the hairpin structure with both pyrenes close enough to form an excimer, which results in a significant shift of the fluorescence emission [81]. Elongation of the aptamer with a sequence characteristic of a peroxidase DNAzyme gives rise to a direct assay where release of complementary strand by target also induces DNAzyme reorganization. In the presence of hemin, DNAzyme shows peroxidase activity, which is monitored spectrophotometrically [82], although other detection systems would also be valid. These are very versatile approaches, with a number of different designs already in place that seem only limited by the imagination of the researcher. However, robust approaches require high-affinity aptamers that allow the target to favourably compete with the complementary strand to be displaced. In some cases, the displacement is only a kinetic question demanding fine optimization of the order in which competition occurs [80].

The ideas aforementioned have been combined with the use of aptamer-conjugated magnetic beads, taking advantage of the easy manipulation of this platform under an external magnetic field, which allows analyte separation from complex matrices. When combined with the use of complementary strands modified with different reporter molecules such as upconversion nanoparticles [83], silver nanoclusters [84] and terbium ions [85] for fluorescent detection, or cadmium and lead ions [86] for electrochemical detection, signal-on assays for diverse analytes have been developed. These assays actually measure the amount of released complementary strand upon target-aptamer binding and can be adapted to multiplex detection of pollutants in food [86]. Given the specificity of the aptamer-based recognition, and the cleanup and preconcentration effect of magnetic particles, they are readily employed in complex samples such as milk, fish or wheat.

3.2 Sandwich Assays

As the name implies, the sandwich format involves three molecular 'in-gredients', ie, a target molecule sandwiched by a capture and a signalling aptamer that recognize two nonoverlapped binding sites on the target, providing a markedly increased selectivity in comparison to single-binding approaches. The factors determining the detection limit of these assays are the affinity constant of the aptamers, the nonspecific binding of the labelled aptamer and the detectability of the tracer bound to the signalling aptamer. Thus the use of amplification systems as labels, such as enzymes or metal nanoparticles, leads to high sensitivity signal-on assays, although it should be noted that binding ability of aptamers as well as detectability of labels should be preserved as fully as possible after conjugation. Another key factor determining the usefulness of this format is the availability of two different aptamer receptors for the same target molecule which limits its applications. For targets with repeated epitopes the use of the same aptamer as primary and secondary receptor allows circumventing this issue [87]. Alternatively, the combination of an aptamer and an antibody has been reported as a convenient and more general option. Since it is easier to label aptamers than antibodies, the use of the aptamer as signal carrier tends to be preferred, but using the aptamer as capturing agent offers reusable sensing phases.

Unfortunately, up to now, none of the relevant targets in food safety control belongs to the pretty exclusive group of molecules with a pair of different aptamer receptors. However, a recently reported array-based platform to obtain these pairs of bioreceptors [88] could prompt change this picture, especially for high-size molecules where multiple interaction sites are easily envisaged.

An appealing variant of the sandwich format is the proximity ligation assay where two specific aptamers are extended with additional sequences at either $3'$ or $5'$ end to tailor a proximity probe pair able to bind two epitopes of the same target. As a result of the dual recognition event, both aptamers are placed close to each other and their free extremities are brought into close proximity allowing revealing the presence of the target by generation of a new amplifiable DNA sequence after enzymatic ligation [89], a self-reporting probe [90] or even by hybridization with surface immobilized oligonucleotides [91]. This approach underpins innovative aptamer-based assays with enhanced specificity and low background signal, since signal generation requires simultaneous binding of two aptamers to the same target molecule, although so far they have only been used for detecting macromolecules such as thrombin and the cytokine platelet-derived growth factor.

Small molecules are not amenable to sandwich assay format because of the lack of two independent binding sites for receptors but can be imitated by splitting the aptamer into two fragments that do not interact unless the target is present. Aptamer reassembling is expected to occur only by target induction. This strategy was originally devised to get rid of loop domains that act as linkers between the two arms of the stem of a hairpin or a three-way junction aptamer and are not essential for binding [92]. Only since 2009, this idea was widely exploited in different heterogeneous assays relying on optical or electrochemical detection but is still restricted to models: cocaine, ATP/adenosine. This fact casts a shadow over the actual applicability. Adenosine aptamer is rather exceptional because target binds to the stem moiety instead of the usual loop, and cocaine is a hydrophobic molecule that binds at the central junction of a T aptamer structure. The latter structure has been considered as privileged for splitting and a general cutting approach was recently demonstrated with four antisteroid aptamers [93]. On the contrary, hairpin structure is easily disturbed by splitting. Beyond the classical sandwich-like assays, some smart approaches have been designed. To improve the low stability of the triplex complex that can dissociate in washing steps, chemical proximity ligation was proposed. Aptamer reassembling through cycloaddition ligation promoted by the target binding to an immobilized azide terminated DNA fragment and a free cyclooctyne carboxylic acid labelled fragment was achieved without interfering with the target. The free segment was also tagged with biotin to allow a classical spectrophotometric readout after labelling with a streptavidin-HRP conjugate [94]. Alternatively, homogeneous approaches, which can be benefited from absence of rinses and steric hindrance issues, can be employed. Inclusion of end-tagged fluorophores in cyclodextrins enhances the fluorescence or excimer emission, and facilitates and stabilizes the formation of the ternary complex [95].

Labelling a DNA with a DNAzyme is straightforward and can be also benefited from splitting. The two halves of DNAzyme are allowed to assemble in the absence of the target, while the target-induced conformational change disrupts or hinders that assembly precluding the enzymatic activity. This strategy was applied to the food relevant mycotoxin aflatoxin B1 in a target induced displacement assay where the DNAzyme subunits, containing a tail partially complementary to the aptamer, are functionally assembled by hybridization but dissociated upon target binding to the aptamer strand [96].

3.3 Competitive Assays

Competitive assays are particularly suitable to detect small molecules lacking multiple epitopes and triggering small conformational changes after binding their specific aptamer. In one of its formats, unlabelled analyte (the native target) present in the sample could be determined by its ability to compete with its labelled counterpart for the aptamer recognition site. Accordingly, the amount of labelled target recorded is inversely related to the analyte in the sample. This strategy allowed the determination of oxytetracycline in different food matrices using oxytetracycline-peroxidase conjugates as a tracer, with suitable correlation with the HPLC-based official method [97].

However, tagging the wide variety of relevant food targets and maintaining their molecular recognition properties is not always feasible and affordable. In response to this challenge, inhibition competitive format involving labelled aptamer receptors has been developed. Specifically, competition between the free target in solution and its immobilized analogue by a limited amount of labelled aptamer occurs. Hence, the presence of the analyte in the sample prevents the binding of the immobilized target to the labelled-aptamer leading to a decrease in the signal [98]. One of the most frequently used labels are enzymes, conjugated to the aptamer through biotin—streptavidin or antigen—antibody fragment binding. The use of a monovalent labelling by means of antibody Fab fragments to introduce the enzyme significantly improves the sensitivity of the assays when compared with the widely used biotin—streptavidin system [99]. Its implementation by using magnetic microparticles as a scaffold for aptamer immobilization and subsequent electrochemical transduction of the biorecognition event has given rise to fast, highly sensitive and easier-to-use methods for the detection of antibiotics [99] and mycotoxins [100]. Remarkably, the competitive format has turned out to be the strategy of choice to reliably determine the presence of food allergenic proteins [101] in processed foods, heat-treated and hydrolysed one, where target structure has been altered while toxicity remains.

Meanwhile, our group has devised a displacement assay that allows label-free detection of small molecules via EIS [102] and SPR [103] approaches. The target molecule is immobilized on the solid surface and the specific aptamer is attached via affinity recognition. The target present in the sample displaces the aptamer from the surface leading to changes in solution—surface interface, which could be sensitively recorded by EIS or SPR. This pioneered work has demonstrated its usefulness in the control of neomycin in milk samples. Likewise, it has been subsequently extended to determine other small molecules such as brevetoxin-2, a marine neurotoxin, in spiked shellfish extracts [17].

3.4 Improved Sensitivity Exploiting Nucleic Acid Amplification Approaches

Due to the oligonucleotide nature of these synthetic receptors, amplification techniques originally conceived only for nucleic acids can be extended to the detection of other molecules of great interest specifically recognized by aptamers.

PCR, the gold standard for nucleic acid amplification and quantification involving thermal cycling, has been successfully combined with different aptamer-based assays pushing down the detection limit of nonnucleic acid targets. In this technology coined as apta-PCR, the aptamer itself or a surrogate, depending on whether the assay format is sandwich [104], competitive [105] or target-induced strand displacement [106,107], serves as a template to be exponentially amplified by PCR. To take full advantage of the excellent amplification power of the PCR, this is performed in solution with real-time monitoring and preceded, when appropriate, by thermal elution of the bound aptamer. Apta-PCR is an extension of immuno-PCR technique which uses antibodies conjugated with specific DNA sequences for PCR amplification. While offering great sensitivity, the antibody-DNA coupling is somewhat laborious and lacks control of DNA tags per antibody, leading to results not as reproducible as desired. These limitations can be overcome by replacing the chimeric conjugate by a readily extended aptamer which, in turn, acts both as biorecognition molecule and as reporter. Likewise, PCR has been combined with other nucleic acid-based strategies such as (exo)nuclease protection assays [108]. In this case, PCR is used for quantifying target-aptamer complexes taking advantage of the protection of aptamer from exonuclease degradation through binding.

Isothermal nucleic acid amplification-based technologies have emerged as thermal cycle-free alternatives to PCR, facilitating the development of miniaturized devices and theirs applicability at the point of need. In this field, extensive research effort has been focused on coupling aptamer-based assays with rolling circle amplification (RCA). Inspired by the mechanism of replication of a viral circular DNA, RCA is an isothermal enzymatic process catalysed by certain DNA and RNA polymerases able to extend a primer annealed to a nucleic acid circular template to synthesize long ssDNA molecules with multiple (tens to hundreds) repeated units. Amplification products can be detected using different methods. The most frequently applied to the development of aptasensors are (1) direct labelling of the RCA product by incorporating modified dNTPs during amplification reaction and (2) hybridization with complementary oligonucleotides tethered with functional moieties. Both of them have been combined with a variety of signal readout techniques: fluorescence, colorimetric detection, electrochemistry and electrochemiluminescence, with a demonstrated improvement in sensitivity.

While homogeneous aptamer-based assays combined with RCA reaction have been achieved by target binding-induced generation of circular template, it demanded a carefully design not so easy to be generalized. Therefore, special attention has been paid to heterogeneous assays, particularly those coupled to electrochemical detection, which is characterized by high sensitivity, inherent equipment simplicity, facilitating on site analysis. RCA implementation in electrochemical platforms is not trivial because it depends on the distance between the redox marker attached to the elongated DNA and the electrode surface. Alternatively, it has been reported the incorporation of nanoparticle tags that after dissolution in acidic pH are conveniently quantified by anodic stripping voltammetry, or the incorporation of multiple redox enzymes capable of generating in solution an electroactive enzymatic product.

Regarding the aptassay format, different RCA-based approaches have been reported. When a heterogeneous sandwich assay (dual-aptamer recognition) is viable, secondary aptamer can be conjugated to RCA primer through standard DNA synthesis, and RCA reaction can be implemented without the need of an intricate design [109]. Likewise, extra sensitivity can be achieved by using bio-bar-code AuNPs functionalized with the secondary aptamer along with a large number of RCA primers (Fig. 4). These are hybridized with the DNA circular template acting as bar codes to generate an enhanced signal [110].

Whether a single aptamer is available for the target, it may act as both recognition element to sense the target and RCA primer to form the aptamer-circular template duplex (Fig. 5A and B). In that case, the amplification is initiated by the aptamer itself in the absence of its cognate ligand. Conversely, in its presence, the isothermal replication does not occur, giving rise to a signal-off approach [111,112]. It should be noted that signal-off architectures entail intrinsic high susceptibility to false positives, so signal-on counterparts are highly desirable; it can be achieved by designing aptamer and RCA primer

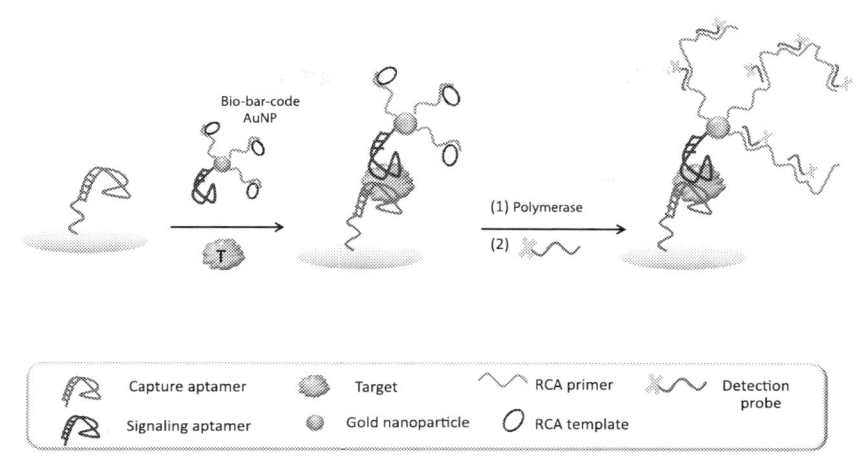

FIGURE 4   Sandwich aptasensor coupled with bio-bar-code DNA and heterogeneous rolling circle amplification (RCA). The secondary aptamer is attached onto gold nanoparticles along with the RCA primer annealed with the RCA template for subsequent isothermal elongation.

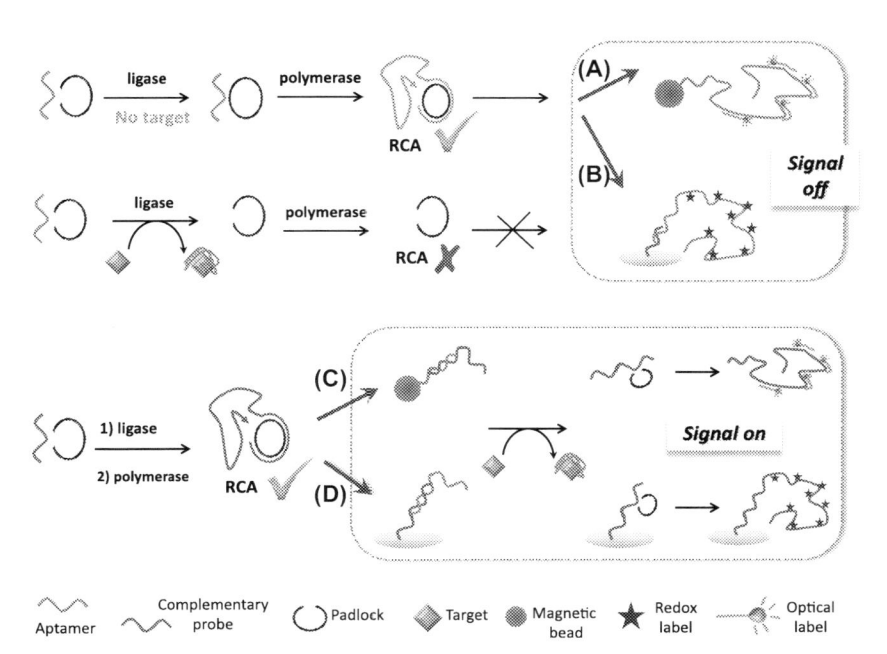

FIGURE 5   Target-induced displacement strategies combined with rolling circle amplification (RCA) for aptamer-based assays. Aptamer elongation on a surface (A) or in solution (B), resulting in signal-off approaches. Complementary probe elongation in solution (C) or on a surface (D), leading to signal-on approaches.

as two separated entities/probes (Fig. 5C and D). This way, aptamer can be immobilized onto the surface through partial hybridization with a complementary capture DNA probe (Fig. 5D). Since aptamer affinity for its cognate target is higher than that for the complementary oligonucleotide, affinity complex induces the displacement of the aptamer from the surface. Then, the free capture probe hybridizes with the RCA template and, serving as RCA primer, it allows the solid-phase isothermal extension [113]. Alternatively, the attachment of the aptamer-RCA primer duplex through the aptamer itself onto magnetic beads (Fig. 5C), followed by the target-induced RCA primer release, leads to a homogeneous RCA reaction as well as a reduced background signal stemmed from the magnetic separation [114].

To minimize the steps and concomitantly the analysis time, the padlock (linear precursor of the circular RCA template) can be designed to be complementary to the aptamer and the RCA primer, which is attached to a solid support giving rise to a ternary duplex-type sensing phase. The target-aptamer binding promotes aptamer release and padlock circularization in the presence of a DNA ligase and subsequent RCA reaction on the magnetic beads. Anodic stripping voltammetry of CdS quantum dots incorporated to the RCA products via hybridization allowed the detection of ultralow levels of toxins [115].

Unlike PCR, conventional RCA proceeds according to a linear kinetics, ie, a single-binding event is amplified in a linear manner. This shortcoming can be circumvented by using an exponential variant, the so-called hyperbranched RCA (HRCA) or ramification amplification, where the linear RCA product is employed as template for further amplification with additional sets of primers. Coupling HRCA with aptasensors allowed dramatically increased signals and consequently lower detection limits [116].

Other powerful nucleic acid-based isothermal amplification techniques start to appear in the aptasensing landscape to detect proteins and small molecules. However to date they have only been challenged with few targets of interest in food safety field. With exponential kinetics, Loop-mediated isothermal AMPlification (LAMP) technique has been recently integrated with electrochemical [117] and electrochemiluminescent [118] signal-on readout to determine OTA, and easily tailored to other molecular targets. A dsDNA composed by the aptamer and its complementary probe is immobilized on the electrode. Upon target binding, the affinity complex leaves off the surface and the remaining aptamer plays as one of the four primers involved in the solid-phase replication. Likewise, an aptamer against cytochrome-c has been used as template for isothermal recombinase polymerase amplification (RPA) after thermal release from a sandwich complex with the target and a recognition antibody immobilized onto magnetic beads [119]. In spite of the fact that the versatility of this scheme is restricted to targets with two or more binding sites, RPA could be combined with other assay formats such as competitive or target-induced displacement, among others.

The aforementioned strategies can be denoted as signal-based amplification strategies, where a catalytic process is used to increase the signal that results from a single-binding event. Consequently, the boost in sensitivity will be limited by the binding affinity, which is maximized during the selection process of the synthetic receptor (aptamer) and through subsequent modifications thereof. By contrast, target-based amplification could be considered superior, as these methods rely on target recycling in which each target molecule takes part in multiple affinity events to accomplish a greatly enhanced sensitivity by circumventing the affinity reaction stoichiometry limitation. Nonnucleic target recycling can be achieved with DNA modifying enzymes — mainly nucleases and polymerases — acting at constant temperature on the aptamer sequence, which has been previously engineered in some extension.

Several aptassays where target recycling is assisted by nicking endonucleases have been reported. Such endonucleases, considered as molecular scalpels, recognize specific nucleotide sequences in dsDNA but cleave only one strand of the duplex. However, their sequence-specificity restricts their scope of application. To overcome this limitation, sequence-independent nucleases such as exonucleases can be used. They catalyse the stepwise removal of mononucleotides at either 3'- or 5'-terminus from ssDNA or dsDNA irrespective of the sequence, giving rise to more general schemes. Most of the exonuclease-aided target recycling aptamer-based assays are homogeneous [120] probably motivated by a recent trend to develop single-step mix-and-read assay formats. However, even if homogeneous assays are performed without the need for washing or separation steps, simplifying assay procedure, they turn out to be challenging to deal with an often complicated sample matrix. In this sense, heterogeneous assays (eg, electrode- or nanoparticle-based formats) might result more appealing, although the enzyme activity with oligonucleotide substrates tethered to a support could be a limitation. To circumvent this drawback, an electrochemical aptasensor has been developed for the detection of OTA wherein the enzymatic DNA digestion leading to the target recycling takes place in solution [121]. A DNA strand complementary to the aptamer, ferrocene-tagged in its 5'-end, is attached to an electrode surface at the 3' extremity and hybridized to the aptamer to generate the sensing phase (Fig. 6A). After the recognition event, the aptamer-OTA affinity complex is released from the sensing platform and then the distance between the electroactive label (ferrocene) and the electrode surface decreases due to a stem-loop structure adopted by the immobilized DNA strand, allowing the electron transfer to occur. An exonuclease selective for ssDNA catalyses the hydrolysis of the aptamer sequence in the complex, untying the target for a new catalytic cycle.

Polymerases, which mediate DNA extension instead of strand cleavage, can be also employed to achieve target-recycling amplification. Polymerase-assisted target recycling (PATR) has been applied in homogeneous

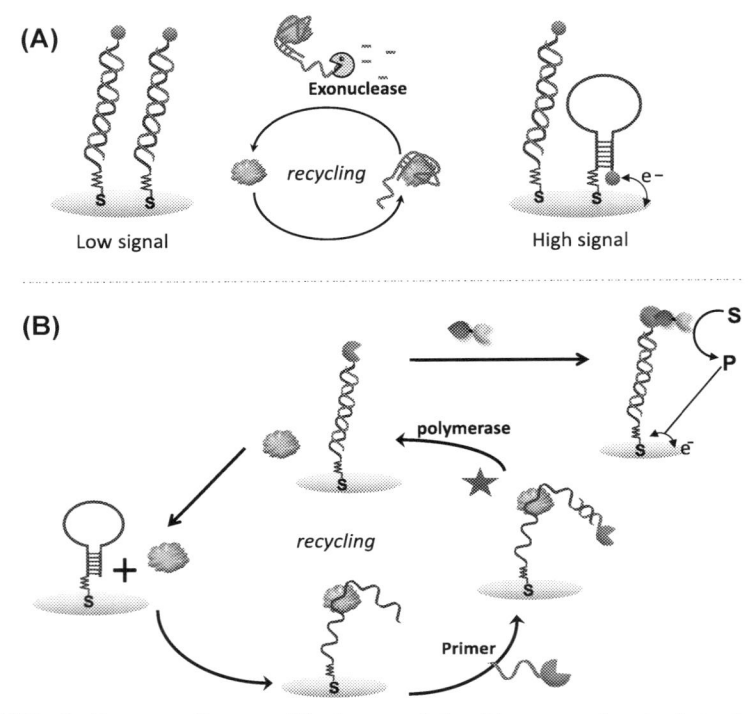

FIGURE 6 Target-recycling amplification coupled with aptamer-based electrochemical platforms: exonuclease-assisted (A) and polymerase-assisted (B) strategy.

fluorescent aptamer-based assays; however, analytical performance was compromised by the blank fluorescence signal [122]. Alternatively, a combination of this strategy with strand displacement amplification and heterogeneous electrochemical detection led to sensitive detection of ampicillin [123]. On the other hand, improved sensitivity has been reported by performing PATR with an immobilized hairpin-like tailored aptamer and enzyme-amplified electrochemical readout resulting in powerful dual amplification [124]. Target binding induces annealing and extension of a biotin-tagged primer with concomitant release of the target to bind a new aptamer. The biotinylated DNA duplex generated from each binding event was subsequently labelled with streptavidin—alkaline phosphatase and the enzyme-generated product was detected by differential pulse voltammetry (Fig. 6B). Enzyme-assisted target recycling is a particularly sensitive amplification route that pushes the thermodynamics of the aptamer—target affinity reaction in favour of the bound target. In consequence, sensitivity of these assays is not limited by the affinity constant of the synthetic receptor towards its cognate target.

4. APPLICATIONS IN FOOD ANALYSIS

There is a growing interest to apply aptamer-based approaches for food analysis in response to most significant challenges posed by food industry and new regulations. In 2013, we reviewed the main analytical strategies developed using aptamers for the control of analytes related to food quality and safety [125], noticing a rather limited number of published papers dealing with a reduced group of suited aptamers against model targets as a proof of concept. After only two years a significant growth is apparent. The targeted analytes include toxins, pathogens, allergens, drug residues and chemical contaminants. In the previous section, we have summarized the main design strategies adopted. Therefore, in this section we will point out the major trends, focussing on those approaches with particularly interesting concepts.

Toxins of importance in food safety include those produced by fungi, bacteria and algae. These are the group of compounds having a greater number of aptamer-based analytical assays by far. Most approaches are still limited to OTA, considered as a model target among mycotoxins. In a recent revision on OTA devices [126], it was noted that only three additional mycotoxins have their own aptamers: aflatoxin M1 [57], aflatoxin B1 [96,107], and zearalenone [126]. Two years later the list includes: aflatoxin B2 [21], fumonisin B1 [46] and versicolorin [127] a precursor of aflatoxin B1. All the strategies discussed previously have been applied to mycotoxin detection, from the direct SPR [50], impedimetric [57] or colorimetric [65] assays, to complicated schemes using DNAzyme [82,96] or fluorescent target induced structure-switching assays [79], and to the most sophisticated experimental approaches based on nucleic acid isothermal amplifications [111,112,115—118] or target recycling [121], in a variety of liquid (wine, beer, peanut oil) or solid matrices (cereals). Marine toxins have appeared as novel targets for aptamers [17] while other toxins (ricin, botulism) are increasing their tool box for improved detection [49,77].

Pathogens are the most important threat in food safety that justifies the second position in the ranking of aptamer-based approaches. Beyond the classical detection of the whole pathogen irrespective of the actual aptamer target, employed, for example, in the direct impedimetric [54], or potentiometric [60] detection of *Escherichia coli* and potentiometric quantification of *Salmonella typhy* [59], the detection of enterotoxins produced by the pathogens has increased with evolution of novel aptamers. Some approaches developed by the same group rely on GO-assisted fluorescent assays [14], but target recycling was also exploited [120]. Aptamer magnetic separation is gaining interest due to equal or even superior recoveries than immunoseparation [128]. Since aptamers are not able to ensure identical selectivity for all strains, capture is usually followed by quantitative PCR detection of genetic material of the pathogen for unambiguous identification and quantitation purposes. This is the approach used for detection of human noroviruses [20] and *Listeria* spp. [128]. Aptamers against spores from bacteria that overcome pasteurization treatments, such as *Bacillus cereus*, are now available also for capturing [129].

The incidence of allergies is unstoppably increasing worldwide. The only way of ensuring the safety of sensitized population is through the accurate labelling of food, and producers, retailers and consumers demand simple and rapid analytical assays for allergens aimed to certify healthy food for all. Novel aptamers for gliadin that causes celiac disease [15,101], concanavalin A, a lectin present in legumes and tubers [18], Ara h 1 protein, one of the most important peanut allergens [27], and β-conglutin from lupine legume, an emerging allergen [105], are the most noticeable recent advances. In all cases, the availability of new receptors allowed the development of assays with performance comparable to those of the existing immunoassays, although only in some cases a wide variety of samples have been tested [101].

Another important field in which aptasensors have been increasingly applied is testing veterinary drug residues. Accumulation of antibiotics in animal-derived food products including milk, meat, eggs and honey has serious implications for human health, triggering the emergence of antibiotic-resistant bacteria. Many countries have set maximum safety levels for these drugs in human foodstuffs. By using aptamers as specific recognition elements, fast methods have been developed for the control of aminoglycoside antibiotics [45,52,71,102], oxytetracycline [44,64,97,98] and chloramphenicol [83] in different food matrices, with a generally low degree of cross-reactivity, after suitable sample pretreatment. Within the group of antibiotics that can persist in food after veterinary use, a generalized use of magnetic micro- and nanoparticles has been detected due to the easiness of handling and cleanup of samples. Besides the trend to develop DNA aptamers for aminoglycosides that have only old RNA aptamers [71], the most remarkable achievements are the kanamycin assay making use of catalytic properties of AuNPs [62], the ultrasensitivity achieved with polymerase-assisted target-recycling amplification combined with strand displacement and electrochemical detection for ampicillin [123] and the good correlation of an oxytetracycline magnetoassay with an HPLC method tested in 40 samples [97].

Nonintentionally added substances in food include those released from food contact materials. These chemical contaminants constitute other potential targets for aptamers, only represented so far by bisphenol A, which can migrate into food from resin coating on cans or from polycarbonate food containers. Aptamers against bisphenol A have been applied to direct [66,68] and sandwich [87] rapid detection of the chemical with detectability in the range of pg/mL, but only tested in water samples. Other chemical contaminants such as pesticides [58] or PCBs [86] are less amenable to be detected by aptamers due to their great variety and similar chemical structure. For them, aptamers evolved to have a broad cross-reactivity have been described [80].

In food analysis it is of outmost importance to demonstrate the applicability of the method in as many matrices as possible. It is fair to say that most methods are tested in at least one though in a very limited number of samples, and sometimes, in easy matrices. In solid samples, it is very common the need for time-consuming and laborious sample pretreatment. All samples had to be enriched with the analyte, but the practice of spiking the final extract before measurement, though more prone to be successful, is not a realistic approach. Sensitivity and selectivity are in general adequate, and attention needs to be turned to demonstrate the ability to detect more analytes, at lower cost and greater speed with capacity to handle many diverse samples.

5. CONCLUDING REMARKS AND FUTURE DIRECTIONS

The arrival of aptamers onto the scene 25 years ago has offered enormous opportunities to biosensing in different fields including food analysis. These receptors of a striking chemical simplicity and production have shown to behave as efficiently as antibodies giving rise to versatile and readily adapted approaches for detecting many of the targets of interest for ensuring food safety and quality. Despite the progress described in this chapter, many challenges still remain.

Great effort has been made in selecting an increasing repertoire of aptamers against targets of paramount importance in the food safety field as well as in making SELEX process easier, faster and more generalizable. However, aptamers recognizing many other compounds of significance in food analysis have yet to be selected. On the other hand, aptasensing tool box is wealthy in design strategies and the nucleic acid nature of the aptamers makes them unique to implement amplification approaches, although most of them are devoted to a few model targets whose aptamers possess well-known structures, and they are tested in simulated media often distant from the complex food samples. In consequence, the future of the aptamer-based sensing for food analysis strongly depends on the effectiveness of the reported proof-of-concept designs in food matrices, which still requires further investigation. In this regard, stringent specificity and sensitivity are key requirements to achieve products capable of passing AOAC certification.

Innovation is continual, and there is a clear trend to use nanomaterials for simplifying the assays or improving their analytical performance. In this sense, the use of GO, or in some cases carbon nanotubes, as a quenching or protecting agent from nucleases is gaining popularity in food analysis approaches due to the commercial availability. Likewise gold nanoparticles are often used due to their cargo capacity not only as a solid support but also as amplification tool [62]. These methods have not been applied to real samples yet, and there is huge untapped potential in the combination of new nanomaterials and aptamers for developing fast and easy-to-use analytical tests.

In addition, food industry is demanding analytical tools able to cover more analytes at a time. There is a faint but determined move to multiplexing, especially for pathogens and toxins, taking advantage of colour-coded fluorescent nanostructured labelling [72] or metal-coded electrochemical assays [86]. However real multiplexing power has only be shown in dual approaches so far, which most likely serve as the basis for future work in multiplexed platforms.

Agro-food industry has special needs with regard to compliance with maximum residue levels of contaminants (pesticides, antibiotics, mycotoxins and pathogens) or labelling regulations (allergens or genetically modified organisms). Development of semiquantitative low-cost methods for rapid and *in situ* assays performed by untrained personnel is starting to find their market. Aptamer-based lateral flow assays are lagged with respect to immunoassays but some examples have appeared recently. A competitive device for aflatoxin B1 with fluorescent detection has been tested in corn using a modified extraction method compatible with the dipstick [130]. This format has the disadvantage of a signal-off readout that implies difficulties to establish the threshold for a positive result. Signal-on devices are more convenient. Pathogen capture from the matrix through apta-magneto separation is becoming commonplace and has been combined with a sandwich format where the secondary aptamer is subsequently isothermally amplified by strand displacement amplification. These strands are the target for the DNA sandwich assay carried out in the lateral flow device [131]. For these and other previously discussed methods reach the commercial market, considerable advantages over conventional methods need to be consistently demonstrated. In fact, a cumbersome issue in the implementation of the aptamer technology is the industrial commitment to antibodies and the reluctance to change well-established methods as well as ignorance about aptamers and their performance. Aptamer-based assays should be validated and demonstrate their affordability, especially for those analytes for which other suitable tests are not available, and in doing so earn their place in food analysis market.

ACKNOWLEDGEMENTS

This work was sponsored by Principado de Asturias government and FICYT under a CLARIN postdoctoral contract (RMC) and project FC15-GRUPIN14-025. Financial support from the Spanish Ministerio de Economía y Competitividad (Project CTQ2012-31157) and FEDER funds are also acknowledged.

REFERENCES

[1] United Nations, Department of Economic and Social Affairs, Population Division, Working Paper No. ESA/P/WP.241, 2015.
[2] M. Zourob, Recognition Receptors in Biosensors, Springer, New York, 2010.

[3]   F.G. Banica, Chemical Sensors and Biosensors: Fundamentals and Applications, Wiley, Chichester, 2012.

[4]   M. Mascini, I. Palchetti, S. Tombelli, Angew. Chem. Int. Ed. Engl. 51 (2012) 1316—1332.

[5]   C. Tuerk, L. Gold, Science 249 (1990) 505—510.

[6]   A.D. Ellington, J.W. Szostak, Nature 346 (1990) 818—822.

[7]   M. Blind, M. Blank, Mol. Ther. Nucleic Acids 4 (2015) e223.

[8]   E. Ouellet, E.T. Lagally, K.C. Cheung, C.A. Haynes, Biotechnol. Bioeng. 111 (2014) 2265—2279.

[9]   W. Pan, G.A. Clawson, Molecules 14 (2009) 1353—1369.

[10] E. Ouellet, J.H. Foley, E.M. Conway, C. Haynes, Biotechnol. Bioeng. 112 (2015) 1506—1522.

[11] N. Derbyshire, S.J. White, D.H.J. Bunka, L. Song, S. Stead, J. Tarbin, M. Sharman, D. Zhou, P.G. Stockley, Anal. Chem. 84 (2012) 6595—6602.

[12] J.G. Bruno, M.P. Carrillo, T. Phillips, C.J. Andrews, J. Fluoresc. 20 (2010) 1211—1223.

[13] Y.K. Huang, X.J. Chen, Y. Xia, S.J. Wu, N. Duan, X.Y. Ma, Z.P. Wang, Anal. Methods 6 (2014) 690—697.

[14] Y.K. Huang, X.J. Chen, N. Duan, S. Wu, Z. Wang, X. Wei, Y. Wang, Food Chem. 166 (2015) 623—629.

[15] S. Amaya-González, N. de-los-Santos-Álvarez, A.J. Miranda-Ordieres, M.J. Lobo-Castañón, Anal. Chem. 86 (2014) 2733—2739.

[16] L. Civit, A. Fragoso, C.K. O'Sullivan, Anal. Biochem. 431 (2012) 132—138.

[17] S. Eissa, M. Siaj, M. Zourob, Biosens. Bioelectron. 69 (2015) 148—154.

[18] R. Ahirwar, P. Nahar, J. Agric. Food Chem. 63 (2015) 4104—4111.

[19] R. Stoltenburg, C. Reinemann, B. Strehlitz, Anal. Bioanal. Chem. 383 (2005) 83—91.

[20] B.I. Escudero-Abarca, S.H. Suh, M.D. Moore, H.P. Dwivedi, L.A. Jaykus, PLoS One 9 (2014) e106805.

[21] X.Y. Ma, W.F. Wang, X.J. Chen, Y. Xia, N. Duan, S.J. Wu, Z.P. Wang, Food Control 47 (2015) 545—551.

[22] R. Stoltenburg, N. Nikolaus, B. Strehlitz, J. Anal. Methods Chem. (2012). ID. 415697.

[23] J.A. He, Y.A. Liu, M.T. Fan, X.J. Liu, J. Agric. Food Chem. 59 (2011) 1582—1586.

[24] F.M. Spiga, P. Maietta, C. Guiducci, ACS Comb. Sci. 17 (2015) 326—333.

[25] S.D. Mendonsa, M.T. Bowser, J. Am. Chem. Soc. 126 (2004) 20—21.

[26] J.J. Tang, J.W. Xie, N.S. Shao, Y. Yan, Electrophoresis 27 (2006) 1303—1311.

[27] D.T. Tran, K. Knez, K.P. Janssen, J. Pollet, D. Spasic, J. Lammertyn, Biosens. Bioelectron. 43 (2013) 245—251.

[28] J.W. Park, R. Tatavarty, D.W. Kim, H.T. Jung, M.B. Gu, Chem. Commun. 48 (2012) 2071—2073.

[29] M.U. Musheev, S.N. Krylov, Anal. Chim. Acta 564 (2006) 91—96.

[30] A. Nitsche, A. Kurth, A. Dunkhorst, O. Panke, H. Sielaff, W. Junge, D. Muth, F. Scheller, W. Stocklein, C. Dahmen, G. Pauli, A. Kage, BMC Biotechnol. 7 (2007) 48.

[31] M.V. Berezovski, M.U. Musheev, A.P. Drabovich, J.V. Jitkova, S.N. Krylov, Nat. Protoc. 1 (2006) 1359—1369.

[32] S.E. Kim, W. Su, M. Cho, Y. Lee, W.S. Choe, Anal. Biochem. 424 (2012) 12—20.

[33] Y.M. Liu, C. Wang, F. Li, S.W. Shen, D.L.J. Tyrrell, X.C. Le, X.F. Li, Anal. Chem. 84 (2012) 7603—7606.

[34] C.J. Huang, H.I. Lin, S.C. Shiesh, G.B. Lee, Biosens. Bioelectron. 35 (2012) 50—55.

[35] K.M. Ahmad, S.S. Oh, S. Kim, F.M. McClellen, Y. Xiao, H.T. Soh, PLoS One 6 (2011) e27051.

[36] T. Hunniger, H. Wessels, C. Fischer, A. Paschke-Kratzin, M. Fischer, Anal. Chem. 86 (2014) 10940−10947.

[37] W.Y. Zhang, W.H. Zhang, Z.Y. Liu, C. Li, Z. Zhu, C.J. Yang, Anal. Chem. 84 (2012) 350−355.

[38] Z. Zhu, Y.L. Song, C. Li, Y. Zou, L. Zhu, Y. An, C.J. Yang, Anal. Chem. 86 (2014) 5881−5888.

[39] T. Schutze, B. Wilhelm, N. Greiner, H. Braun, F. Peter, M. Morl, V.A. Erdmann, H. Lehrach, Z. Konthur, M. Menger, P.F. Arndt, J. Glokler, PLoS One 6 (2011) e29604.

[40] N. de-los-Santos-Álvarez, M.J. Lobo-Castañón, A.J. Miranda-Ordieres, P. Tuñón-Blanco, Trends Anal. Chem. 27 (2008) 437−446.

[41] V.C. Ozalp, Analyst 136 (2011) 5046−5050.

[42] A. Osypova, D. Thakar, J. Dejeu, H. Bonnet, A. Van der Heyden, G.V. Dubacheva, R.P. Richter, E. Defrancq, N. Spinelli, L. Coche-Guérente, P. Labbé, Anal. Chem. 87 (2015) 7566−7574.

[43] M.D. Schlensog, T.M.A. Gronewold, M. Tewes, M. Famulok, E. Quandt, Sens. Actuat. B Chem. 101 (2004) 308−315.

[44] H. Hou, X. Bai, C. Xing, N. Gu, B. Zhang, J. Tang, Anal. Chem. 85 (2013) 2010−2014.

[45] X. Bai, H. Hou, B. Zhang, J. Tang, Biosens. Bioelectron. 56 (2014) 112−116.

[46] X. Chen, X. Bai, H. Li, B. Zhang, RSC Adv. 5 (2015) 35448−35452.

[47] I. Mihai, A. Vezeanu, C. Polonschii, C. Albu, G.L. Radu, A. Vasilescu, Sens. Actuat. B Chem. 206 (2015) 198−204.

[48] P. Subramanian, A. Lesniewski, I. Kaminska, A. Vlandas, A. Vasilescu, J. Niedziolka-Jonsson, E. Pichonat, H. Happy, R. Boukherroub, S. Szunerits, Biosens. Bioelectron. 50 (2013) 239−243.

[49] B. Wang, Z. Lou, B. Park, Y. Known, H. Zhang, B. Xu, Phys. Chem. Chem. Phys. 17 (2015) 307−314.

[50] Z. Zhu, M. Feng, L. Zuo, Z. Zhu, F. Wang, L. Chen, J. Li, G. Shan, S.Z. Luo, Biosens. Bioelectron. 65 (2015) 320−326.

[51] Y. Shevchenko, T.J. Francis, D.A.D. Blair, R. Walsh, M.C. DeRosa, J. Albert, Anal. Chem. 83 (2011) 7027−7034.

[52] G. Cappi, F.M. Spiga, Y. Moncada, A. Ferreti, M. Beyeler, M. Binachessi, L. Decosterd, T. Buclin, C. Guiducci, Anal. Chem. 87 (2015) 5278−5285.

[53] F. Rohrbac, H. Karadeniz, A. Erdem, M. Famulok, G. Mayer, Anal. Biochem. 421 (2012) 454−459.

[54] R.B. Queirós, N. de-los-Santos-Álvarez, J.P. Noronha, M.G.F. Sales, Sens. Actuat. B Chem. 181 (2013) 766−772.

[55] C. Ocaña, E. Arcay, M. del Valle, Sens. Actuat. B Chem. 191 (2014) 860−865.

[56] S. Khezrian, A. Salimi, H. Teymourian, R. Hallaj, Biosens. Bioelectron. 43 (2013) 218−225.

[57] E. Dinçkaya, Ö. Kınık, M.K. Sezgintürk, Ç. Altuğ, A. Akkoca, Biosens. Bioelectron. 26 (2011) 3806−3811.

[58] L. Fan, G. Zhao, H. Shi, M. Liu, Z. Li, Biosens. Bioelectron. 43 (2013) 12−18.

[59] G.A. Zelada-Guillén, J. Riu, A. Duezguen, F.X. Rius, Angew. Chem. Int. Ed. 48 (2009) 7334−7337.

[60] G.A. Zelada-Guillén, S.V. Bhosale, J. Riu, F.X. Rius, Anal. Chem. 82 (2010) 9254−9260.

[61] K. Maehashi, T. Katsura, K. Kerman, Y. Takamura, K. Matsumoto, E. Tamiya, Anal. Chem. 79 (2007) 782−787.

[62] T.K. Sharma, R. Ramanathan, P. Weerathunge, M. Mohammdtaheri, H.K. Daima, R. Shukla, V. Bansal, Chem. Commun. 50 (2014) 15856−15859.

[63] T. Hermann, D.J. Patel, Science 287 (2000) 820−825.

[64] Y.S. Kim, J.H. Kim, I.A. Kim, S.J. Lee, J. Jurng, M.B. Gu, Biosens. Bioelectron. 26 (2010) 1644−1649.

[65] C. Yang, Y. Wang, J.L. Marty, X. Yang, Biosens. Bioelectron. 26 (2011) 2724−2727.

[66] Z. Mei, H. Chu, W. Chen, F. Xue, J. Liu, H. Xu, R. Zhang, L. Zheng, Biosens. Bioelectron. 39 (2013) 26−30.

[67] W. Yun, H. Li, S. Chen, D. Tu, W. Xie, Y. Huang, Eur. Food Res. Technol. 238 (2014) 989−995.

[68] K.V. Ragavan, L.S. Selvakumar, M.S. Thakur, Chem. Commun. 49 (2013) 5960−5962.

[69] L. Sheng, J. Ren, Y. Miao, J. Wang, E. Wang, Biosens. Bioelectron. 26 (2011) 3494−3499.

[70] Y. Zhang, X. Sun, Chem. Commun. 47 (2011) 3927−3929.

[71] J. Cheng, Z. Li, J. Ge, R. Yang, L. Zhang, L. Qu, H. Wang, L. Zhang, Talanta 139 (2015) 226−232.

[72] S. Wu, N. Duan, X. Ma, Y. Xia, H. Wang, Z. Wang, Q. Zhang, Anal. Chem. 84 (2012) 6263−6270.

[73] A.A. Lubin, K.W. Plaxco, Acc. Chem. Res. 43 (2010) 496−505.

[74] R. Miranda-Castro, N. de-los-Santos-Álvarez, M.J. Lobo-Castañón, A.J. Miranda-Ordieres, P. Tuñón-Blanco, Electroanalysis 21 (2009) 2077−2090.

[75] B. Prieto-Simón, J. Samitier, Anal. Chem. 86 (2014) 1437−1444.

[76] R.J. White, A.A. Rowe, K.W. Plaxco, Analyst 135 (2010) 589−594.

[77] L. Fetter, J. Richards, J. Daniel, L. Roon, T.J. Rowland, A.J. Bonham, Chem. Commun. 51 (2015) 15137−15140.

[78] R. Xiao, D. Wang, Z. Lin, B. Qiu, M. Liu, L. Guo, G. Chen, Anal. Methods 7 (2015) 842−845.

[79] J. Chen, Z. Fang, J. Liu, L. Zeng, Food Control 25 (2012) 555−560.

[80] C. Zhang, L. Wang, Z. Tu, X. Sun, Q. He, Z. Lei, C. Xu, Y. Liu, X. Zhang, Y. Yang, X. Liu, Y. Xu, Biosens. Bioelectron. 55 (2014) 216−219.

[81] J. Huang, Z. Zhu, S. Bamrungsap, G. Zhu, M. You, X. He, K. Wang, W. Tan, Anal. Chem. 82 (2010) 10158−10163.

[82] C. Yang, V. Lates, B. Prieto-Simón, J.L. Marty, X. Yang, Talanta 116 (2013) 520−526.

[83] S. Wu, H. Zhang, Z. Shi, N. Duan, C. Fang, S. Dai, Z. Wang, Food Control 50 (2015) 597−604.

[84] J. Chen, X. Zhang, S. Cai, D. Wu, M. Chen, S. Wang, J. Zhang, Biosens. Bioelectron. 57 (2014) 226−231.

[85] J. Zhang, X. Zhang, G. Yang, J. Chen, S. Wang, Biosens. Bioelectron. 41 (2013) 704−709.

[86] Z. Yan, N. Gan, D. Wang, Y. Cao, M. Chen, T. Li, Y. Chen, Biosens. Bioelectron. 74 (2015) 718−724.

[87] J. Lee, M. Jo, T.H. Kim, J.Y. Ahn, D.K. Lee, S. Kim, S. Hong, Lab Chip 11 (2011) 52−56.

[88] M. Cho, S.S. Oh, J. Nie, R. Stewart, M.J. Radeke, M. Eisenstein, P.J. Coffey, J.A. Thomson, H.T. Soh, Anal. Chem. 87 (2015) 821−828.

[89] S. Fredriksson, M. Gullberg, J. Jarvius, C. Olsson, K. Pietras, S.M. Gustafsdottir, A. Ostman, U. Landegren, Nat. Biotech. 20 (2002) 473−477.

[90] C. Zong, J. Wu, M. Liu, L. Yang, L. Liu, F. Yan, F. Ju, Anal. Chem. 86 (2014) 5573−5578.

[91] Y.L. Zhang, Y. Huang, J.H. Jiang, G.L. Shen, R.Q. Yu, J. Am. Chem. Soc. 129 (2007) 15448−15449.

[92] M.N. Stojanovic, P. de Prada, D.W. Landry, J. Am. Chem. Soc. 122 (2000) 11547−11548.

[93] A.D. Kent, N.G. Spiropulos, J.M. Heemstra, Anal. Chem. 85 (2013) 9916−9923.

[94] A.K. Sharma, A.D. Kent, J.M. Heemstra, Anal. Chem. 84 (2012) 6104−6109.

[95] C. Yang, N. Spinelli, S. Perrier, E. Defrancq, E. Peyrin, Anal. Chem. 87 (2015) 3139−3143.

[96] Y. Seok, J.Y. Byun, W.B. Shim, M.G. Kim, Anal. Chim. Acta 886 (2015) 182−187.

[97] C. Lu, Z. Tang, C. Liu, L. Kang, F. Sun, Anal. Bioanal. Chem. 407 (2015) 4155−4163.

[98] C.H. Kim, L.P. Lee, J.R. Min, M.W. Lim, S.H. Jeong, Biosens. Bioelectron. 51 (2014) 426−430.

[99] E. González-Fernández, N. de-los-Santos-Álvarez, A.J. Miranda-Ordieres, M.J. Lobo-Castañón, Sensor Actuat. B Chem. 182 (2013) 668−674.

[100] A. Hayat, R. Mishra, G. Catanante, J. Marty, Anal. Bioanal. Chem. 407 (2015) 7815−7822.

[101] S. Amaya-González, N. de-los-Santos-Álvarez, A.J. Miranda-Ordieres, M.J. Lobo-Castañón, Anal. Bioanal. Chem. 407 (2015) 6021−6029.

[102] N. de-los-Santos-Álvarez, M.J. Lobo-Castañón, A.J. Miranda-Ordieres, P. Tuñón-Blanco, J. Am. Chem. Soc. 129 (2007) 3808−3809.

[103] N. de-los-Santos-Álvarez, M.J. Lobo-Castañón, A.J. Miranda-Ordieres, P. Tuñón-Blanco, Biosens. Bioelectron. 24 (2009) 2547−2553.

[104] A. Csordas, A.E. Gerdon, J.D. Adams, J. Qian, S.S. Oh, Y. Xiao, H.T. Soh, Angew. Chem. Int. Ed. 49 (2010) 355−358.

[105] M. Svobodova, T. Mairal, P. Nadal, M.C. Bermudo, C.K. O'Sullivan, Food Chem. 165 (2014) 419−423.

[106] W. Ma, H. Yin, L. Xu, Z. Xu, H. Kuang, L. Wang, C. Xu, Biosens. Bioelectron. 42 (2013) 545−549.

[107] X. Guo, F. Wen, N. Zheng, Q. Luo, H. Wang, H. Wang, S. Li, J. Wang, Biosens. Bioelectron. 56 (2014) 340−344.

[108] X.L. Wang, F. Li, Y.H. Su, X. Sun, X.B. Li, H.J. Schluesener, F. Tang, S.Q. Xu, Anal. Chem. 76 (2004) 5605−5610.

[109] J. Lee, K. Icoz, A. Roberts, A.D. Ellington, C.A. Savran, Anal. Chem. 82 (2010) 197−202.

[110] S. Bi, B. Ji, Z. Zhang, S. Zhang, Chem. Commun. 49 (2013) 3452−3454.

[111] L. Huang, J. Wu, L. Zheng, H. Qian, F. Xue, Y. Wu, D. Pan, S.B. Adeloju, W. Chen, Anal. Chem. 85 (2013) 10842−10849.

[112] L. Yao, Y. Chen, J. Teng, W. Zheng, J. Wu, S.B. Adeloju, D. Pan, W. Chen, Biosens. Bioelectron. 74 (2015) 534−538.

[113] Q. Wang, H. Zheng, X. Gao, Z. Lin, G. Chen, Chem. Commun. 49 (2013) 11418−11420.

[114] C. Ma, W. Wang, Q. Yang, C. Shi, L. Cao, Biosens. Bioelectron. 26 (2011) 3309−3312.

[115] P. Tong, W.W. Zhao, L. Zhang, J.J. Xu, H.Y. Chen, Biosens. Bioelectron. 33 (2012) 146−151.

[116] Y. Zhang, L. Yang, C. Lin, L. Guo, B. Qiu, Z. Lin, G. Chen, Anal. Methods 7 (2015) 6109−6113.

[117] S. Xie, Y. Chai, Y. Yuan, L. Bai, R. Yuan, Biosens. Bioelectron. 55 (2014) 324−329.

[118] Y. Yuan, S. Wei, G. Liu, S. Xie, Y. Chai, R. Yuan, Anal. Chim. Acta 811 (2014) 70−75.

[119] J.F.C. Loo, P.M. Lau, H.P. Ho, S.K. Kong, Talanta 115 (2013) 159−165.

[120] S. Wu, N. Duan, X. Ma, Y. Xia, H. Wang, Z. Wang, Anal. Chim. Acta 782 (2013) 59−66.

[121] P. Tong, L. Zhang, J.J. Xu, H.Y. Chen, Biosens. Bioelectron. 29 (2011) 97−101.

[122] L.P. Qiu, Z.S. Wu, G.L. Shen, R.Q. Yu, Anal. Chem. 83 (2011) 3050−3057.

[123] H. Wang, Y. Wang, S. Liu, J. Yu, W. Xu, Y. Guo, J. Huang, Chem. Commun. 51 (2015) 8377−8380.

[124] W. Cheng, W. Ding, Q. Li, T. Yu, Y. Yin, H. Ju, G. Ren, Biosens. Bioelectron. 36 (2012) 12−17.

[125] S. Amaya-González, N. de-los-Santos-Álvarez, A.J. Miranda-Ordieres, M.J. Lobo-Castañón, Sensors 13 (2013) 16292–16311.

[126] A. Rhouati, C. Yang, A. Hayat, J.L. Marty, Toxins 5 (2013) 1988–2008.

[127] H.L. Jiang, X.Y. Liu, Y.X. Qiu, D.S. Yao, C.F. Xie, D.L. Liu, Food Control 56 (2015) 202–210.

[128] S.H. Suh, L.A. Jaykus, J. Bictechnol. 167 (2013) 454–461.

[129] C. Fischer, T. Hunniger, J.H. Jarck, E. Frohnmeyer, C. Kallinich, I. Haase, U. Hahn, M. Fischer, J. Agric. Food Chem. 63 (2015) 8050–8057.

[130] W.B. Shim, M.J. Kim, H. Mun, M.G. Kim, Biosens. Bioelectron. 62 (2014) 288–294.

[131] W. Wu, S. Zhao, Y. Mao, Z. Fang, X. Lu, L. Zeng, Anal. Chim. Acta 861 (2015) 62–68.

New Technologies Improving Biosensor Efficacy

Chapter 7

Emerging Nanomaterials for Analytical Detection

G. Alarcon-Angeles,[1] G.A. Álvarez-Romero[2] and A. Merkoçi[3,4,*]

[1]*Universidad Autónoma Metropolitana-Xochimilco, Mexico City, México;* [2]*Universidad Autónoma del Estado de Hidalgo, Hidalgo, México;* [3]*Catalan Institute of Nanoscience and Nanotechnology (ICN2), CSIC and The Barcelona Institute of Science and Technology, Barcelona, Spain;* [4]*ICREA — Catalan Institution for Research and Advanced Studies, Barcelona, Spain*
Corresponding author: E-mail: arben.merkoci@icn.cat

Chapter Outline

Comprehensive Analytical Chemistry, Vol. 74. http://dx.doi.org/10.1016/bs.coac.2016.03.022

1. ANALYTICAL CHEMISTRY ROLE IN FOOD ANALYSIS

Quality control in food and beverage industry is important to ensure that product characteristics are acceptable to consumers. Chemical analysis is essential for monitoring food attributes such as composition, structure, sensory attributes, microbiological status and nutritional characteristics. In addition, chemical analysis is fundamental for the rational identification of main factors and their interrelations possibly affecting the properties of food. Thus, analytical chemistry has an important role in the development and application of analytical methods for the study of the properties, content and food contaminants.

Organic and inorganic compounds, heavy metals and pathogenic microorganisms can be toxic for the consumer; so they must be identified in order to prevent diseases and even death. The assurance of food quality is an important challenge for the production and distribution of food. According to the WHO, the quality assurance of food requires security measures in all from harvesting, production, transport, process, storage, marketing and food preparation areas [1].

New analytical methods and techniques are required for a reliable, fast, highly sensible, selective, quantitative, easy handling, low cost and real time analysis. Chromatography methodologies are the typical methods used for food analysis; however, colourimetric and electrochemical methods have arisen as quick detection methods. Among techniques with high specificity, the bioanalytical techniques are one of the most promising; these are based on the use of biological recognition agents. In recent decades, optical and electrochemical techniques have made use of biosensors, which substantially improve the analytical response. In this chapter, we summarize some important aspects of biosensors.

IUPAC states that 'a chemical sensor is a device that transforms chemical information, ranging from the concentration of a specific sample component to total composition analysis, into an analytically useful signal. Chemical sensors usually contain two basic components connected in series: a chemical (molecular) recognition system (receptor) and a physicochemical transducer'. Biosensors are 'chemical sensors in which the recognition system uses a biochemical mechanism'. Likewise, 'an electrochemical biosensor is a self-contained integrated device, which is capable of providing specific quantitative or semiquantitative analytical information using a biological recognition element (biochemical receptor) which is retained in direct space contact with an electrochemical transduction element' [2].

Furthermore, the use of nanomaterials (NMs) and nanotechnology represents a new trend in the development of new analytical methods. Nanotechnology is already recognized as an important tool for the construction of new devices, such as biosensors, and also brings improvements in the performance of chromatography, lab-on-a-chip devices and other analytical platforms. Moreover, NMs have been employed as labels to track a signal, either optical or electrochemical. In the next section, we describe the different NMs, their characteristics and their main applications in quality assurance of food and beverages.

1.1 Opportunities of Using Nanomaterials

The development of new analytical methods based on nanotechnology has been a great success. The use of NMs as nanoparticles (NPs), nanotubes (NTs), nanorods, nanowires, nanodiamonds and quantum dots (QDs) has contributed to the creation of new powerful tools applicable in different areas of the food industry, such as the development of 'intelligent' packaging materials, food preservation and quality assurance.

NMs have numerous applications in food industry including their use as antioxidants, masking-flavour agents, nutrient encapsulation control, release control of nutraceuticals, vitamins or flavours and the construction of nanosensors for the detection of pathogens and chemical contaminants [3]. In the scope of food microbiology, nanobiosensors have great relevance since they reduce the pathogen's analysis time from days to hours or even minutes. Nanobiosensors are based on the use of NMs for the immobilization of enzymes, antigens and nucleic acids over a transducer surface in order to promote direct electron transference, causing an amplification of the analytical signal generated by the biorecognition events. In addition, NMs can be also used as signalling tools (labels) in nanobiosensors such as those based on electrical, optical, mass change and other kind of responses.

The success of the NMs used as analytical tool lies in their electrical, mechanical, conductive, magnetic and optical properties, which largely depend on their synthesis procedure [4,5]. Certain properties such as size, shape, surface area, aggregation states, charge, chemistry and reactivity of NMs are essential and necessary when evaluating their potential, toxicity, stability and physicochemical behaviour for applications in qualitative and quantitative analysis. Different NMs have been used for these purposes, being the most reported: gold NPs, carbon nanotubes (CNTs), QDs and graphene (GR) (Fig. 1). The most important characteristics of these materials are described in the following section.

1.1.1 Gold Nanoparticles

Gold nanoparticles (AuNPs) are the most used and are reported by Michael Faraday since the 18th century. AuNPs can easily be synthesized and functionalized, they have high chemical stability and low toxicity and excellent

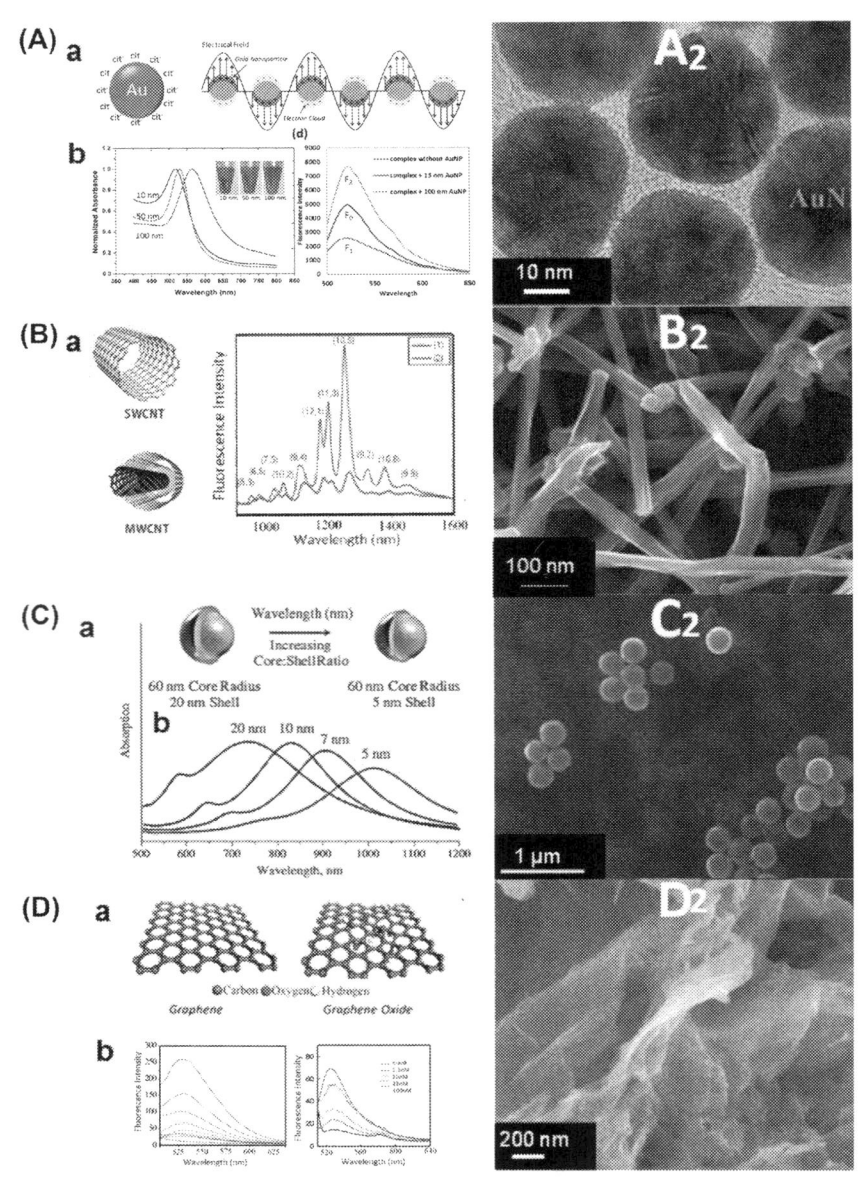

FIGURE 1   (A). (a) Absorption spectra correspond to the nanoparticles. *(From L. Sutarlie, K.M. Moh, M.G.L. Lim, S. Lukman, E. Cheung, X. Su, Plasmonics 9 (2014) 753 with permission from Springer.)* (A_2) Transmission electron microscopy images of synthesized AuNPs with average size of 15 nm. *(From S. Halldorsson, E. Lucumi, R. Gomez-Sjoberg, R.M.T. Fleming, Biosens. Bioelectron. 67 (2015) 595−600 with permission from Elsevier.)* (B) Illustration of CNT quenching: fluorescence spectra. *(From B.C. Satishkumar, L.O. Brown, Y. Gao, C.-C. Wang, H.-L. Wang, S.K. Doorn, Nat. Nanotechnol. 2 (2007) 506 with permission from Nature Publishing Group.)* (B_2) SEM

electronic, redox and optical properties [14,15]. Optical properties of these NPs can be related to their size [6], so the transmission electron microscopy is an essential tool when characterizing this material in terms of its size, morphology and dispersion (Fig. 1A).

Synthesis of AuNPs can generally involve a 'top-down' approach or a 'bottom-up' approach. The top-down synthesis is based on reducing the size of an initial material. It can involve physical or chemical methods such as mechanical milling/ball milling, chemical etching, thermal ablation/laser ablation, sputtering and others. Top-down synthesis has the disadvantage of producing defects on the structure's surface, which is a major limitation due to the fact that physicochemical properties are associated with the surface structure. On the other hand, with the bottom-up approach the AuNPs obtained are more homogeneous [16].

Due to its high surface area, AuNPs are very reactive and aggregation is favoured, producing an undesired passivation of the surface, so pretreatments are important to prevent this aggregation or precipitation, for example, protecting AuNPs surface. There are many methods reported in literature to protect AuNPs such as the use of citrates, functionalization with thiolated compounds, encapsulation with microemulsions and dispersion in polymer matrixes. From the synthesis process, it is possible to control some characteristics of NPs: size, monodispersion, morphology and surface chemistry. The latter is particularly important when considering the applications of AuNPs for the construction of catalysts, nanosensors and biosensors whose activity strongly depends on the AuNPs chemistry surface [17]. Likewise, properties of NPs not only depend on size, but also on the shape; and physicochemical, electric and optical properties along with catalytic activity and selectivity are highly dependent one of each other.

Nowadays, among the most important applications of AuNPs is the design of electrochemical and optical biosensors, due to the improved electron's transference they present, the high ultraviolet—visible (UV—Vis) absorption coefficient, the small size (up to 100 nm) and consequently large surface area per unit of mass and biocompatibility compared to other particles [18,19]. Integration of AuNPs in analytical methods and techniques represents an

---

of MWCNT. *(From I. Cesarino, F.-C. Moraes, M.L.V. Lanza, S. Machado, Food Chem. 135 (2012) 873—879 with permission from Elsevier.)* (C) Optical resonances of gold shell-silica core nanoshells as a function of their core/shell ratio. (C$_2$) SEM images of nanoshells. *(From S.J. Oldenburg, R.D. Averitt, S.L. Westcott, N. Halas, J. Chem. Phys. Lett. 288: 243—247, 1998 with permission from Elsevier.)* (D) Illustration of GRO quenching: fluorescence spectra GR *from S. He, B. Song, D. Li, C. Zhu, W. Qi, Y. Wen, L. Wang, S. Song, H. Fang, C. Fan, Adv. Funct. Mater. 20 (2010) 453—459 with permission from WILEY* and fluorescence spectra graphene oxide *from R. Yang, Z. Tang, J. Yan, H. Kang, Y. Kim, Z. Zhu, W. Tan, Anal. Chem. 80 (2008) 7408 with permission from American Chemical Society.* (D$_2$) SEM image of MRGO. *(From S. Yan, T.-T. Qi, D.-W. Chen, Z. Li, X.-J. Li, S.-Y. Pan, J. Chromatogr. A 1347 (2014) 30—38 with permission from Elsevier.)* MWCNT, multiwalled carbon nanotubes; *SWCNT*, single-walled carbon nanotubes.

important strategy to increase sensitivity and selectivity, which represents a great technological improvement for the in situ monitoring and real-time analysis of chemical compounds and pathogenic contaminants.

1.1.2 Quantum Dots

QDs are conductive or semiconductive NPs with diameters between 1 and 20 nm and contain 100 to 100,000 atoms per NP [20]. The most reported QDs are based on CdSe, CdTe, ZnSe, CdS, CdSe@ZnS (core shell QDs) and InP. Nowadays, new carbon-based QDs have been proposed which, in contrast to traditional QDs, are safer and nontoxic [21].

As all NMs, QD properties such as size, shape and surface chemistry depend on the synthetic route. Among the different physical and chemical synthetic processes, the colloidal chemistry is the best method for the synthesis of nanocrystals with appropriate surfaces to provide high luminescence efficiency, narrow size distribution and ability to interact with specific species. The surface modification with suitable ligands can help to prevent the agglomeration of QDs, which controls some properties of the nanocrystals in solution, including size, distribution, stability as well as their crystalline structure.

QDs present optical, electrical, thermal and mechanical characteristics, some of these include high quantum yield (>20%), high molar extinction coefficients, broad absorption spectra, narrow and symmetric emission bands [30−50 nm], generally long photoluminescence (PL) decay times (often >10 ns), large effective stokes shifts and high resistance to photobleaching and chemical degradation [22]. They are also thermally and chemically stable and have good biocompatibility. The properties of these NMs are attributed to their small size and large electron confinement, which means that they can be used for many different applications including photovoltaic, thermal, development of light emitting diodes, fluorescence linked immunoassays, fluorescence resonance energy transfer assays, immunosensors, DNA probes and enzymatic labels or isotopic labels [23].

In general, the use of QDs is attractive as an analytical tool for the detection and quantification of a great variety of analytes. Because of their unique fluorescent properties, QDs can be used as labels to improve detection limits, sensitivity and specificity in optical and electrochemical methods.

1.1.3 Carbon Nanotubes

CNTs were discovered in 1991 by Iijima [24]. These generally present micrometre lengths and diameters below 100 nm, with surface areas in the range of 150−1500 m^2/g [25]. CNTs can be considered as sheets of GR rolled into a cylinder and covered by fullerene structures, whose properties can change depending on the synthesis route, and particularly on the rolling direction [26]. CNTs can be single-walled (SWCNTs), double-walled or multiwalled (MWCNTs) and can be easily functionalized with different organic or inorganic molecules to achieve greater selectivity in the analysis of different chemical species [27−29].

CNTs usually offer certain interesting advantages compared to spherical NPs for biotechnological applications. For instance, NTs have large internal volume (in relation to the NT dimensions), which can be filled with any chemical substance or biochemical species wanted, from small molecules to proteins. Furthermore, NTs have different internal as well as external surfaces, which may be chemically or biochemically functionalized [30].

CNTs have high mechanical strength, high elasticity, high thermal conductivity, and high chemical and thermal stability and can present semiconductor, metal or superconductor properties [31]. In addition, they present unique optical properties as small band-gaps, and PL in the near-infrared region (NIR) [27]. In the electroanalysis field, their unique electrochemical properties are used to improve the analytical parameters of quantification methodologies.

Nowadays, CNTs have found many applications, such as DNA sequencing, drug delivery, batteries and fuel cells, nanoreactors and also the development of new sensors and biosensors for monitoring of different analytes; this last has had great success over the past decades. CNTs have many advantages when used for the construction of biosensors, for example, due to their small size and large surface area, they exhibit an excellent electronic transference. Also, immobilization of proteins is improved in order to retain its biological activity.

CNTs have also been used as tools for adsorption and separation of various compounds. Analytical techniques such as capillary electrophoresis, microchip capillary electrophoresis and some chromatographic techniques have employed CNTs as stationary phase with excellent results [27].

1.1.4 Graphene and Other Nanomaterials

GR is the carbon structure with sp^2 hybridization in two dimensions (Fig. 1D). It is also known as a single carbon sheet and is the main component of other important allotropes. These carbon sheets can be stacked to form a 3-D graphite or rolled to form NTs or fullerenes. Novoselov reported for the first time the synthesis of a single layer of GR [32]. This material exhibited high stability and crystallographic quality with excellent thermal, electrical and mechanical properties, due to the π conjugation [33]. GR has an analogous structure to benzene and polycyclic aromatic hydrocarbons, so the chemistry of these compounds is similar; but a discriminative factor in GR is the formation and/or breaking of conjugated C−C bonds in the basal plane and the C−H bonds on the edges.

The characterization of GR properties is achieved using scanning probe microscopy (used to study thickness), atomic force microscopy (used to measure mechanical properties) and scanning electron microscopy (employed for morphological studies, Fig. 1. D_2). Raman spectroscopy has been recently used to determine the thickness of the sheets obtained by mechanical exfoliation [34].

In general, the main properties of GR are great surface area (2630 m^2/g), high electron mobility (15,000 cm^2/v s), high thermal conductivity: 5000 W/m K, transmittance of about 97.7%, high conductivity ($\sim 104 \, \Omega^{-1} cm^{-1}$). GR can support a current density about six times higher than that of copper; the electronic characteristics of this material are mainly due to its topology, because of the size of the films at atomic levels, the electron transport can be ballistic at sub-micrometre distances [35–37].

Another kind of NMs is core shell NPs (Fig. 1C). These NMs commonly consist of magnetic NPs as core and metal/metal oxide as shell which have attracted extensive attention because of their unique properties such as super magnetism, nontoxicity, biocompatibility, ease synthesis, high specific surface area, chemical stability, electrochemical activity and high electron mobility (115–155 cm^2/V s), which are beneficial for the kinetics of electron transport [38]. The most popular is $Fe_3O_4@ZnO$ where coprecipitation deposition is involved in its synthesis, and also solvothermal methods for Fe_3O_4 magnetic NPs.

Magnetic nanoparticles (MNPs) have a size of about 50 nm and supra-paramagnetic properties. Their ability to separate target molecules from other compounds by the simple use of a permanent magnetic field makes them attractive materials for the fabrication of sensors. MNPs commonly consist of magnetic elements such as iron, cobalt and nickel and their chemical compounds. As stated before, Fe_3O_4 NPs are the most prevalent materials because they have no toxicity, good biocompatibility and do not retain residual magnetism after the removal of external magnetic field. MNPs can be functionalized by groups, eg, -OH, -COOH, $-NH_2$ [39], suitable for further modifications by the attachment of various bioactive molecules for different applications, eg, antibodies or aptamers.

MNPs are usually integrated with detection techniques for food analysis, as in the polymerase chain reaction (PCR), high-performance liquid chromatography (HPLC), liquid chromatographic-mass spectrometric (LC-MS), electrochemical and optical biosensors. The use of MNPs into conventional detection techniques can make them simple, rapid, highly selective and sensitive [40–44].

2. BIOSENSING TECHNOLOGIES

Among the various applications of NMs, their use in the construction of biosensors has been one of the most important. NMs are used for modifying the transducer's surface (with favourable effects on the sensitivity) improving the catalytic effect as well as for an efficient functionalization and immobilization of biological materials [45,46].

Biosensor technology has certain advantages due to the possibility of massive production of these devices (and thus reducing the analysis costs), use of sample volumes on the order of microlitres or even nanolitres, short analysis time and multiple analysis (detection of several analytes simultaneously).

Analytical success of biosensors lies on both, the proper selection of biological material (enzyme, antibodies, DNA, aptamers, cells, bacteria and even tissues) as well as the strategy employed for the immobilization, and not just the biological material but the used NM, in order to prevail its physico-chemical characteristics. In particular, when it comes to electrochemical biosensors, an incorrect strategy could affect the catalytic properties of NMs, producing low selectivity due to a high imposed potential.

There are various classifications for biosensors such as DNA sensors, aptasensors and immunosensors among others (Fig. 2), in which NMs have been used as labels in the detection of a specific analyte, either a chemical contaminant or a pathogen agent [46,47]. The advantage of using NMs in biosensing technology stands in general improvements of analytical performance of the fabricated devices.

Usually, in routine food analysis it is necessary to evaluate chemical substances like sugars, alcohols, amino acids, flavourings, colourants, pesticides, as well as to control pathogen microorganisms and mycotoxins [48]. The interest in quality control related to pathogens is due to cases of food-borne diseases. For example, it has been reported that *Salmonella* causes 31% of the deaths related to food, followed by *Listeria* (28%), *Campylobacter* (5%) and *Escherichia coli* O157:H7 (3%) [49]. In this context, biosensor technology represents fast and simple analytical methods enabling food monitoring along the entire food production chain (until delivery to the consumer), and helping to prevent and avoid the spread of diseases and even death.

Thus biosensing technology and integration of NMs represent in food industry a great alternative considering the analysis required for control and quality assurance of food. Development of biosensors has evolved in order to construct fast, sensitive, reliable and inexpensive devices, so the integration of new materials as NPs, carbon NTs and QDs, among other NMs, is a breakthrough in food diagnostics using specific and efficient transducers. In the following sections, we present a review of these NMs applied to the development of biosensors useful in the quality assurance of foods. Table 1 summarizes the favourable properties, roles of each NM in transducer-based biosensors, as well as the challenges that need to be taken care of for sensor design [50].

2.1 Batch Devices

2.1.1 Gold Nanoparticle-Based Biosensors

Electrochemical biosensor's response is essentially based on the change in current, voltage, potential difference or impedance due to oxidation and/or reduction of chemical or biological molecules; this change can be measured with the aid of an appropriate two or three electronic system.

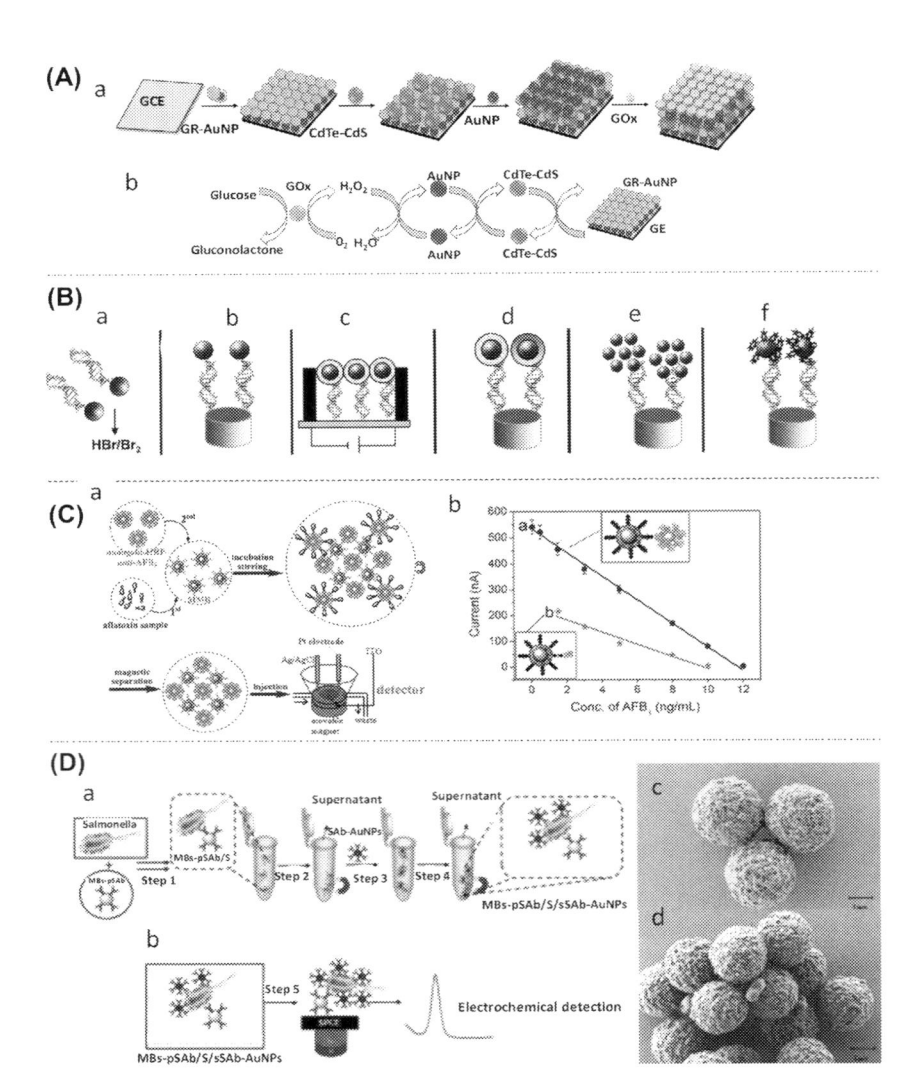

FIGURE 2 (A) Procedure for fabrication of (a) GOx/AuNP/CdTe-CdS/G-AuNP/GE, (b) enzymatic reaction. *(From G. Zhiguo, Y. Shuping, L. Zaijun, S. Xiulan, W. Guangli, F. Yinjun, L. Junkang, Electrochim. Acta 56 (2011) 9162−9167 with permission from Elsevier.)* (B) Schematic procedure of the different strategies used for the integration of gold nanoparticles (AuNPs) into DNA sensing systems: (a) previous dissolving of AuNPs by using HBr/Br$_2$ mixture followed by Au (III) ions detection; (b) direct detection of AuNPs anchored onto the surface of the genosensor; (c) conductometric detection; (d) enhancement with silver or gold followed by detection; (e) AuNPs as carriers of other AuNPs; (f) AuNPs as carriers of other electroactive labels. *(From M.T. Castañeda, S. Alegret, A. Merkoçi, Electroanalysis 19 (2007) 743−753 with permission from Wiley-VCH.)* (C) Schematic illustration of the MMB and the nanogold-HRP-anti-AFB1, and (b) measurement process of the competitive immunoassay method linear curves of the developed immunoassay toward AFB1 standards by using various detection antibodies: (a) nanogold-HRP-anti-AFB1 and (b) HRP-labelled anti-AFB1. *(From D. Tang, Z. Zhong, R. Niessner, D. Knoop, Analyst 134 (2009) 1554−1560 with permission from Royal Society of Chemistry.)* (D) Schematic (not in scale) of Salmonella (S) detection. (a) Principle of the assay. (b) Electrochemical detection of MBs-pSAb/S/sSAb-AuNPs onto screen-printed carbon electrode captured with a magnetic field (step 5) by using DPV technique. SEM images of MBs-pSAb: before (c) and after (d) incubation with 10^5 cells/mL of Salmonella. *(From A. Afonso, B. Pérez-López, R. Faria, L. Mattoso, M. Hernández-Herrero, A. Roig-Sagués, M. Maltezda Costa, A. Merkoçi, Biosens. Bioelectron. 40 (2013) 121−126 with permission from Elsevier.)*

TABLE 1 Nanomaterials Used for Constructing Dual-Transducer Biosensors

Nanomaterial	Favourable Properties	Major roles in Design	Challenges	Other Remarks
Quantum dots	Tunable fluorescence	FRET donor (more preferential)	Cytotoxicity	Surface functionalization is required to increase aqueous solubility
	Photostability	FRET acceptor	Fluorescence blinking	Biocompatibility
	Emission brightness		No standard synthetic protocol	
Metal NP	LSPR absorption	Fluorescence quencher/enhancer	Nonspecific aggregation (eg, during particle surface modification or assaying procedures)	Size-dependent optical behaviour (eg, absorption dominates for AuNPs <80 nm; scattering for any larger ones)
	Light scattering	Fluorescence polarization amplifier		Various other particle shapes (eg, rods, plates, triangles, stars) to provide tunable absorption spectra
	Facile surface modification (eg, Au—thiol interaction)	Carrier for DNA barcode or chemiluminescence (CL) reaction catalysers		
	Intrinsic affinity to ssDNA and proteins through coordination chemistry			

Continued

TABLE 1 Nanomaterials Used for Constructing Dual-Transducer Biosensors—cont'd

Nanomaterial	Favourable Properties	Major roles in Design	Challenges	Other Remarks
Graphene, GO, CNT	Optical quenching	Fluorescence quencher	Nonspecific adsorption	GO is more commonly used than intact graphene
	Different binding affinity with ds- and ssDNA	Enzyme mimicking	CL reaction catalyser	
Carbon nanodot	Tunable fluorescence Facile synthesis Low fabrication cost	FRET donor	Lack of thorough understanding on physical and chemical properties	Usually need further surface modification (eg, ligand attachment or solid capping layers) for efficient emission

$AuNP$, gold nanoparticles; CNT, carbon nanotubes; $FRET$, fluorescent resonance energy transfer; $LSPR$, localized surface plasmon resonance; NP, nanoparticles.
From N. Li; X. Su; Y. Lu, Analyst 140 (2015) 2916—2943. © The Royal Society of Chemistry.

In particular, the use of AuNPs for the modification of electrodes generates a microenvironment similar to the natural system of redox proteins, giving these molecules better orientation. Also AuNPs reduce the insulating effect of the protein and allow direct electron transfer.

The construction of a biosensor can include several steps such as the electrodeposition of AuNPs, followed by physical adsorption of biological material [51], a polymer can sometimes be used for encapsulating the biological material used. However, physical adsorption may lead to low reproducibility, so it is necessary to use other strategies.

Among reported strategies employed to integrate the AuNPs-protein (enzyme) conjugate (Fig. 2A), are the following: self-assembly technology [52], layer-by-layer (LBL) assembly technique (based on electrostatic interactions [53] and polymer-nanoparticle composites [54]. A feature of these composites is that the polymer helps to preserve bioactivity in extreme conditions, in these circumstances, the AuNPs exhibit good affinity for the enzyme and a very stable composite is obtained. Special applications have been reported by the inclusion of AuNPs in carbon paste electrodes, where the NM creates a microenvironment suitable for the complete and direct transfer of electrons of different redox proteins [55,56].

AuNPs have been used as labels in the detection of pathogens, using as recognition agent DNA or antibodies [34]. The strategy in the construction of DNA biosensor is based on nucleotide hybridization using a capture probe immobilized over the sensor surface (Fig. 2B), where the analyte of interest can be detected due to a hybridization event, the complementary probe is usually labelled with AuNPs, enabling electrochemical detection [57]. AuNPs have the same use in aptasensors, where aptamers are used as target-recognition element and NPs as transducer or signal amplifier [58].

Some biosensors used for the detection of chemicals important in food quality are summarized below. Antioxidants are important in food industry because they retard oxidation processes in food; usually in food manufacturing. Synthetic antioxidants are used for this purpose [59]; however, an excess of these chemicals can be harmful to health. Synthetic antioxidant analysis is generally performed by conventional techniques, but with disadvantages of time-consuming analyses and requirement of sample pretreatment. Development of electrochemical sensors based on AuNPs provides a competitive alternative for analysis of antioxidants. A proper design in the construction of these sensors has allowed the simultaneous detection of up to three synthetic antioxidants with lower detection limit (LOD) in the order of ng/mL (Table 2). These results are comparable with traditional techniques such as HPLC-time-of-flight mass spectrometry (HPLC-TOFMS) and gas chromatography—mass spectrometry (GC—MS). This type of sensor was used for the detection of synthetic antioxidants in commercial oil samples and validated against the HPLC methodology [60]. The success of this sensor is strongly associated with the strategy used for the immobilization, in this sense, electrodeposition results

TABLE 2 AuNP-based Biosensors for the Detection of Some Compounds of Interest in Food Quality

Analyte	Technique	Biosensing Surface	Sample	Lower Detection Limit	References
Part I					
Butylated hydroxyanisole (BHA) Butylated hydroxytoluene (BHT) Butylated hydroquinone (TBHQ)	Linear sweep voltammetry (LSV)	AuNPs	Oil	BHA 0.039 µg/mL BHT 0.080 µg/mL TBHQ 0.079 µg/mL	[60]
Methyl parathion	—	AuNPs-MWCNT/methyl parathion hydrolase	Garlic	0.3 ng/mL	[61]
Phenol Catechol Caffeic acid Chlorogenic acid Gallic acid Protocatechualdehyde	Amperometry	AuNPs/Tyr	Wine	2.1×10^{-7} M 1.5×10^{-7} M 6.6×10^{-7} M 6.2×10^{-7} M 70×10^{-7} M 20×10^{-7} M	[63]
Pesticide formetanate hydrochloride (FMT)	—	AuNPs/Lac	Mango Grapes	9.5×10^{-8}	[72]
Malathion Chlorpyrifos Monocrotophos Carbofuran	—	AuNPs/AChE	Cabbage Lettuce Leek Pakchoi	0.05 nM 0.05 nM 0.1 M 2.5 nM	[74]

Antioxidants Isoflavone glycosides	HPLC	AuNPs	Soya supplements Boiled soya bean Tofu	125 ng/mL	[77]
Phenol	Amperometry	AuNPs/Lac	Wine	0.006 mM	[78]
Phenol total	Amperometry	AuNPs/Lac	Tea Alcoholic beverages	0.1 µm	[79]
Ochratoxin A (OTA)		AuNPs	—	20 nM	[47]
	Impedimetry	AuNPs/aptamer	Beer	0.02 nM	[80]
	Cyclic voltammetry	AuNPs	—	30 pg/mL	[81]
Part II					
Atrazine	Conductimetry	AuNPs/Ag-Ab	Red wine	—	[75]
Cyanides	Piezoelectric quartz crystal (PQC)	AgNPs	Drinking water	2.2 µg/L	[82]
Sugar	Amperometry	AuNPs/GOx/graphite		0.1 mM	[64]
Glucose		AuNPs/GOx	Beer	—	[83]
		SNS-NH$_2$/AuNP/GOx	Fizzy with orange Coke Lemonade Ice tea Peach juice	0.21 mM	[71]
		AuNPs-silica	Soft drink	360 nM	[84]
	Chronoamperometry	AuNPs-PANI	—	2.5 µM	[85]

Continued

TABLE 2 AuNP-based Biosensors for the Detection of Some Compounds of Interest in Food Quality—cont'd

Analyte	Technique	Biosensing Surface	Sample	Lower Detection Limit	References
Salmonella typhi	Nano test-strips	AuNPs	Rice, corn and wheat	2 μg/mL	[86]
Salmonella	Amperometry	MBs-pSAb/S/sSAb-AuNPs	Milk	143 cells/mL	[67]
Salmonella enteritidis	EIS	Self-assembly onto gold NPs	—	600 CFU/mL	[87]
Bacillus cereus	Voltametry	Bc-AuNPs-chit-GCE	—	10 CFU/mL	[76]
Escherichia coli O157:H7	Chronoamperometry	MBs—pECAb/AuNPs—sECAb	Meat Water	10^3 CFU/mL 10^4 CFU/mL	[88]
Aflatoxins	Impedance	GO/Py(DPB)/AuNPs	Peanuts, rice, milk, soya bean	1 fM	[89]
		AnAb/AuNP-Lcys	Soya fermentation	18 CFU/mL	[90]
Ochratoxin	DPV	OTA-BSA-AuNPs	—	0.86 ng/mL	[91]

AuNP, gold nanoparticles; *BSA*, albumin from bovine serum; *MWCNT*, multiwalled carbon nanotubes.

to be the best technique since it allows the control of size and dispersion. Deposition time and the used potential are crucial for obtaining homogeneous NPs size. This strategy has been implemented for the analysis of different compounds [61,62]. Another commonly used technique is the physical adsorption, but problems have been reported in the reproducibility, uniformity and size of the AuNPs deposited layer. Antioxidants such as phenols (phenol, catechol, caffeic acid, chlorogenic acid, gallic acid, protocatechuic aldehyde) have been analysed with a variety of electrochemical biosensors based primarily on AuNPs with enzymes such as tyrosine (Tyr) and lacasse (Lac) [63]. It has been reported that the electrodeposition of AuNPs followed by immobilization of the enzyme via cross-linking with glutaraldehyde results in an efficient biosensor which is able to detect and quantify up to six different types of phenols in different commercial wines. Biosensor stability and sensitivity are associated with the presence of the NPs, which act as nanoscale electrodes electrically communicated with the enzyme and the transducer. These biosensors have demonstrated to be useful in the analysis of antioxidants present in foods like juices, wines, vinegar, oil, tea, among others (Table 2) [63].

On the other hand, modifications in food composition during storage is an important aspect in food industry, especially when considering the main carbohydrate constituents, such as glucose and fructose, which could be responsible of food-browning processes. For this reason, glucose monitoring is important as it is an indicator of food freshness. Considering this, biosensors based on nanocomposites have been reported for glucose analysis (Fig. 2A); generally, a combination of AuNPs and glucose oxidase (GOx) is used [54,55,64]. The composite may contain nafion or teflon polymers to encapsulate and protect the enzyme, these polymers have a permeability characteristic that allows good interaction with the substrate. It has been reported that the NP's size is a contributing factor on the orientation of the enzyme: generally, a size of 5−50 nm for gold particles allows free orientation of the enzyme, which promotes easy electron transfer between the electrode and the enzyme [68]. Moreover, it has been demonstrated that the use of 10 nm AuNPs enhances amperometric response [69,70]. The application of AuNPs could increase the rate of mediated electron transfer, providing an improved sensitivity with a low LOD.

These biosensors have shown an LOD of 17 μmol/L with high stability (lifetime of about 3 months) due to the presence of AuNPs. The analytical results were validated with traditional methods such as spectrophotometry. AuNP-based biosensors have been applied for the detection of glucose in a variety of food products such as juice, cola drinks and tea [71].

Other important chemicals concerning food quality are pesticides, since their presence in foods can cause toxic effects in human health. Pesticides that are considered strong contaminants are organophosphates (OPs), N-methyl carbamates, triazines, chloroacetanilides, pyrethrins, pyrethroids. These pesticides are used in agriculture, so they are absorbed by fruits and plants that will later be used as food.

Pesticides such as carbamate, formetanate hydrochloride are usually used in fruits like grapes, mango, potato, onion, citrus, bean, watermelon, pepper and tomato. As an alternative to ultraperformance liquid chromatography—tandem mass spectrometry (UPLC—MS/MS) used in the detection of this pesticide, a biosensor as detector was developed based on the immobilization of laccase on an electrode modified with AuNPs. This strategy allowed the detection of pesticide residues in fruits such as mango and grapes [72].

Among the most commonly used pesticides are the organophosphorus. To quantify them, various biosensors based on acetylcholinesterase (AChE) have been proposed. It was found that the immobilization of this enzyme on AuNP electrodeposits combined with chitosan (chit)-calcium carbonate provides a favourable and biocompatible microenvironment enhancing the detection. Moreover, it has been found that electrodeposition and a combination of CNTs/ NPs [61,73] improves the electrocatalytic activity and help to stabilize the enzyme. With this architecture, response times and LODs are improved [74].

Atrazine is another important contaminant present in fruits such as grapes. For the detection of this herbicide, immunosensors based on antibodies using AuNPs as labels have been developed. Detection of very low concentrations of atrazine is possible because of the signal amplification caused by the NM. The method is attractive because it is inexpensive, has no matrix effects and presents low analysis times (48 samples can be analysed in 5 h), this biosensor has been used for the detection of atrazine in wine [75].

Trends in food quality control related to the detection of pathogens are oriented to minimize analysis times. Traditional methods for microorganism's analysis such as morphological and biochemical identification are performed in 2—7 days until confirmation. Other methods such as RAPD (random amplified polymorphic DNA-PCR) require expensive instruments and consumables. In this sense, biosensors can generate quick responses when predicting the existence of microorganisms in real time analyses.

To date, there have been reported different strategies involving biosensors for pathogens such as *E. coli, Staphylococcus aureus, Bacillus, Salmonella* [67], *Vibrio parahaemolyticus, Campylobacter jejuni* and *Listeria monocytogenes* (Table 2) [67,86—88,91].

Thus affinity biosensors are the most employed for the detection of microorganisms, eg, *Bacillus cereus (Bc)* bacteria occurring in various foods such as vanilla sauce, custards, soups, rice, pasta, meat, vegetables and dairy products. The immunosensor used to determine these bacteria are based on a monoclonal antibody as recognition agent. A glassy carbon electrode (GCE) is used in which a monolayer of CS is placed on its surface, then AuNPs are electrostatically adsorbed on it and finally the antibody is placed to generate Bc-AuNPs-Chit-GCE. The reported LOD for the bacteria is 10 CFU/mL, demonstrating that this biosensor can be used in quality assurance of foods [76].

The development of an immunosensor capable of detecting *Salmonella* in milk has also been reported. Biosensor construction is based on the incubation of *Salmonella* (S) with magnetic particles (MBs) modified with the specific primary antibody related to the bacteria (PSAb), generating the conjugate MBs-PSAb. During this step, the bacteria is captured and retained in this conjugate. Subsequently, the system MBs-PSAB/S is captured by applying a magnetic field. Finally, the MBs-PSAB/S is incubated with AuNPs modified with the secondary antigen (SSAB-AuNPs). The bacteria are then detected by differential pulse voltammetry using a screen-printed electrode and a magnet to collect the conjugated MBs-PSAB S/SSAB-AuNPs (Fig. 2C) [67]. The analysis time is estimated in 1:30 h.

Escherichia coli O157:H7 has also been analysed in meat and water samples. Detection has been carried out by using an anti-*E. coli* O157—magnetic beads conjugate (MBs—pECAb) as a capture platform and sandwiching it afterwards with AuNPs modified with secondary antibodies (AuNPs—sECAb), the detection is achieved by using an electroanalytical technique with screen-printed carbon electrodes (SPCEs). LODs of 148, 457 and 309 CFU/mL were obtained for a synthetic solution, minced beef and tap water samples, respectively. When compared with a commercial lateral-flow kit, electrochemical immunoassay performance presented clear advantages such as specificity and reproducibility [88].

Immunosensors represent a very good strategy for the detection of mycotoxins [72,73,92], it has been reported that about 25—50% of crops worldwide are contaminated by aflatoxigenic fungi [93], so early analysis is a key to ensure food safety. A strategy for mycotoxin detection is based on the incorporation of AuNPs and polyclonal antibody to extracellular antigen from *Aspergillus parasiticus*. For the immobilization of these components in the biosensor, an AuNPs/L-cysteine (Lcys) monolayer is prepared (Fig. 2B). The developed immunosensor has high sensitivity, good recovery and reproducibility for the direct detection of aflatoxigenic Aspergilli during fermentation of soya bean meal [90].

2.1.2 Quantum Dot-Based Biosensors

The combination of QDs with biomolecules provided new opportunities for the design of biosensors useful in food analysis, due to the characteristics of these NMs, whose chemical, electrical and optical properties can be manipulated and adapted by changing their size, shape and composition.

In the design of biosensors, QDs can be used as platform for the immobilization of the biomolecule or as recognition label material.

QDs promote electron transfer and stability of biological materials when used as surface modifiers for transducers, examples of these materials are enzymes, DNA, aptamers and antibodies. To date, QDs have been tested as a platform for the immobilization of enzymes such as GOx, Lac, horseradish peroxidase (HRP), among others.

To evaluate the advantages of QDs in the architecture of biosensors, Jian Du et al. developed three devices for the detection of glucose using AuNPs and QDs (CdS, ZnS). In this research, variables such as NP size, conductivity, rate constant (Ks) and sensitivity were analysed. The values obtained for the conductivity corroborated the NP's nature. It was determined that AuNPs have better conductivity when compared with CdS or ZnS, although the latter showed better electron transference and the highest sensitivity [94]. Electrochemical and photoelectrochemical techniques were compared and the analysis indicated that the increase in sensitivity is evident when this last technique is used.

Another biosensor based on a ZnS-QDs mixture, which is used as platform for the immobilization of GOx, exhibits a low LOD (3 μM) using the method with a GOx/Chit/ZnS-CdS/PGE architecture.

In general, QDs can be used for the immobilization of enzymes taking advantage of different strategies, eg, self-assembled monolayer (Table 3). This strategy is used in the detection of ascorbic acid (one of the most common antioxidants) and is based on the modification of an electrode with a cysteine self-assembled monolayer and coated with functionalized QDs. With this biosensor (Lac/CdTe/Cys/Au), very low LOD are obtained (1.4 μM), furthermore, the biosensor showed good repeatability and stability [95]. Among the most recent strategies reported, the use of QDs in combination with GR stands out; graphene quantum dots (GRQDs) have been used for the determination of total polyphenols in red wine [molybdenum disulphide (MoS_2)] [96].

Moreover, due to its composition and surface, QDs can be easily modified, immobilized and bound to molecules (eg, DNA, oligonucleotides, aptamers and antibodies).

At present, QDs are a good alternative for multianalyte detection [97−103]. The multiscreen strategy based on QDs employs a biorecognition agent (Fig. 3A and C), usually antibodies, and QDs of different compositions [97]; the most used are CdS, CuS and PbS [98]. For example, in the case of bacteria, multiscreening is based on marking each bacterium (antibody) with a QD constituted by a different metal (Cd, Cu, Pb), each specie will have a defined and particular analytical signal. In the case of electrochemical immunosensors, discrimination between signals is possible with the use of a voltammetric technique, where each metal has a particular redox potential. *Escherichia coli*, O157:H7 *Campylobacter* and *Salmonella* have been detected with this type of immunosensor, the antibodies are labelled with CdS, PbS, CuS [99] and the analysis is performed using square-wave voltammetry. *Escherichia coli* is associated with the Cd reduction potential (-0.7 V), *Campylobacter* with PbS reduction potential (-0.5 V) and *Salmonella* with the reduction potential of Cu (-0.7 V), the construction of the immunosensor is described in Fig. 3B. LOD for *Salmonella*, *Campylobacter* and *E. coli* were 400, 400 and 800 cells/mL, respectively. It was demonstrated that the inclusion of CNTs significantly increases sensitivity. With this immunosensor, simultaneous detection of these

TABLE 3 QD-Based Biosensors for the Detection of Some Chemical Species of Interest in Food Quality

Analyte	Technique	Biosensing Surface	Sample	Lower Detection Limit	References
Ascorbic acid	Amperometry	Lac/CdTe/Cys/Au	—	1.4 μM	[95]
Escherichia coli Campylobacter Salmonella	SWASV	MWCNT-PAH/SPE E. coli-CdS Campylobacter-PbS salmonella-CuS	Milk	400 cells/mL 400 cells/mL 800 cells/mL	[99]
Salmonella Typhimurium Salmonella	Fluorescence	QDs	Milk Chicken Carcass wash water	13 cells/mL 10^3 cells/mL	[101] [103]
E. coli	Fluorescence	SiO$_2$NPs	Water	3—5 cells	[105]
E. coli S. Typhimurium	Fluorescence	QDs	—	10^4 CFU/mL	[106]
Chloramphenicol	Electrochemical	QDs	Milk	0.0054	[102]
Polyphenol	Electrochemical	MoS$_2$ nanoflakes/GRQDs/Lac	Red wine	0.32 μM	[96]
	Fluorescence	QDs	—	—	[104]

QD, quantum dot; SWASV, square-wave anodic stripping voltammetry.

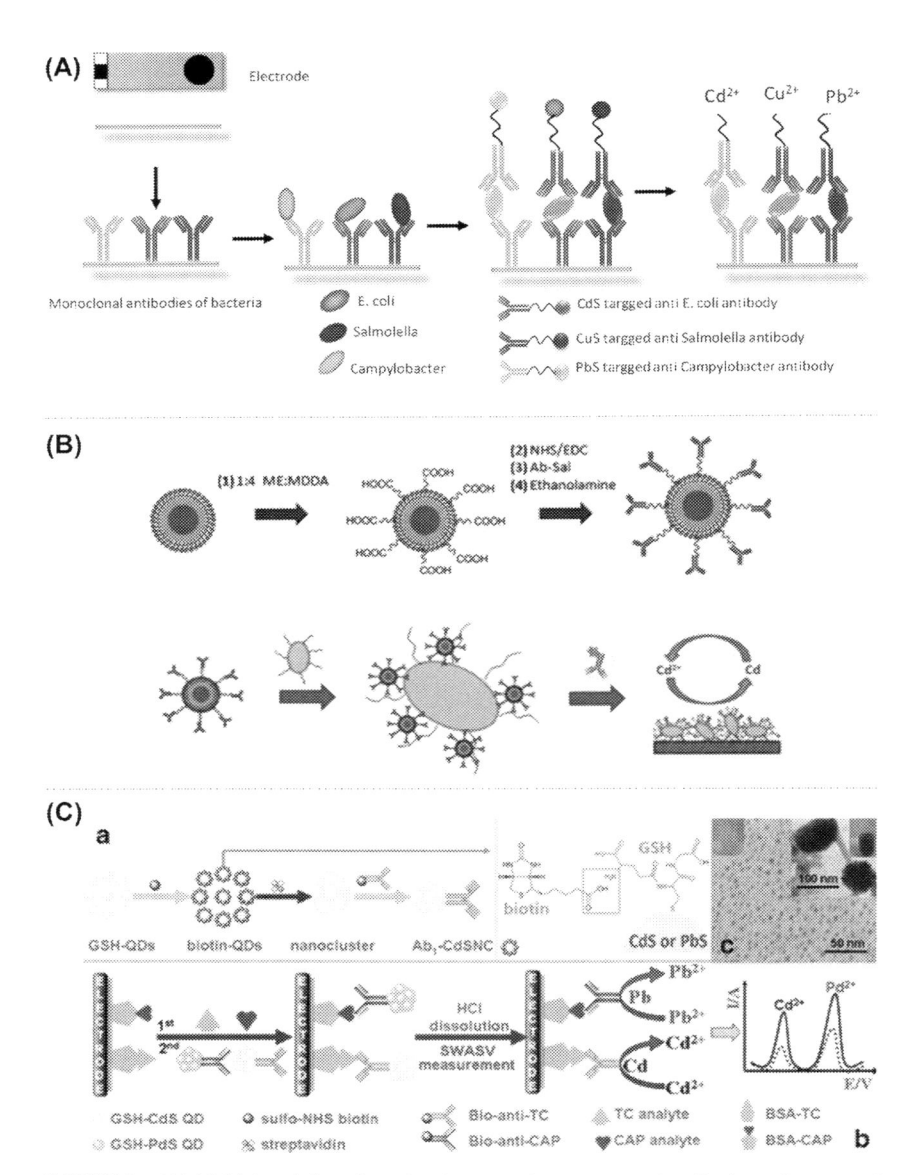

FIGURE 3 (A) Multiplexed detection of pathogens using nanocrystal antibody conjugates and SWCNT-PAH/SPE. *(From S. Viswanathan, C. Rani, J. Ho, Talanta 94 (2012) 315−319 with permission from Elsevier.)* (B) (a) Formation of the polyallylamine (ME-MDDA) SAM-functionalized nanoparticles linked to anti-*Salmonella* antibodies. (b) Representation of labelled electrochemical determination of *Salmonella typhimurium*. *(From M. Freitas, S. Viswanathan, H.P.A. Nouws, M.B.P.P. Oliviera, C. Delerue-Matos, Biosens. Bioelectron. 51 (2014) 195−200 with permission from Elsevier.)* (C) (a) Schematic illustration of synthesis and bioconjugation of metal sulfide nanoclusters (CdS used as an example), (b) electrochemical measurement of multiplex immunoassay, and (c) transmission electron microscopy image of metal sulfide quantum dots (CdS used as an example) *(Insets*: TEM image of CdS nanoclusters, and the corresponding samples). *(From B. Liu, B. Zhang, G. Chen, D. Tang, Microchim. Acta 181 (2014) 257−262 with permission from Springer.)*

three pathogens in milk is possible, demonstrating its potential for quality control in food industry (Table 3).

Detection of pathogens using QDs and Fe@AuNPs stands out (Fig. 3C), because of the very low detection limit obtained (13 cells/mL). The inclusion of these NPs (13 nm) with magnetic properties improves sensitivity and analysis times ($\sim$ 1 h). This immunosensor is functional for the detection of *Salmonella* in milk [101].

Antibiotic residues in dairy products represent a constant concern in food quality control; usually these chemicals are analysed by techniques such as HPLC, ELISA and GC−MS. As an alternative, it has been proposed that an immunoassay using QDs for the analysis of tetracycline (TC) and chloramphenicol (CAP), which consist on the immobilization of TC and CAP conjugates with bovine serum albumin on the surface of a GCE, modified with AuNPs. Anti-TC and anti-CAP monoclonal antibodies are functionalized with CdS and PbS. Detection is performed by square wave anodic stripping voltammetry, the construction of the immunosensor is described in Fig. 3D.

2.1.3 Carbon Nanotube-Based Biosensors

Because of the electrical and electrochemical properties of CNTs, these materials are ideal for the construction of transducers for biosensors. CNTs as part of electrochemical biosensors have been implemented in different architectures as described in the following sections.

1. Carbon paste electrodes and nanotubes (CNTPEs). CNTPEs are constructed with a mixture of graphite, CNT and a mineral oil [107]. Biological material is added to the mixture too (usually enzymes) [108,109]. These composite electrodes take advantage of the improvement on the electron transference due to the presence of CNT.
2. Modified electrodes with CNTs. These electrodes are associated to the modification of conventional carbon or metal electrodes (Au, Pt, ITO, etc.). The most common technique used for the modification is physical adsorption. CNT adsorption sometimes involves the mixture with a polymer such as poly(glycidyl methacrylate-co-vinylferrocene), Chit, polypirrole, polyaniline (PANY), among others. In general, polymers help the CNT dispersion promoting the electron transference, sensitivity and stability. CNTs have also been used for immobilization of biomolecules; however, the most effective method is the growth of CNTs on the metal substrate, achieving an arrangement of the CNTs which substantially improve the electronic transference. Moving from randomly aligned mixtures of NTs to well-ordered arrays can enhance their electrical conductivity [110]. To improve the catalytic activity of CNTs, it has been reported the use of metal or semiconductor NPs in the electrode's modifications.
3. LBL self-assembling modification. This technique involves alternate adsorption on the electrode's surface of polyelectrolites with opposite

charge, through electrostatic interactions. LBL self-assembling modification is not only used for the modification of electrodes with CNTs, but for immobilization of biomolecules as well. Among the advantages of this technique are its simplicity, the minimum quantity of reagents used, and minimization of protein denaturation.

Different strategies for the construction of CNT-based biosensors using enzymes as recognition agents have been reported, enzyme immobilization is a crucial step for the proper functioning of the biosensor. Some of the reported strategies include the following:

1. Physical adsorption. CNTs are dispersed on the surface of conventional electrodes, usually glassy carbon (GCE), and then a solution containing the enzyme is used to cover the electrode's surface and is allowed to dry. In spite of being a simple and fast technique, this strategy presents some disadvantages: low amount of adsorbed enzyme, leaching, lack of homogeneity of the enzyme on the electrode surface and low reproducibility. The incorporation of CNTs has solved some of these problems since they favour enzyme's adsorption. The adsorption technique has been employed for the detection of different analytes including phenol compounds [111] and OPs [112].
2. Covalent binding. This technique allows direct attachment of the enzyme on the electrode via covalent bonds, which enables the direct electron transference to the enzyme's active site [113].
3. Electropolymerization. This is an attractive and well-controlled method for immobilizing enzymes on the surface of electrodes. This technique consists on mixing the enzyme with a monomer and applying an optimum potential to induce the polymerization. The incorporation of the enzyme in the matrix is often promoted through electrostatic interactions. The benefits of this technique lie in the control of the chemical system; to produce thin layers, polypyrrole is commonly used for these purposes [114,115].
4. Encapsulation. This technique allows the immobilization of the enzyme through the use of hydrogel or sol—gel materials, it can be carried out from a modified electrode with CNTs followed by the deposition of the hydrogel containing the enzyme. Alternatively, CNTs can be incubated with enzymes and then incorporated into the hydrogel or sol—gel matrix and finally deposited on the electrode [116].

Recently, new developments involving CNTs for the detection of proteins, antibodies and DNA have been reported [117]. CNTs play a double role as transducers and recognition agents, amplifying the analytical response. In food analysis, these sensors are essential in determining the presence of contaminants or pathogens.

Due to their size and excellent electrochemical properties, CNTs continue attracting interest in the field of biosensors development. It has been demonstrated that CNT-based biosensors have electrochemical properties superior to most conventional electrodes, their use for the determination of chemical species such as, sulphites, glucose, pesticides, polyphenols and microorganisms is of great significance for food analysis. This section points out the advances reported using these CNT-modified biosensors to detect different compounds associated with the quality control of food (Table 4).

Sulphites are one of the most common additives in food and beverages technology. They help to preserve colour and flavour during preparation, storage and distribution. The properties of sulphites are mainly associated with their reducing power; they can also inhibit bacterial growth. The most common sulphites used in foods are sulphur dioxide, sodium sulphite, sodium and potassium bisulphite, sodium and potassium metabisulphite. However, there have been reports of allergic reactions and intolerance to foods containing this additive, so the quantification of sulphite is a parameter that must comply with the regulations of the FDA (Food and Drug Administration). The quantitative method recommended by the Association of Analytical Chemists is based on an acid—base titration [118], which despite being a low-cost method, has some limitations such as the high analysis times and errors associated to interferences. Electrochemical methods arise as very good alternatives because they do not require sample pretreatment even if it is a suspension or it is coloured. Generally, the method is based on electrochemical oxidations or reductions using conventional electrodes, although there are some drawbacks such as poor sensitivity and high errors due to interferences. A new alternative based on the use of carbon paste electrodes chemically modified with MWCNT has been highly successful in the quantification of sulphite in water, vinegar, juices, alcoholic beverages and wines [115,119—124].

Different configurations involving polymers, CNTs and NPs have been reported, with LODs for sulphite in the order of nanomolar concentrations [121]. The inclusion of hybrids (Fig. 4A) containing different NMs (CNT-AuNPs, CNT-QDs, CNT-GR) highly improves sensitivity [122], where CNTs favour the electrocatalytic activity and reduces fouling of the electrode's surface. To date, few biosensors based on CNT have been developed for the detection of sulphite, in spite of its better selectivity. Sulphites are indirectly quantified when using these biosensors, the role of CNTs in this device is related to improved half-life time (90 days), response time (3 s) and low detection limit.

Carbohydrate analysis is a great concern in food industry; a clear example is the quantification of glucose, fructose and maltose which are very important in the fermentation of wine and beer. To date, diverse configurations of biosensors for the detection of glucose have been proposed, CNTs have been used in a great variety and complex configurations [125—129]. An interesting configuration is the double-stranded DNA (dsDNA); its aim is to support the CNTs and to maintain the bioaffinity of the recognition agent even after being exposed to extreme conditions [125]. The biosensor is constructed from the dispersion of

TABLE 4 Carbon Nanotubes in Food Analysis

Analyte	Technique	Biosensing Surface	Sample	Lower Detection Limit	References
Part I					
Sulphite	SWV	CP/MWCNT	Red Grape juice or red wine samples	16 μM	[119]
	SWV	GCE/PAH/MWCNTs	Vinegar, pickle water, coconut water and shredded coconut	4.2 μM	[120]
	SWV	BFCNPE	Drinking water	90 nM	[121]
	Amperometry	CNTs-PDDA-AuNPs/GC	Fruit juices and red wines	0.03 mg/L	[122]
	DPV	PPO-AuNPs-MWCNTs-PAH/GCE	White and red wine	0.4 μmol/L	[123]
	Amperometry	SOx/AuNPs/Chit/MWCNT/PANI/Au	Juices and alcoholic beverages	0.5 μM	[124]
Glucose	Amperometry	GCE/bCNT-dsDNA/[PDDA/GOx]n	Beverages	50 μM	[125]
		ITO/PB/(chit/MWCNT/GOx]$_6$	—	0.05 mM	[126]
		GOx/amino terminated-MWCNT	—	8.0 μM	[127]
		GCE/[MWCNT-Polyhys/GOx]5/Naf	Milk	2.2 μM	[128]
	Chronoamperometry	Composite GOx/meso-HAP/meso-TiO$_2$/MWCNT/GCE	Fruit juice	2 μM	[129]
Maltose	Amperometric	AG/PyOx/Chit—MWCNT	Beer	—	[130]
Total phenolics	DPV	Polyeugenol/MWNT/GCE	Red and white wine	0.21 μM	[131]

Gallic acid	Chronoamperometry	MWCNT/GCE	Cognac and brandy	0.19 μM	[132]
Tannic acid	DPV	SWNTs/GC	Tea Beer	8 nM	[133]

Part II

Carbaryl	Chronoamperometry	GC/MWCNT/PANI/AChE	Cabbage, broccoli and apple	1.4 μM	[9]
Cethomyl				0.95 μM	
Organophosphorus	Amperometry	f-MWCNT/poly(SNS-NH2)/AChE	–	0.09 mM	[135]
Carbofuran	Amperometry	GS-PEI-Au/MWCNTs/GC	Vegetables	0.03 ng/mL	[136]
Organophosphate	Amperometry	PDDA/AChE/PDDA/CNT/GC	–	0.4 pM	[137]
Histamine	Amperometry	Polysulfone/MWCNT/ferrocene	Sardines	1.7×10^{-7} M	[138]
Xanthine	Amperometry	P(GMA-Co-vfc)/MWCNT/XO	Fish, meat	0.12 μm	[139]
Ochratoxin A	Amperometry	SWCNTS		24.1 nM	[140]
Mycotoxin	DPV	MWCNTS	Beer and wine	1.7 μg/L	[141]
Aflatoxin b	DPV	AFB1-BSA-SWNTS/CS/GCE	Corn powder	3.5 pg/mL	[142]
	Potentiometry	Covalent binding to SWCNTs	Milk	5 CFU	[143]
	DPV	MWCNT-CHI-BI/fDNA-tDNA-sDNA-CDsNPS	Beef	1.97×10^{-14} M	[146]
Staphylococcus aureus	DPV	MWCNTCHI-BI/fDNA–tDNA–cDNA–PBS	Beef	3.17×10^{-14} M	[147]
Salmonella spp.	Amperometry	DWCNT	–	8.9 CFU/mL	[148]

AG, α-glucosidase; *BF*, benzoylferrocene; *cDNA*, capture DNA; *DPV*, differential-pulse voltammetric; *DWCNT*, double wall carbon nanotube; *fDNA*, fixing DNA; *tDNA*, target DNA; *PAH*, poly(allylamine hydrochloride); *PDDA*, poly(diallyldimethylammonium chloride); *PyOx*, pyranose oxidase; *SWV*, square-wave voltammetry.

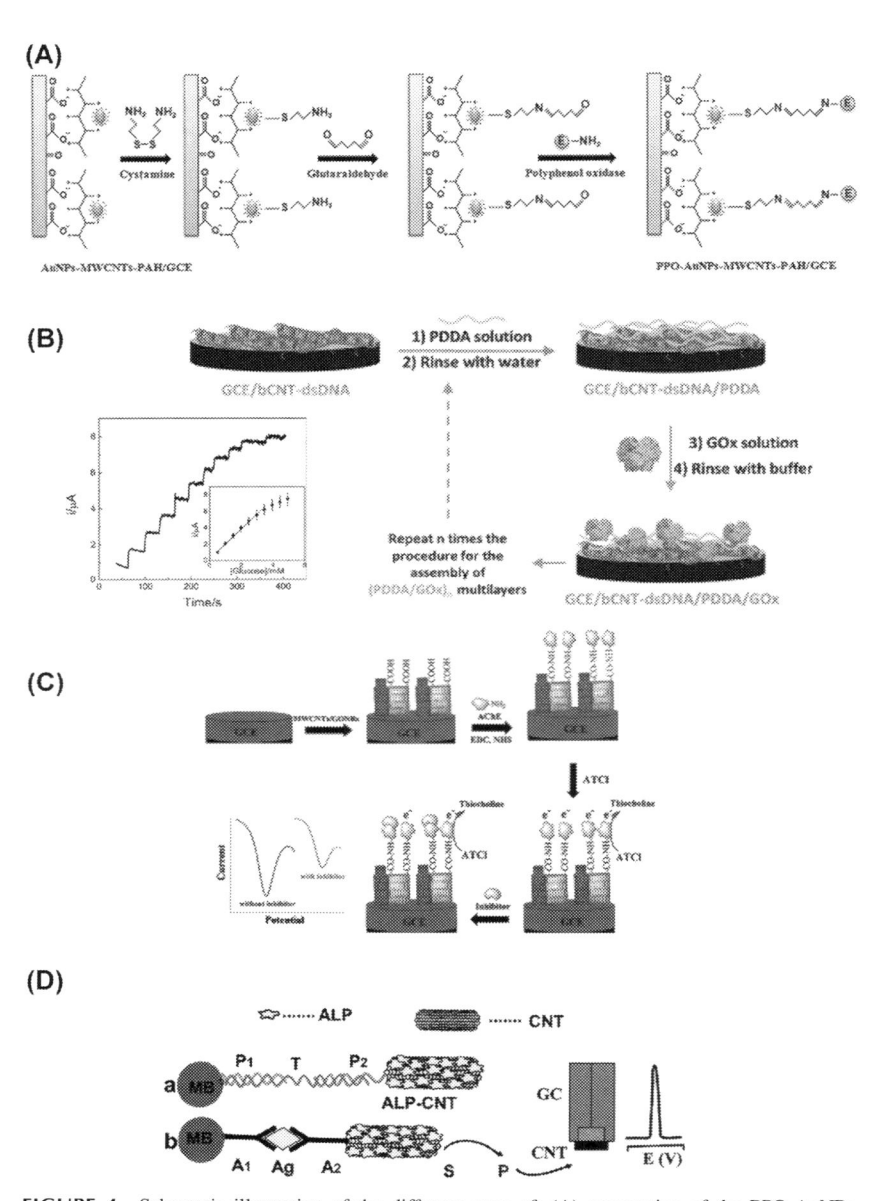

FIGURE 4   Schematic illustration of the different steps of: (A) preparation of the PPO-AuNPs-MWCNTs-PAH/GCE biosensor for sulfite. Not drawn in to scale. *(From E. Sartori, F. Vicentini, O. Fatibello-Filho, Talanta 87 (2011) 235—242 with permission from Elsevier.)* (B) Glucose biosensor using bamboo-like MWCN dispersed in double stranded calf-thymus DNA as a new analytical platform for building layer-by-layer based biosensors. *(From E. Primo, F. Gutierrez, M. Rubianes, G. Rivas, Electrochim. Acta 182 (2015) 391—397 with permission from Elsevier.)* (C) Stepwise amperometric biosensor fabrication and principle for pesticide determination. *(From Q. Liu, A. Fei, J. Huan, H. Mao,*

CNTs with dsDNA, which is placed on the electrode's surface, then the electrode is immersed into a solution of polydiallyldimethylammonium (PDDA) followed by immersion in a solution of the enzyme GOx, until multiple layers are obtained. The technique used for the development of the dsDNA-CNT biosensor is shown in Fig. 4B.

The challenge in the construction of such biosensors is to find the optimum immersion time and number of layers that allow the highest sensitivity and stability of the biosensor. The use of hybrid composites with NMs is a good strategy to improve detection limits when quantifying glucose [129], as demonstrated when used in fruit juices, milk and other beverages (Table 4).

Maltose is a component of food, such as commercial sugar, beer and milk. Enzyme sensors can accomplish this and several studies have been made in order to develop maltose biosensors. Usually two enzymes are used to construct maltose sensors, one hydrolyses maltose and the other oxidizes glucose, as alternative a new bi-enzymatic system was designed by coimmobilization of α-glucosidase (AG) and pyranose oxidase (PyOx) and the integration of CNT for maltose analysis. This biosensor is suitable for maltose detection in beer [130].

Other chemical species important in preserving food quality are polyphenols [130–133], which have also been analysed with CNT-modified electrodes. Gallic and tannic acids are some natural polyphenolic compounds usually used in the production of beer, wine, candies and meat products because of their antioxidant properties. These can be analysed according to its oxidation signal with an electrode modified with SWCNT, the very low detection limit (8 nM) and the simplicity of this device makes the method comparable with the official HPLC one [133].

Pesticide detection and quantification has motivated the development of biosensors based on CNT [134–137] in combination with enzymes such as AChE (Fig. 4C) [134], butyrylcholinesterase, laccase and methyl parathion hydrolase. In the construction of these biosensors, the immobilization of enzymes is achieved using Chit because of its low toxicity and biocompatibility. Polymers have been used along with NMs such as gold or iron NPs, which have been included using the layer by layer adsorption or electrodeposition techniques. Moreover, the construction of CNT-based immunosensors for the detection of pesticides has been a great success because of its high specificity.

K. Wang, J. Electroanal. Chem. 740 (2015) 8–13 with permission from Elsevier.) (D) Analytical protocol: (a) Capture of the ALP-loaded CNT tags to the streptavidin-modified magnetic beads by a (1) sandwich DNA hybridization or Ab-Ag-Ab (2) interaction enzymatic reaction. (b) Electrochemical detection of the product of the enzymatic reaction at the CNT/GCE; Magnetic beads (MB), DNA probe (P1); DNA target (T1); DNA probe (P2) first antibody (A1); antigen (Ag); secondary antibody (A2); S and P, substrate and product, respectively, of the enzymatic reaction; GCE; CNT, layer. *(From J. Wang, G. Liu, M. Jan, J. Am. Chem. Soc. 126 (2004) 3010–3011 with permission from American Chemical Society. Publications.) CNT,* carbon nanotubes; *GCE,* glassy carbon electrode; *MWCNT,* multiwalled carbon nanotubes.

So far, immunosensors for the detection and quantification of carbofurans and atrazine have been developed. Biosensors based on hybrid nanocomposites have improved the detection limits, which have been reported in the order of pg/mL.

Biosensors based on CNTs have been successfully used to quantify pesticides as carbaryl [9], monocrotophos, chlorpyrifos, dichlorvos, carbomate, carbofuran, parathion, methyl parathion (MP), triazophos, pirimicarb and atrazine in fruits (grape, apple), vegetables (cabbage, broccoli, tomato) and potable water (Table 4).

In addition to the parameters already discussed, the effect of the methods used for the synthesis of CNTs used in pesticide biosensors has been studied, particularly MWCNT for the detection of p-nitrophenol. It was demonstrated that MWCNTs obtained by vapour deposition have better response and greater stability with respect to those obtained by arc discharge. Here a GCE is modified with MWCNT and OPH enzyme. The construction consists of two layers, the first of MWCNTs dispersed in nafion and the second of the enzyme dissolved in nafion. Amperometry is used to quantify paraoxon and MP, both LOD and sensitivity are better for the first pesticide due to the selectivity of the enzyme [112].

Biosensors based on nanobiomaterial hybrids have had great success for the detection of pesticides such as carbaryl [9], a carbamate pesticide, which is highly effective in controlling insects, consequently, highly employed in agriculture. A concern about this pesticide is the long-term accumulation in the biosphere, since it represents serious risks to human health due to its high toxicity related to the inhibition of AChE, a key enzyme for many functions in the central nervous system. A biosensor has been reported for the quantification of this pesticide based on the modification of glassy carbon by an electropolymerization process with PANI and MWCNT; the immobilized enzyme is AChE. This device exhibits a detection limit in accordance with the established standard regulations in Brazil, and is used in the analysis of fruits and vegetables (Table 4), the results are comparable with the traditional HPLC method [9].

As already mentioned, a very important matter in food is the analysis of pathogens; biosensors based on CNT, DNA, antibodies; thus aptamers have resulted a very good alternative for the detection of these harmful species.

To date, genosensors have arisen as rapid methods for the detection of pathogens. Such methods are based on the use of a DNA probe (capture probe), which is functionalized with thiol groups (SH) for its immobilization on the electrode's surface. Moreover, the probe is functionalized with a label which may be an NP with redox properties (Fig. 4D). When the electrode's surface is modified with CNTs, genosensor conductivity and sensitivity are favoured [143−148].

The analytical response and stability of the genosensor depends on the materials used for the modification of the electrode's surface with CNT, for this purpose, polymers as Chit have been used. This design can detect *E. coli* and *S. aureus* using the so-called 'sandwich structure', which is based on the hybridization of the capture probe with a specific fragment of the pathogen's gene (fDNA-tDNA-sDNA-NPs) [145,146].

Thus, the use of CNTs as a platform for the detection of pathogens has been very versatile in combination with nafion, teflon, epoxy polymers, Chit, polypirrol (Py) and poly(ethylenedioxythiophene). In order to improve and facilitate the immobilization of biocomponents in some polymers, $-NH_2$ functional groups are used. Another advantage of using CNT-polymer composites is the homogeneity of the surface which is related to the stability and biocompatibility of the recognition agent [144].

Advantages of CNTs in biosensor's architecture are related with the improvement of detection limits when determining pathogens, eg, a bioassay based on CNTs and the immobilization of LBL for the detection of *E. coli* O157:H7 achieved a detection limit of 250 CFU/mL (Table 4). This biosensor presents no interference when used for the analysis of milk samples, with no significant differences when compared and validated with the ELISA methodology [144].

Recently, aptamers have been reported for the recognition of pathogens such as *E. coli* and *Salmonella*, due to its high specificity not only to proteins but also to metals and organic molecules.

2.1.4 Graphene and Other Nanomaterials

Recently, biosensors have been considered for the production of novel systems using NMs as GR, carbon-QD (CQDs) and core shell NPs.

GR has been used in the construction of electrodes and transducers in biosensors. GR has great electrochemical properties, some of them are chemical stability, low cost, wide potential window, electrochemically inert and their electrocatalytic activity for several redox reactions.

Surfactants have been used in some research regarding GR and sensors due to their ability of changing the electrical properties of the electrode's interface and so, the electrochemical processes. This is due by the surfactant's adsorption on the electrode's surface and/or its aggregation into supramolecular structures [147].

Polymers have also been used with GR to improve some material properties, eg, nafion can serve as an antifouling [148], Chitosan is used in combination with GR to improve permeability and adhesive strength [149–153].

Modification of electrode's surface with GR and biorecognition elements is usually done through composites. It has been reported that the use of GR in sensor construction is inexpensive and can be produced on a large scale in comparison with other carbon NMs such as CNTs [154].

It has been demonstrated that GR-based biosensors have electrochemical properties that are equal or even superior compared to other electrodes, therefore its use for food quality. In subsequent sections, advances reported in literature regarding the use of these GR-based biosensors for food content and composition are presented.

Enzymes as recognition agents for biosensors based on GR have been proposed considering different construction strategies and the fact that enzyme immobilization is a crucial step for the appropriate response of biosensors. As discussed in the past sections, the most important immobilization techniques are adsorption technique, covalent binding, electropolymerization, LBL.

Phenolic compounds have been determined using tyrosinase-based biosensors. A novel GR-nanosheets matrix for biosensor's construction was proposed, where the enzyme is immobilized with glutaraldehyde. LOD where obtained compared to those reported with other methods based on GR sensors [155].

Among the different strategies to improve the quantification of phenolic compounds like chlorophenols, an interesting proposal is the combination of carbon, GR and β-cyclodextrin (β-CD) (cD/GRs/CPE); quantification of 2-chlorophenol (2-CP) and 3-chlorophenol (3-CP) can be easily achieved with detection limits of 0.2 y and 0.09 μM, respectively. This sensor was successfully used in real water samples [156]. Its success is based on the presence of the β-CD due to its nanocavities and supramolecular recognition, which in combination with GR increases the surface area. The success of this configuration depends on the formation of inclusion complexes with the pesticide.

Interesting proposals are reported in literature using biosensors for pesticide determination, where no enzyme is used. A GR/CNT/CS sensor has been reported to detect OPs like MP, the detection limit obtained was 0.5 ng/mL, much lower compared to sensors using enzymes. Reproducibility and stability also improved and the interference analysis resulted satisfactory. In general, the sensor presented high sensitivity and fast response times [157]. In other report, a cobalt (II) oxide (CoO)-reduced graphene oxide (rGO) based sensor for the detection of carbofuran (CBF) and carbaryl (CBR) is developed. This sensor has excellent electrocatalytic behaviour and it was successfully applied for the simultaneous detection of CBF and CBR in fruits and vegetables [158]. Other reports in literature conclude that β-CD acts like an adsorbent for the preconcentration and electrochemical detection of MP. The analytical characteristics of this sensor are comparable to others based on enzyme structures, β-CD reduces significantly GR aggregation and keeps an effective surface area, so the strong π-electrons de-location can strongly interact with π electrons of MP. This induces superconductivity and a fast electron-transference resulting in an excellent electrochemical response [147].

Combination of GR with metal NPs has proved to improve detection limits with values ranging picomolar concentrations [154]. Composites are used to construct these biosensors, where chemically reduced GR sheets (RGR) and AuNP with poly-(diallyldimethylammonium chloride) (PDDA) are used to produce a nano-hybrid AuNP/RGR.

Recently, new sensors based on carbon QDs, also called C-Dots, have been reported. A great advantage of C-Dots is its low toxicity, making them a very attractive material for the development of sensors. These NMs have been used to detect one of the most common pesticides: MP, where C-Dots with zirconia (ZrO_2) are electrodeposited. The detection limit resulted to be 0.056 ng/mL and it was used for detection of this pesticide in rice samples [159].

Actually, different GR-based biosensors have been reported for *E. coli* detection, through the GR functionalization. The proposed biosensor is a combination of GR/CNT [160] and GR oxide nanosheets, which improves the capture of the bacteria compared to the GR alone [161,162].

Other strategy of biosensors use a sulphate-reducing bacteria (SRB) which is sensitive through the anti-(SRB) antibody (Ab) [163,164] or also, aptamers can be used, as reported for the detection of *S. aureus* [165]. This sensor is particularly based on GRO and AuNPs linked by single-stranded DNA (rGOssDNA-AuNPs) which works as an amplification system to improve LOD.

Trends in several analyses are focussed on the development of composite biosensors based on NMs and enzymes as recognition agents.

2.2 Lab-on-a-Chip (Microfluidics)

In the late 1980s, with the evolution of several microfluidic structures, the 'micro total analysis systems' (μTAS) (also called lab-on-a-chip) emerged. These microsystems integrate injection, reaction, separation and analyte detection in a unique device. The advantages of these microsystems in chemical analysis are portability, small sample volumes (nanolitres), high sensitivity, high stability, short analysis times, low cost per analysis and less laboratory space.

There are different types of lab-on-a-chip devices [166−174]; one of the most important is the capillary electrophoresis microchip (CE microchip), which consists on a reservoir (where the sample is loaded, injector) and a microchannel (where the electrophoretic separation of the analytes is performed). Generally, the microchannel and reservoirs are manufactured using photolithography. In a typical configuration of a CE microchip (both simple cross-T and a twin-T injectors) channels vary from 10 to 100 mm, with a separation length in the range of 3−10 cm (Fig. 5A). Usually microchips can be constructed with glass, quartz or polymers such as poly(methyl methacrylate) (PMMA), polycarbonate (PC), and poly(dimethylsiloxane) (PDMS) which is the most used. Recently, paper has been reported as substrate. Diverse applications have demonstrated the functionality of electrochemical sensors coupled to these microsystems.

Nowadays, microfabrication technologies and nanotechnology offer great compatibility, so intelligent integrated systems can be constructed with the inclusion of electrochemical sensors. The use of NMs in these microsystems allows the improvement of separation stages, eg, using pseudostationary

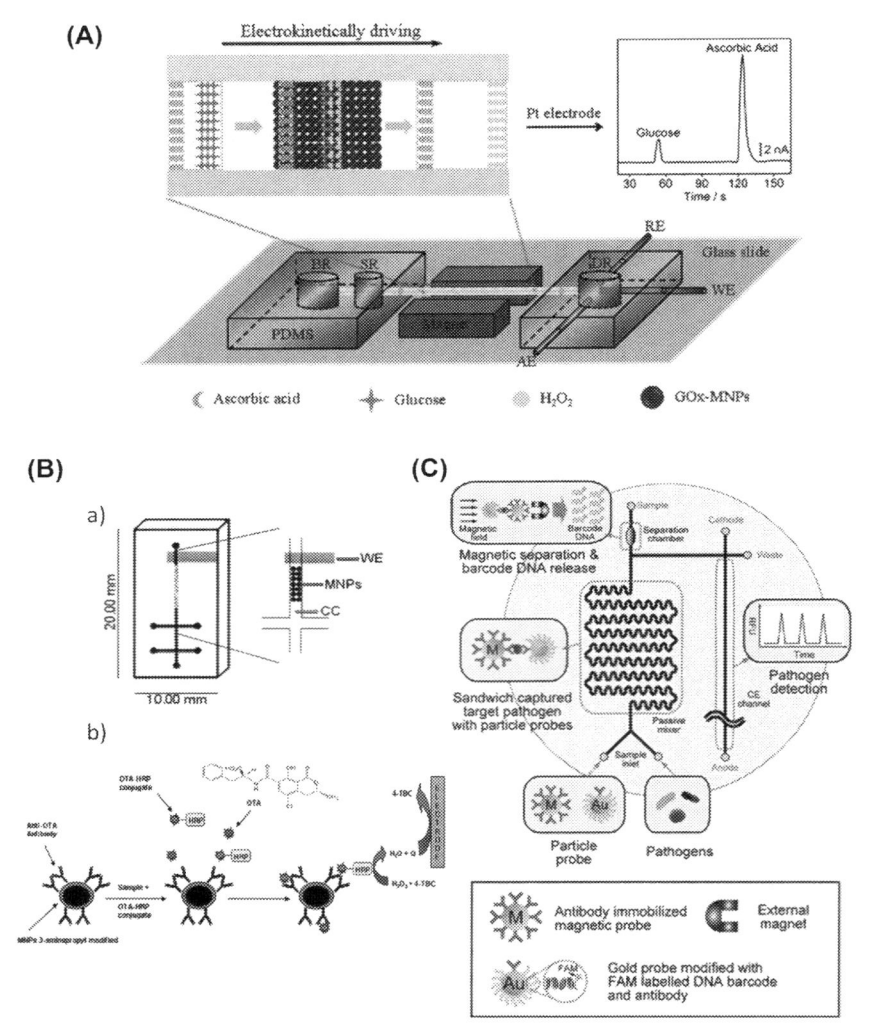

FIGURE 5 (A). Scheme representation of the construction and analytical procedure of GOx–MNPs. *AE*, auxiliary electrode; *BR*, Buffer reservoir; *DR*, detection reservoir; *RE*, reference electrode; *SR*, sample reservoir; *WE*, working electrode. *(From J. Sheng, L. Zhang, L. Jianping, H. Ju, Anal. Chim. Acta 709 (2012) 41–46 with permission from Elsevier.)* (B) (a) Schematic representation of microfluidic immunosensor. *CC*, central channel; *WE*: Au working electrode. All measurements are given in millimetres, (b) representation of the biochemical immunoreaction. *(From M. Fernández-Baldo, F. Bertolino, G. Fernández, G. Messina, M. Sanz, J. Raba, Analyst 136 (2011) 2756–2762 with permission from Royal Society of Chemistry.)* (C) Schematic diagram of an integrated PMMS-CE microdevice which performs a DNA barcode assay and a capillary electrophoretic separation for multiplex pathogen detection. The PMMS-CE microdevice consists of a passive mixer for target capture with particle probes, a magnetic separation chamber for isolating and purifying the target complexes, and a capillary electrophoretic microchannel for detecting barcode DNA strands to identify target pathogens. *(From J. Jung, G. Kim, T. Seo, Lab. Chip 11 (2011) 3465–3470 with permission from Royal Society of Chemistry).*

phases in the microchip-based capillary electrochromatography, besides these can be used to improve detection and sensitivity.

2.2.1 Nanoparticle-Based Lab-on-a-Chip

Several NMs can be used to modify the microchannels in the lab-on-a-chip systems, the most used are AuNPs and CNTs [167]. In particular, NPs have several functions as part of microchips, these can be used as support or for concentrating and/or separating the analyte. These devices can act as biological and molecular recognition systems improving the analysis methodologies.

AuNPs used as stationary phase on microchips was first proposed by Pumera et al. [168]. The use of NMs allows a significant improvement in mobility, resolution and selectivity. It has been reported that the number of theoretical plates can be doubled compared to systems without NPs. Several investigations are based on this principle, which involves coating the walls of microfluidic networks with polymer and adding NMs that adsorb on such modified walls. This method has been applied to DNA analysis [169].

Moreover, MBs have been used to generate bioreactors in the microchannel. This bioreactor is based on the immobilization of an enzyme through covalent bonds on the MBs, and then these particles are introduced into the microfluidics channel. Care must be taken when functionalizing the microchannel with MBs because aggregates can block the channel and interfere with the detection.

Some advantages of using MBs are handling the bioreactor zone in the microchannel through the magnet and regeneration of the bioreactor by replacing the MBs in order to renew them. Based on this principle, MBs have been used in the construction of a bioreactor for detecting carbofuran [170]. AChE is immobilized on the MBs, incubating for 24 h, and then the last ones are introduced into the microchannel and then the electrochemical detection is performed. Indirect quantitation of the pesticide is made through the inhibition of the enzyme. The main advantage of this microsystem is the low detection limit that can be achieved ppb.

Another interesting strategy employed is the use of an on-chip sandwich immunoassay strategy based on the functionalization of NPs with antibodies. This strategy has been used for detection of some mycotoxins, eg, Ochratoxin A in apples with a detection limit lower than that obtained with the traditional ELISA methodology [174]. Fig. 5B shows the representative diagram of lab-on-chip sandwich configuration for immunoassay.

Recently, a new barcode DNA assay has been proposed for the detection of target proteins or nucleic acids with high sensitivity. This analysis is based on the use of thiolated barcode DNAs, and two particle probes: magnetic microparticles (MMP) with the recognition element and AuNP with a second recognition element to form a sandwich structured immuno-complex (MMP-target-AuNP); then the dehybridization of the barcode DNA from the NP probes is done. Fig. 5C shows a schematic representation of the steps involved in the DNA barcode assay for the detection of pathogens like *S. aureus*, *E. coli* and *Salmonella typhimurium*. The application of DNA barcode assay substantially increases sensitivity and presents detection limits in the range of attomolar concentrations.

Because of their optical properties, QDs have been used as labelling biomolecules in microfluidic systems. Typically, QDs can be attached to protein assemblies for protein detection via a streptavidin-biotin link or a DNA link bridge. QDs are used to enhance the performance and analytical characteristics of fluorescence, optical and electrochemical detectors, but there are few real applications on lab-on-a-chip based on QDs for food analysis. Table 5 summarizes some analysis based on lab-on-a-chip.

2.2.2 Carbon Nanomaterial-Based Lab-on-a-Chip

CNTs have demonstrated great advantages when used in food analysis, eg, increasing the analytical signal (due to its active surface area) [175], improving selectivity (due to their electrocatalytic properties) and high reproducibility.

CNTs are used in the lab-on-a-chip devices as stationary phase. This NM improves the separation of various compounds such as DNA fragments, polycyclic aromatic hydrocarbons and phenolic compounds. The separation is associated with $\pi-\pi$ interactions between the aromatic rings of the solute and the sp^2 hybridized aromatic rings of the CNT, as well as with oxygen-containing groups.

CNTs are used with hydrogel polymers to produce a stationary phase. The durability of the hydrogel as well as the separation efficiency is favoured especially with SWCNT. In general, CNTs have different surface areas and impurities such as metal particles, amorphous carbon and different functional groups from sample to sample. That is an inconvenience for stationary phase development.

The integration of CNT-based electrochemical detector in microfluidic separation systems is a promising alternative to their traditional sensitivity problems that arise from the extremely low volume samples introduced.

Thus microfluidic devices may use different CNTs as SWCNT or MWCNT. Antioxidants such as vitamins, isoflavones and vanilla have been analysed using a lab-on-a-chip coupled to a detector with CNTs. Analytical peaks presented high resolution and using MWCNT, high sensitivity is achieved compared to other systems without CNTs, making this system suitable for food analysis (Table 5).

3. CONVENTIONAL ANALYTICAL TECHNIQUES

Chemical analysis has evolved with the arrival of NMs, as they have been used in different areas of analytical chemistry, including spectroscopy, electronic detection and separation techniques beside sensors and biosensors.

TABLE 5 NMs-Based Lab-on-a-Chip Used in Food Analysis

Technique	Nanomaterial	Analyte	Sample	Lower Detection Limit	References
Electrochemical	MWCNT	Polyphenols Vitamins Vanilla flavours Total isoflavones	Vitamin C Polyphenols Apple	2 µM 12 µM 11 µM 0.8 µM	[176]
Electrochemical	MWCNT/SPE	Polyphenol	Fruits	7.1 µM	[177]
Optical	CNT	Staphylococcal enterotoxin B	Soya milk	0.1 ng/mL	[178]
Electrochemical	Fe_3O_4/MWCNTs/β-CD	Hypoxanthine	Meat	0.3×10^{-8} M	[179]
Electrochemical	MNPs	Ochratoxin A	Apples	0.05 µg/kg	[174]
Electrochemical	MWCNT/PEI/GCE	Herbicides	ND	9 µM	[180]
Electrochemical	MWCNT/PEI/GCE	Phenol	ND	4.2 µM	[181]
Electrochemical	MWCNT/NiNP	Sugar	Herb	0.32 µM	[182]
Electrochemical	rGRO/CuNP	Carbohydrates	Food	1.64 µM	[183]
Electrochemical	MWCNTs/PMMA	Lignans	Herb	0.26 µM	[184]
Electrochemical	MWCNT/PEI/GCE	Antioxidants, acids	Wine	3.2 µM	[185]
Electrochemical	MWCNT/EVA	Coumarin derivatives	Herb	0.27 µM	[186]
Electrochemical	MWCNT/PET	Flavonoids	Peanut hulls	4.2 µM	[187]
Electrochemical	SW and MWCNT/SPE	Antioxidants	Fruit	12 µM	[188]

CNT, carbon nanotubes; GCE, glassy carbon electrode; MNPs, magnetic nanoparticles; MWCNT, multiwalled carbon nanotubes; ND, not detected; PMMA, polymethyl meth-acrylate; SPE, screen-printed electrodes; PEI, polyethylenimine; EVA, Polyethylene-co-vinyl acetate); PET, poly(ethylene terephthalate); SW, single-walled nanotube.

3.1 Nanomaterial-Based Chromatography

The chromatography is a physical separation technique for mixture resolution relying on the distribution of the analyte between a stationary phase and mobile phase. In a single step process, it can separate a mixture into its individual components and simultaneously provide a quantitative estimate of each constituent. The analytical efficiency of a chromatography method depends not only on adequate separation but also on suitable detection.

NMs offer several advantages for separations due to their unique properties, in particular a high surface-to-volume ratio, which can facilitate mass transfer and increase separation efficiency. Compared to silica, the most commonly used packing material, NPs exhibit chemical stability over a wide pH range. NMs consisting of metal oxides, polymers, CNTs, QDs and fullerene NMs are either used directly as the packing material or to modify this last in columns for applications in gas and liquid chromatography, capillary electrophoresis, and electrochromatography. NMs can also act as a pseudostationary phase.

Recently, monolith columns in combination with NMs have been used for separations. Nanosized metal organic frameworks are also used for separating different analytes. Different magnetic NPs can be used for concentration of analyte and separations, where the analyte detection usually occurs by some type of spectroscopic or electrochemical technique.

QDs have been implemented as part of fluoroimmunochromatographic technique for MP determination at picogram level by using thiol-stabilized CdTe-QDs as fluorescent probes in a flow-injection analysis system comprising an immunoreactor column packed with immobilized anti-MP IgY antibodies [189].

On the other hand, SWCNT-hydroxypropyl-β-cyclodextrin (HP-β-CD) was used as stationary phase for the liquid chromatographic separation of organic compounds, the $\pi-\pi$ interaction and the hydrophobic interaction in the stationary phase is improved by introducing the SWCNTs [187]. The SWCNTs make it possible to extend the application range on the newly prepared stationary phases for HPLC.

Currently, the sensor array used to distinguish between different compounds is the gas sensor array, also called 'electronic nose', which is modelled directly on the olfactory system. It can detect the odour of molecules as vinegar to form a 'fingerprint' database through the use of an array of broadly cross reactive sensors in conjunction with pattern recognition methods. This method has been established by chemiluminescence liquid sensor array (electronic tongue) based on NMs for the discrimination of eight vinegars.

3.2 Nanomaterial-Based Optical Techniques

So far, the NMs have been used in fluorescence spectroscopy, surface plasmon resonance (SPR) spectroscopy, localized surface plasmon resonance (LSPR) spectroscopy, surface-enhanced Raman spectroscopy (SERS), and mass spectrometry, IR, NPs in nuclear magnetic resonance (NMR) spectroscopy, magnetic resonance imaging (MRI), and infrared spectroscopy. In this section, only NM's applications for SPR and fluorescence techniques are discussed.

3.2.1 Surface Plasmon Resonance Spectroscopy

SPR spectroscopy is based on detecting changes in refractive index that occurs in thin metal films, usually Au, as the result of chemical reactions or a recognition event. When a small change is detected in the refractive index on the analyte—metal interface, information about molecular interactions can be obtained by measuring an optical property (intensity, phase and polarization of light). In the early 1980s SPR was first proposed for the direct monitoring of specific antigen—antibody interactions.

In SPR sensors, changes in plasmonic resonance signals in the metal thin film are strongly dependent of the refractive index of the medium. The sensitivity of the method is limited by the change amount in the refractive index at the metal surface and the minimum SPR displacement is detected due to the recognition event between the receptor and the analyte.

The SPR technique has limitations when the analyte is present in very low concentrations and when analysing small molecules. NMs, particularly metal NPs, can help to amplify the analytical signal. Coupling between the localized surface plasmon of metal NPs and the surface plasmon wave of the metal film was found to increase the shift in the SPR spectrum.

A sandwich assay strategy has been proposed, where a molecular receptor is on a surface. The analyte binds to the surface, and then the receptor-functionalized metal AuNPs can also bind to the analyte at the surface, this means that the AuNP will not bind to the surface if the analyte is not present (Fig. 6). Absorbance/scattering or visual colour of the AuNP is directly correlated with the amount of analyte present in the solution. In general, all analytes can be analysed based on this principle; however, there are several strategies for amplifying the signal to improve the detection limits [190,191].

3.2.2 Spectroscopy Using Gold Nanoparticles

NPs and gold nanorods are common NMs used to increase the SPR signal due to their optical properties such as the refractive index, this property depends on the size, shape and chemical composition of the NP (Table 6).

Modification of the metal film with AuNPs is a good strategy to amplify the signal. NPs can be deposited directly onto the surface to form a nano-composite, or chemically by cross-link.

AuNPs can be used to form a biomolecular conjugate using a secondary biomolecular probe sandwich assay with the metal surface, in order to increase the sensitivity and detection limits of SPR sensors [192]. The detection limit is enhanced 10^6 times when using AuNPs [193] whereas using nanorods leads to an

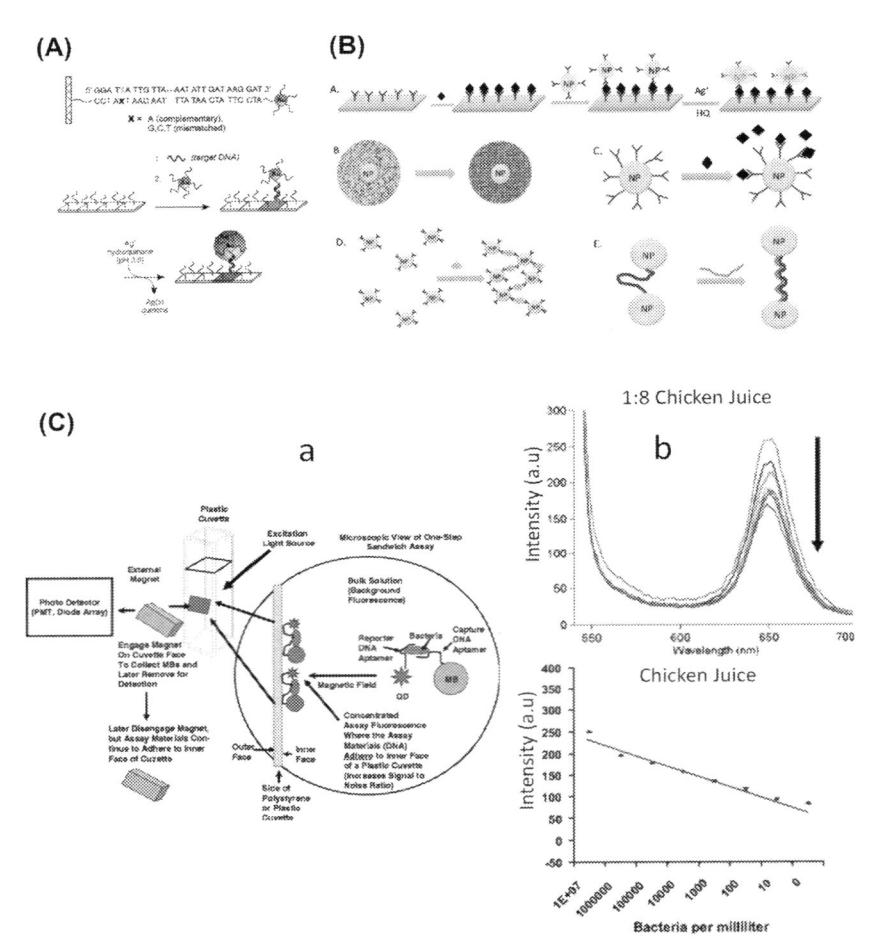

FIGURE 6 (A) Schematic representation of different NP LSPR-based sensing mechanisms. *(From T. Taton, C. Mirkin, R. Letsinger, Science 289 (2000) 1757—1760 with permission from AAAS.)* (B) (a) DNA assay, (b) AuNPs as labels with Ag deposition using hydroquinone (HQ) for signal amplification, (c) LSPR bulk refractive index (RI) sensing, (d) LSPR sensing of local binding events, (e) LSPR aggregation based sensing. *(From F. Zamborini, L. Bao, R. Dasari, Anal. Chem. 84 (2012) 541—576 with permission from American Chemical Society Publications.)* (C), (a) Schematic representation of the self-assembling plastic adherent aptamer-magnetic bead plus aptamer quantum dot sandwich assay and magnetic collection system. (b) Spectrofluorometric response of adherent *Campylobacter* assays in chicken juice. *(From J.G. Bruno, T. Phillips, M.P. Carillo, R. Crowell, J. Fluores. 19 (2009) 427—435 with permission from Springer.)* LSPR, localized surface plasmon resonance.

improvement of about 10^3 times. However, sensitivity of the system decreases as the size of the probe becomes smaller than the target molecule [194].

Due to its high extinction coefficient, NPs are used as labels for visual, light scattering or absorbance detection. The most used NPs for this purpose

TABLE 6 Summary of Nanoparticles and Their Refractive Index Sensitivities

Particle	Type	λ_{max} (nm)	Sensitivity (nm/RIU)
Au nanoring	Ensemble	1300	691
Au nanostar	Single	725	218
Au nanodisk	Ensemble	820	327
Au nanotube	Ensemble	650	225
Au triangle@SiO$_2$	Ensemble	1250	727
SiO$_2$@cushell	Ensemble	610	67.8
Ag triangle	Ensemble	1093	1096
AuNP dimer	Ensemble	650	471.15

From F. Zamborini, L. Bao, R. Dasari, Anal. Chem. 84 (2012) 541–576 with permission from ACS Publications.

are AuNPs and AgNPs because they present absorption in the visible region and can be easily functionalized.

A strategy for DNA signal amplification is the 'scanometric' DNA microarray [190]. To amplify the signal of the AuNPs hybridized at the surface, silver ions are reduced to metallic silver with hydroquinone at the surface of AuNPs, this process increases the signal by a factor of 10^5. Based on this, *E. coli, Enterobacter cloacae, Bacillus subtilis, S. aureus* have been detected using AuNPs as labels [195].

3.2.3 Spectroscopy Using Quantum Dots

Use of QDs have a lot of advantages due to their optical properties eg, the absorption spectrum shows the simplicity to perform multiplex analysis, since it can be excited with a single light source to multiple wavelengths, compared with dyes which are limited to a small region. Multiple analysis allows the evaluation of several parameters in a single run, which reduces the time and cost of the analysis.

Fluorescence detection based on QDs has been used for biological detection, through the conjugation of QDs with specific DNA. In addition, QDs have been used in the detection of heavy metals using the 'turn off' mode, as it can reduce the risk of a false positive.

Fluorescence is one of the most popular analysis methods due to its high sensitivity and very low detection limits. However, the main disadvantage of this technique is that fluorophores can undergo photobleaching and have a short stability times. Using NMs with this technique is desirable due to its optical properties, better when compared with organic fluorescent dyes. NPs tend to have greater stability and higher luminescent intensity. The most important feature is that emission is limited and symmetrical, and NPs are also tunable in size, shape and composition, so they can be used in the detection of biomolecules and heavy metals [140,196−212].

There are two strategies for 'turn-on' and 'turn-off' detection, when initially fluorescence detection is deactivated turn-on detection occurs, usually through the process of fluorescent resonance energy transfer (FRET), and returns through the interaction with the analyte, which separates the fluorophore and the excitation agent. FRET is a process involving the distance dependent energy transfer from a donor molecule to an acceptor molecule through dipole−dipole interactions.

In the case of NPs, these are used as FRET donors and the dye serves as an acceptor molecule. The proximity between the actor molecule and the NPs decreases the emission intensity of these last ones and increases the intensity of the acceptor. In 'turn-on' sensors, fluorescence of NPs is deactivated by the interaction with the acceptor molecule or the proximity to the surface, so the fluorescence loss is proportional to the analyte concentration. The most used NMs are QDs and metal NPs, and less frequently used are CNTs.

It must be remembered that the semiconductor QDs have emission in the NIR. Most commercially available QDs have a semiconductor core, usually a mixture of Cd, Zn, Se, Te, In, P and/or As, the size of these NPs is usually $2-10$ nm in diameter. This core is surrounded by a cover, usually another semiconductor material and may include a polymer outer layer.

Recently, a new type of nanosized carbon particulates like CQDs, with similar size to conventional quantum has been reported [196]. Their use has been significantly increased because of their nontoxic nature, good dispersion in water along with competitive fluorescent properties similar to the toxic metal-based semiconductor QDs. This NM is a new option for *E. coli* and cholesterol detection.

Integration of QDs and DNA aptamers were developed against $MgCl_2$-extracted surface proteins from *C. jejuni*. The two highest affinity aptamers were selected for use in a magnetic bead (MB) and QD-based sandwich assay scheme (Fig. 6D). The assay exhibits specificity, sensitivity and detection limits of $10-250$ UFC colony forming unit (CFU). This strategy may help to prevent major foodborne pathogen outbreaks.

3.2.4 Spectroscopy Using Carbon Nanotubes

Immunosensing based on CNT was designed for the analysis of a marine (palytoxin, PlTX) toxin, it is usually detected in seafood. The electro-chemiluminescent device was developed by preparing a ruthenium-based complex linked to an anti-PlTX polyclonal antibody (pAb). The quantitative detection of PlTX is performed measuring the intensity of the emitted light under excitation conditions. This immunosensor shows a detection limit within subnanograms per millilitre of solution, which is better than reported by HPLC coupled to mass spectrometry or fluorescence polarization. This device is suitable for detection of mussels and microalgae [213].

3.2.5 Surface-Enhanced Raman Spectroscopy

Recently, a nano-enabled SERS technique is finding its way into food-related applications. It is a technique capable of measuring samples within minutes, with minimal sample preparation. So, the development of NMs has significantly contributed to Raman spectroscopy techniques. The low sensitivity of SERS can be overcome if a suitable nanostructured surface is used as a substrate to perform SERS [197,214—216]. In the presence of suitable nanostructures, SERS can improve the technique sensitivity to 10^{15} times, leading the potential of this technique for single molecule or single cell detection.

In Table 7 a review of NM-based optical techniques for some analytes of interest in food science and industry is presented.

4. CONCLUSIONS AND PERSPECTIVES

Quality control of food draws the attention of both the institutions responsible for regulating and researchers because of their intimate relationship with human health; hence, the need to generate fast, accurate and reliable information on the quality and safety of foodstuffs.

Due to these requirements, conventional analytical techniques in recent years have been adapted with new strategies to improve stability, sensitivity, detection limit and the selectivity of the method.

With the arrival of NMs to the analytical chemistry, chemical and biological analysis techniques have found more opportunity areas. Inclusion of NMs in spectrophotometric techniques and electrochemical separation has improved substantially these techniques when considering sensitivity, specificity and detection limits where femtomolar concentrations can be achieved. In biochemical analysis, detection of pathogens can be made in minutes, thanks to the integration of NMs in biosensors.

In this sense, the design and development of sensitive devices, which can detect contaminants in food and/or food chain is a vital issue. Their low toxicity and low cost make NMs ideal for the development of new analytical devices with potential applications in food science and industry, eg, carbon QDs.

Biosensors based on NMs are considered forefront devices representing a great alternative to conventional analytical methods such as HPLC, UV—Vis, fluorescence, electrochemical, among others, as they are able to provide solutions to analytical problems easily, quickly and inexpensively, although this depends on the food area and the analyte considered.

The unique properties of NMs have enabled their application with great success in microfluidics, promoting the multiscreen for detection of analytes in complex matrixes and with similar structures and even for the detection of pathogens.

TABLE 7 Nanomaterials-Based Optical Techniques

Technique	Nanomaterial	Analyte	Sample	Lower Detection Limit	References
ECL	TiNT	Choline	Milk	0.01 µM	[198]
	CdS	Hypoxanthine	Fish	5 nM	[199]
FL	CdTe/CdS	Glucose	—	1.8 µM	[200]
	QD	*Campylobacter*	Chicken	2.5 CFU	[201]
	SiO$_2$	*Salmonella*	Bacteria	3 fM	[202]
	SWNTs	Ochratoxin A	Beer	24.1 nM	[140]
	CdSe/ZnS	Diquat herbicide	Oat grains	0.01 mg/kg	[203]
	β-CD−CdSe/ZnS	Vanillin	Sugar Milk	0.99 mg/mL	[204]
	CdSe/ZnS	Melamine	Milk	0.15 mg/kg	[205]
	CdTe	Melamine		0.02 mg/L	[206]
		Choline		0.1 µM	[207]
		Acetylcholine		10 µM	[207]
	Silica-NPs	Soya protein	Soya milk	0.05 mg/mL	[208]
	Mn:ZnSe	OPs	Milk	1.3×10^{-11} M	[209]
	AuNPs	Streptomycin		47.6 nM	[210]
Colourimetric	AuNPs	Streptomycin		71.3 nM	[210]

Surface plasmon resonance	Magnetic nanoparticles (MNPs)	Peanut allergen Ara h1	Cocoa	0.09 µg/mL	[211]
	MNPs	Ochratoxin A	Wine	0.94 ng/mL	[212]
Surface-enhanced Raman spectroscopy	MNPs	Ovalbumin	Milk	5 µg/mL	[214]
	AgNPs	*Escherichia coli*		10^3 cells/mL	[215]
	AuNPs	Pesticides	Apple juice and cabbage	2.5 ppm	[216]

AuNPs, gold nanoparticles; *ECL*, electrochemiluminescence; *FL*, Fluorescence; *OPs*, organophosphates; *QD*, quantum dot.

The development of biosensors based on NMs represents a significant opportunity area with the emergence of new materials with mechanical, electrical and optical properties of a higher quality than the current ones. As a result, carbon-based NMs draw special attention in order to create new devices as carbon — QDs and GR. On the other hand, new electrodes have emerged based on AuNPs chains. In this regard, there are no reports indicating their use for detecting food substances, but there are some for glucose detection with optimal results and representing promising opportunities for future developments with applications in food analysis.

The use of nanotechnology supports the development of new analytical devices. A challenge in this regard, is a massive device production without changing the NM properties.

Another challenge in the use of NMs is related to the detection and monitoring of real samples in real time, along with the inclusion of these new devices and methods in already standardized and regulated methods, considering the official legislations.

ACKNOWLEDGEMENTS

This work was supported by MINECO (Spain, MAT2014-52485-P). ICN2 acknowledges support from the Severo Ochoa Program (MINECO, Grant SEV-2013-0295). Nanobiosensors and Bioelectronics Group acknowledges the support from Secretaria d'Universitats i Recerca del Departament d'Economia i Coneixement de la Generalitat de Catalunya (2014 SGR 260).

REFERENCES

[1] World Health Organization: WHO (Online) (reported 01.09.15) http://www.who.int/mediacentre/factsheets/fs399/en/.

[2] D.R. Thevenot, K. Tóth, R.A. Durst, G.S. Wilson, Pure Appl. Chem. 71 (1999) 2333−2348.

[3] N. Sozer, L.J. Kokini, Trends Biotechnol. 27 (2008) 82−89.

[4] A. Tiwari, S.K. Shukla, Advanced Carbon Materials and Technology, Wiley, USA, 2014, pp. 87−128.

[5] D. Tasis, N. Tagmatarchis, A. Bianco, M. Prato, Chem. Rev. 106 (2006) 1105−1136.

[6] L. Sutarlie, K.M. Moh, M.G.L. Lim, S. Lukman, E. Cheung, X. Su, Plasmonics 9 (2014) 753.

[7] S. Halldorsson, E. Lucumi, R. Gomez-Sjoberg, R.M.T. Fleming, Biosens. Bioelectron. 67 (2015) 595−600.

[8] B.C. Satishkumar, L.O. Brown, Y. Gao, C.-C. Wang, H.-L. Wang, S.K. Doorn, Nat. Nanotechnol. 2 (2007) 506.

[9] I. Cesarino, F.-C. Moraes, M.L.V. Lanza, S. Machado, Food Chem. 135 (2012) 873−879.

[10] S.J. Oldenburg, R.D. Averitt, S.L. Westcott, N. Halas, J. Chem. Phys. Lett. 288 (1998) 243−247.

[11] S. He, B. Song, D. Li, C. Zhu, W. Qi, Y. Wen, L. Wang, S. Song, H. Fang, C. Fan, Adv. Funct. Mater. 20 (2010) 453−459.

[12] R. Yang, Z. Tang, J. Yan, H. Kang, Y. Kim, Z. Zhu, W. Tan, Anal. Chem. 80 (2008) 7408.

[13] S. Yan, T.-T. Qi, D.-W. Chen, Z. Li, X.-J. Li, S.-Y. Pan, J. Chromatogr. A 1347 (2014) 30–38.

[14] E. Susie, E.-S. Mostafa, Chem. Soc. Rev. 35 (2006) 209–217.

[15] O. Yuval, S. Bappaditya, R. Vincent, Chem. Soc. Rev. 37 (2008) 1814–1825.

[16] N. Kaushik, M.S. Thakkar, S. Snehit, M.S. Mhatre, Y. Rasesh, M.S. Parikh, Nanomed (London) 6 (2010) 257–262.

[17] S. Guo, E. Wang, Anal. Chim. Acta 598 (2007) 181–192.

[18] S. Gopinath, T. Lakshmipriya, K. Awazu, Biosens. Bioelectron. 51 (2014) 115–123.

[19] E. Ahmad Sarreshtehdar, D. Noor Mohammad Lavaee, P. Ramezani Mohammad, A. Khalil, T. Seyed Mohammad, Food Chem. 190 (2016) 115–121.

[20] S.J. Rosenthal, J. Chang, O. Kovtun, J. McBride, I. Tomlinson, Chem. Biol. 18 (2011) 10–24.

[21] S. Hao, M.X. Xiaojie, J. Wang, Talanta 127 (2014) 68–74.

[22] A. Russ, K. Ulrich, Anal. Chem. 81 (2009) 4113–4120.

[23] F. Esteve-Turrillas, A. Abad-Fuentes, Biosens. Bioelectron. 41 (2013) 12–29.

[24] S. Iijima, Nature 354 (1991) 56–58.

[25] C. Hussain, S. Mitra, Anal. Bioanal. Chem. 399 (2011) 75–89.

[26] Y. Zhang, J. Zhu, Nanomaterials 33 (2002) 523–534.

[27] K. Scienceda, P. Stege, G. Haby, G. Messina, C. García, Anal. Chim. Acta 691 (2011) 6–17.

[28] B. Pérez-López, A. Merkoçi, Microchim. Acta 179 (2012) 1–16.

[29] C. Herrero Latorre, J. Álvarez Méndez, S. Barciela García, S. García Martín, R.M. Peña Crecente, Anal. Chim. 749 (2012) 16–35.

[30] D.T. Mitchell, S.B. Lee, L.T.N. Li, T.K. Nevanen, H. Söderlund, C.R. Martin, J. Am. Chem. Soc. 124 (2002) 11864–11865.

[31] N. Karousis, N. Tagmatarchis, D. Tasis, Chem. Rev. 110 (2010) 5366–5397.

[32] Y. Lin, C. Dimitrakopoulos, K. Jenkins Farmer, D. Chiu, H. Grill, A. Avouris, Science 327 (2010) 662.

[33] K. Novoselov, A. Geim, S. Morozov, D. Jiang, Y. Zhang, S. Dubonos, A. Firsov, Science 306 (2004) 666–669.

[34] J. Park, W. Mitchel, L. Grazulis, H. Smith, K. Eyink, J. Boeckl, D. Tomich, S. Pacley, J. Hoelscher, Adv. Mat. 22 (2010) 4140–4145.

[35] M. Allen, V. Tung, R. Kaner, Chem. Rev. 110 (2010) 132–145.

[36] K. Geim, K. Novoselov, Nature 6 (2007) 183–191.

[37] K. Novoselov, V. Falko, P. Colombo, P. Gellert, M. Schwab, K. Kim, Nature 490 (2012) 192–200.

[38] E.M. Kaidashev, M. Lorenz, H. von Wenckstern, A. Rahm, H.C. Semmelhack, K.H. Han, G. Benndorf, C. Bundesmann, H. Hochmuth, M. Grundmann, Appl. Phys. Lett. 82 (2003) 3901–3903.

[39] C. Berry, C. Curtis, J. Phys. D: App. Phys. 36 (2003) R198–R206.

[40] H.J. Lee, B.C. Kim, K.W. Kim, Kim, Y.K. Kim, Biosens. Bioelectron. 24 (2009) 3550–3555.

[41] X. Wu, J. Hu, B. Zhu, L. Lu, X. Huang, D. Pang, J. Chromatogr. A 1218 (2011) 7341–7346.

[42] Q. Gao, D. Luo, J. Ding, Y. Feng, J. Chromatogr. A 1217 (2010) 5602–5609.

[43] N. Gan, X. Yang, D. Xie, Y. Wu, W. Wen, Sensors 10 (2010) 625–638.

[44] Y. Cheng, Y. Liu, J. Huang, K. Li, W. Zhang, Y. Xian, Talanta 77 (2009) 1332–1336.

[45] A. Merkoçi, A. Ambrosi, A. De la Escosura-Muñiz, B. Pérez-López, M. Guix, M. Maltez, S. Marin, Nanomater. Electroanal. (2010). http://dx.doi.org/10.1002/9780470027318.a9077.

[46] B. Perez-López, A. Merkoçi, Trends Food Sci. Technol. 22 (2011) 625–639.

[47] R. Sharma, K.V. Ragavan, M.S. Thakur, K.S. Raghavara, Biosens. Bioelectron. 74 (2015) 612—627.

[48] M. Granda, C. Valdéz, A. García, M. Díaz-García, Microchim. Acta 166 (2009) 1—19.

[49] IFT, Bacteria Associated with Foodborne Diseases: A Scientific Status Summary of the Institute of Food Technologists, Chicago: S.N, 2004.

[50] N. Li, X. Su, Y. Lu, Analyst 140 (2015) 2916—2943.

[51] L. Zhou, J. Wang, L. Li Yanbin, Food Chem. 162 (2014) 34—40.

[52] J. Jia, B. Wang, A. Wu, G. Cheng, S. Li, S. Dong, Anal. Chem. 74 (2002) 2217—2223.

[53] T. Hoshi, N. Sagae, K. Daikuhara, K. Takahara, J.I. Anza, Talanta 71 (2007) 644—647.

[54] X. Luo, J. Xu, Y. Du, H. Chen, Anal. Biochem. 334 (2004) 284—289.

[55] S. Liu, H. Ju, Biosens. Bioelectron. 19 (2003) 177—183.

[56] L. Agui, J. Manso, P. Yañez-Sedeño, J.M. Pingarrón, Sens. Actuator B 113 (2006) 272—280.

[57] H. Li, B. Xu, D. Wang, Y. Zhou, H. Zhang, W. Xia, S. Xu, Y. Li, J. Biotech. 203 (2015) 97—103.

[58] K.V. Richa Sharma, M.S. Ragavan, K.S. Thakur, M.S. Raghavarao, Biosens. Bioelectron. 74 (2015) 612—627.

[59] R.A. Medeiros, R.C. Rocha-Filho, O. Fatibello-Filho, Food Chem. 123 (2010) 886—891.

[60] X. Lin, Y. Ni, S. Kokot, Anal. Chim. Acta 765 (2013) 54—62.

[61] S. Chen, J. Huang, D. Du, J. Li, H. Tu, D. Liu, A. Zhang, Biosens. Bioelectron. 26 (2011) 4320—4325.

[62] E. Deab, M.S. Ohsaka, Electrochem. Commun. 4 (2002) 288—292.

[63] V. Carralero, M.L. Mena, A. Gonzalez-Cortes, P. Yañez-Sedeno, J.M. Pingarrón, Anal. Chim. Acta 528 (2005) 1—8.

[64] G. Zhiguo, Y. Shuping, L. Zaijun, S. Xiulan, W. Guangli, F. Yinjun, L. Junkang, Electrochim. Acta 56 (2011) 9162—9167.

[65] M.T. Castañeda, S. Alegret, A. Merkoçi, Electroanalysis 19 (2007) 743—753.

[66] D. Tang, Z. Zhong, R. Niessner, D. Knoop, Analyst 134 (2009) 1554—1560.

[67] A. Afonso, B. Pérez-López, R. Faria, L. Mattoso, M. Hernández-Herrero, A. Roig-Sagués, M. Maltez-da Costa, A. Merkoçi, Biosens. Bioelectron. 40 (2013) 121—126.

[68] T. Hoshi, N. Sagae, K. Daikuhara, K. Takahara, J. Anzai, Mater. Sci. Eng. C 27 (2007) 890—894.

[69] S. Liu, D. Leech, H. Ju, Anal. Lett. 36 (2003) 1—19.

[70] N. German, A. Ramanaviciene, J. Voronovic, A. Arunas Ramanavicius, Microchim. Acta 168 (2010) 221—229.

[71] S. Tuncagil, C. Ozdemir, D. Dilek, T. Suna, T. Levent, Food Chem. 127 (2011) 1317—1322.

[72] F. Ribeiro, M. Barroso, S. Morais, S. Viswanathan, P. De Lima-Neto, A. Correia, M. Oliveira, C. Delerue-Matos, Bioelectrochemistry 95 (2014) 7—14.

[73] V. Dhull, N. Dilbaghi, V. Hooda, Int. J. Pharm. Pharm. Sci. 7 (2014) 17—24.

[74] C. Zhai, X. Sun, X. Zhao, Z. Gong, X. Wang, Biosens. Bioelectron. 42 (2013) 124—130.

[75] E. Valera, J. Ramón-Azcón, A. Barranco, B. Alfaro, F. Sánchez-Baeza, M.-P. Marco, Á. Rodríguez, Food Chem. 122 (2010) 888—894.

[76] X. Kang, G. Pang, Q. Chen, X. Liang, Sens. Actuator B 177 (2013) 1010—1016.

[77] S. Sakamoto, G. Yusakul, B. Pongkitwitoon, H. Tanaka, S. Morimoto, Food Chem. 194 (2016) 191—195.

[78] C. Lanzellotto, G. Favero, M.L. Antonelli, C. Tortolini, S. Cannistraro, E. Coppari, F. Mazzei, Biosens. Bioelectron. 55 (2014) 430—437.

[79] R. Rawal, S. Chawla, C. Pundir, Anal. Biochem. 419 (2011) 196—204.

[80] G. Evtugyn, A. Porfireva, V. Stepanova, M. Kutyreva, A. Gataulina, N. Ulakhovich, V. Evtugyn, T. Hianikal, Sensors 13 (2013) 16129−16145.

[81] H. Kuang, W. Chen, D. Xu, L. Xu, Y. Zhu, L. Liu, H. Chu, C. Peng, C. Xu, S. Zhu, Biosens. Bioelectron. 26 (2010) 710−716.

[82] H. Sun, Y.Y. Zhang, S.H. Si, D.R. Zhu, Y.S. Fung, Sens. Actuator B 108 (2005) 925−932.

[83] J. Li, J. Yu, F. Zhao, B. Zeng, Anal. Chim. Acta 587 (2007) 33−40.

[84] A. Boujakhrout, E. Sánchez, P. Díez, A. Sánchez, P. Martínez-Ruiz, C. Parrado, J.M. Pingarrón, R. Villalonga, Chem. Electrochem. 2 (2015) 1735−1741.

[85] Y. Zhang, W. Jia, M. Cu, C. Dong, S. Shuang, Y. Kwan, M.M. Choi, J. Biotechnol. 6 (2011) 491−500.

[86] S. Xiulan, Z. Xiaolian, T. Jian, G. Xiaohong, Z. Jun, F.S. Chu, Food Control 17 (2006) 256−262.

[87] M. Labib, A.S. Zamay, O.S. Kolovskaya, I.T. Reshetneva, G.S. Zamay, R.J. Kibbee, S.A. Sattar, T.N. Zamay, M.V. Berezovski, Anal. Chem. 84 (2012) 8114−8117.

[88] A. Hussein, A. Hassan, A. De la Escosura-Muñiz, A. Merkoçi, Biosens. Bioelectron. 67 (2015) 511−515.

[89] L. Zhou, R. Li, Z. Li, Q. Xia, Y. Fang, J. Liu, Sens. Actuator B 174 (2012) 359−365.

[90] X. Sun, L. Yan, T. Tang, Y. Zhang, Eur. Food Res. Technol. 234 (2012) 1013−1021.

[91] L. Bonel, J.C. Vidal, P. Duato, J.R. Castillo, Anal. Methods 2 (2010) 335−341.

[92] E. Dinçkayaa, Ö. Kınıkb, M.K. Sezgintürkc, Ç. Altuğa, A. Akkocaa, Biosens. Bioelectron. 26 (2011) 3806−3811.

[93] R. Latha, H.K. Manonmani, E.R. Rati, J. Microbiol. 3 (2008) 136−142.

[94] J. Du, X. Yu, J. Di, Biosens. Bioelectron. 37 (2012) 88−93.

[95] Z. Wang, Q. Xu, J.H. Wang, Q. Yang, J.H. Yu, Y.D. Zhao, Microchim. Acta 165 (2009) 387−392.

[96] I. Vasilescu, S. Eremia, M. Kusko, A. Radoi, E. Vasile, G. Radu, Biosens. Bioelectron. 75 (2016) 232−237.

[97] J. Wang, G. Liu, A. Merkoci, J. Am. Chem. Soc. 125 (2003) 3214−3215.

[98] J. Hansen, J. Wang, A.N. Kawde, Y.K. Xiang, G. Collins, J. Am. Chem. Soc. 128 (2006) 2228−2229.

[99] S. Viswanathan, C. Rani, J. Ho, Talanta 94 (2012) 315−319.

[100] M. Masikini, S.N. Mailu, A. Tsegaye, N. Njomo, K.M. Molapo, C.O. Ikpo, C.E. Sunday, C. Rassie, L. Wilson, P.G.L. Baker, E.I. Iwuoha, Sensors (Basel) 15 (2015) 529−546.

[101] M. Freitas, S. Viswanathan, H.P.A. Nouws, M.B.P.P. Oliviera, C. Delerue-Matos, Biosens. Bioelectron. 51 (2014) 195−200.

[102] B. Liu, B. Zhang, G. Chen, D. Tang, Microchim. Acta 181 (2014) 257−262.

[103] L. Yang, L. Yanbin, J. Food Protect. 68 (2005) 1241−1245.

[104] Y. Liu, D. Yao, H. Chang, C. Liu, C. Chen, Biosens. Bioelectron. 24 (2008) 558−565.

[105] S. Kalele, A. Kundu, S. Gosavi, D. Deobagkar, D. Deobagkar, S. Kulkarni, Small 2 (2006) 335−338.

[106] L. Yang, Y. Li, Analyst 131 (2006) 394−401.

[107] M. Rubianes, G. Rivas, Electroanalysis 17 (2005) 73−78.

[108] F. Valentini, A. Amine, S. Orlanducci, M. Terranova, G. Palleschi, Anal. Chem. 75 (2003) 5413−5421.

[109] M. Rubianes, G. Rivas, Electrochem. Commun. 5 (2003) 689−694.

[110] M.A. Pulickel, M.T. James, Mater. Sci. 447 (2007) 1066−1068.

[111] Q. Zhao, L. Guan, Z. Gu, Q. Zhuang, Electroanalysis 17 (2005) 85−88.

[112] R. Deo, J. Wang, I. Block, A. Mulchandani, K.A. Joshi, M. Trojanowicz, F. Scholz, W. Chen, Y. Lin, Anal. Chim. Acta 530 (2005) 185−189.

[113] J. Liu, A. Chou, W. Rahmat, M. Paddon-Row, J. Gooding, J. Electroanal. 17 (2005) 38−46.

[114] P. Bartlett, J. Cooper, J. Electroanal. Chem. 362 (1993) 1−12.

[115] J. Wang, M. Musameh, Anal. Chim. Acta 539 (2005) 209−213.

[116] P. Joshi, S. Merchant e, Y. Wang, D. Schmidtke, Anal. Chem. 77 (2005) 3183−3188.

[117] J. Ye, Y. Wen, W. Zhang, H. Cui, G. Xu, F. Sheu, Electroanalysis 17 (2005) 89−96.

[118] C. Livingstone, A.J. Wain, Trends Anal Chem. 25 (2006) 569−579.

[119] E. Silva, R. Takeuchi, A. Santos, Food Chem. 173 (2015) 763−769.

[120] E. Sartori, H. Takeda, O. Fatibello-Filho, Electroanalysis 23 (2011) 2526−2533.

[121] H. Moghaddam, M. Malakootian, H. Beitollah, P. Biparva, Int. J. Electrochem. Sci. 9 (2014) 327−341.

[122] M. Amatatongchai, W. Sroysee, S. Chairam, D. Nacapricha, Talanta 133 (2015) 134−141.

[123] E. Sartori, F. Vicentini, O. Fatibello-Filho, Talanta 87 (2011) 235−242.

[124] R. Rawal, S. Chawla, C. Pundir, Anal. Bioanal. Chem. 401 (2011) 2599−2608.

[125] E. Primo, F. Gutierrez, M. Rubianes, G. Rivas, Electrochim. Acta 182 (2015) 391−397.

[126] Y. Zou, C. Xiang, L. Sun, F. Xu, Electrochim. Acta 53 (2008) 4089−4095.

[127] Y. Sun, H. Wang, C. Sun, Biosens. Bioelectron. 24 (2008) 22−28.

[128] P. Dalmasso, M. Pedano, G. Rivas, Biosens. Bioelectron. 39 (2013) 76−81.

[129] D. Kuang, Y. Feng, F. Zhang, M. Liu, Microchim. Acta 176 (2012) 73−80.

[130] D. Odaci, A. Telefoncu, S. Timur, Bioelectrochemistry 79 (2010) 108−113.

[131] G. Ziyatdinova, E. Kozlova, H. Budnikov, Electroanalysis 27 (2015) 1660−1668.

[132] G. Ziyatdinova, I. Salikhova, H. Budnikov, Talanta 125 (2014) 378−384.

[133] H. Wan, Q. Zou, R. Yan, F. Zhao, B. Zeng, Microchim. Acta 159 (2007) 109−115.

[134] Q. Liu, A. Fei, J. Huan, H. Mao, K. Wang, J. Electroanal. Chem. 740 (2015) 8−13.

[135] M. Kesik, F.E. Kanik, J. Turan, M. Kolb, S. Timur, M. Bahadir, L. Toppare, Sens. Actuator B: Chem. 205 (2014) 39−49.

[136] Y. Zhu, Y. Cao, X. Sun, X. Wang, Sensors 13 (2013) 5286−5301.

[137] G. Liu, Y. Lin, Anal. Chem. 78 (2006) 835−843.

[138] S. Pérez, J. Bartrolí, E. Fàbregas, Food Chem. 141 (2013) 4066−4072.

[139] M. Dervisevic, E. Custiuc, E. Çevik, M. Şenel, Food Chem. 181 (2015) 277−283.

[140] Z. Guo, J. Ren, J. Wang, E. Wang, Talanta 85 (2011) 2517−2521.

[141] J. Pacheco, M. Castro, S. Machado, M. Barroso, H. Nouws, M. Delerue-Matos, Sens. Actuator B 215 (2015) 107−112.

[142] X. Zhang, C. Li, W. Wang, J. Xue, Y. Huang, X. Yang, B. Tan, X. Zhou, C. Shao, S. Ding, J. Qiu, Food Chem. 192 (2016) 197−202.

[143] J. Wang, G. Liu, M. Jan, J. Am. Chem. Soc. 126 (2004) 3010−3011.

[144] G. Zelada-Guillén, S. Bhosale, J. Riu, F. Rius, Anal. Chem. 82 (2010) 9254−9260.

[145] Y. Li, P. Cheng, J. Gong, L. Fang, J. Deng, W. Zheng, J. Liang, Anal. Biochem. 421 (2012) 227−233.

[146] M. Abdalhai, A. Fernandes, X. Xia, A. Musa, J. Ji, X. Sun, J. Agric. Food Chem. 63 (2015) 5017−5025.

[147] S. Wu, X. Lan, L. Ciu, L. Zhang, S. Tao, H. Wang, M. Han, Z. Liu, C. Meng, Anal. Chim. Acta 699 (2011) 170−176.

[148] R. Xue, T. Kang, L. Lu, S. Cheng, Anal. Lett. 46 (2013) 131−141.

[149] H. Yin, Q. Zhang, Y. Zhou, Q. Ma, T. Liu, L. Zhu, S. Ai, Electrochim. Acta 56 (2011) 2748−2753.

[150] H. Fan, Y. Li, D. Wu, H. Ma, K. Mao, D. Fan, B. Du, Q. Wei, Anal. Chim. Acta 711 (2012) 24−28.

[151] Y. Liu, S. Yang, W. Niu, Colloid Surf. B 108 (2013) 266−270.

[152] Y. Zhou, S. Liu, H. Jiang, H. Yang, H. Chen, Electroanalysis 22 (2010) 1323−1328.

[153] H. Song, Y. Kokot, Anal. Chim. Acta 788 (2013) 24−31.

[154] Y. Wang, S. Zhang, D. Du, Y. Shao, Z. Li, J. Wang, M. Engelhard, J. Li, Y. Lin, J. Mat. Chem. Phys. 21 (2011) 5319−5325.

[155] L. Zaijun, S. Xiulan, X. Qianfanf, L. Ruivi, F. Yinjun, Y. Shuping, L. Junkang, Electrochim. Acta 85 (2012) 42−48.

[156] M. Wei, D. Tian, S. Liu, X. Zheng, S. Duan, S.C. Zhou, Sens. Actuactor B 195 (2014) 452−458.

[157] Y. Chai, X. Niu, Ch Chen, H. Zhao, M. Lan, Anal. Lett. 46 (2013) 803−807.

[158] M. Wang, J. Huang, R. Wang, R. Zhang, J. Chen, Food Chem. 151 (2014) 191−197.

[159] P. Prasada, E. Naidooa, N. Sreedhar Arab, J. Chem. (2015). http://dx.doi.org/10.1016/j.arabjc.2015.02.012.

[160] M. Manoor, H. Tao, J. Clayton, A. Sengupta, D. Kaplan, R. Naik, N. Verma, F. Omenetto, M. Alpine, Nature 3 (2012) 763.

[161] O. Akhavan, E. Ghaderi, A. Esfandiar, J. Phys. Chem. B. 115 (2011) 6279−6288.

[162] Y. Wan, Y. Wang, J. Wu, D. Zhang, Anal. Chem. 50 (2011) 648−653.

[163] Y. Huang, X. Dong, Y. Liu, L. Lic, P. Chen, J. Mater. Chem. 21 (2011) 12358−12362.

[164] P. Basu, D. Indukuri, S. Keshavan, V. Navratna, S. Vanjari, S. Raghavan, N. Bhat, Sens. Actuactor B 190 (2014) 342−347.

[165] Y. Zhu, S. Murali, W. Cai, W. Li, J. SukPotts, R. Ruoff, Adv. Mater. 22 (2010) 3906.

[166] D. Mark, S. Haeberle, G. Roth, F. Stettenz, R. Zengerle, Chem. Soc. Rev. 39 (2010) 1153−1182.

[167] M. Pumera, Chem. Commun. 47 (2011) 5671−5680.

[168] M. Pumera, J. Wang, E. Grushka, R. Polsky, Anal. Chem. 73 (2001) 5625.

[169] Y. Lin, M. Huang, H. Chang, J. Chromatogr. A 1014 (2003) 47−55.

[170] X. Llopis, M. Pumera, S. Alegret, A. Merkoci, Lab. Chip 9 (2009) 213−218.

[171] J. Sheng, L. Zhang, L. Jianping, H. Ju, Anal. Chim. Acta 709 (2012) 41−46.

[172] M. Fernández-Baldo, F. Bertolino, G. Fernández, G. Messina, M. Sanz, J. Raba, Analyst 136 (2011) 2756−2762.

[173] J. Jung, G. Kim, T. Seo, Lab. Chip 11 (2011) 3465−3470.

[174] X. Gao, G. Tang, X. Su, Analyst 136 (2011) 2756−2762.

[175] M. Pumera, A. Merkoçi, S. Alegret, Electrophoresis 28 (2007) 1274−1280.

[176] A. González, M. Ávila, M. Pumera, M.C. González, A. Escarpa, Anal. Chem. 79 (2007) 7408−7415.

[177] A.G. Crevillen, M. Pumera, M.C. González, Lab Chip 9 (2009) 346−353.

[178] M. Yang, S. Sun, Y. Kostova, A. Rasooly, Lab. Chip 10 (2010) 1011−1017.

[179] Y. Wang, L. Wang, T. Tian, G. Yao, X. Hu, C. Yang, Q. Xu, Talanta 99 (2012) 840−845.

[180] M. Chicharro, A. Sanchez, M. Moreno, E. Bermejo, A. Zapardiel, Talanta 74 (2007) 376−386.

[181] A. Sanchez, M. Moreno, E. Bermejo, M. Angeles, Electrophoresis 30 (2009) 3480−3488.

[182] W. Zhang, X. Zhang, L. Zhang, G. Chen, Sens. Actuator B 192 (2014) 459−466.

[183] Q. Chen, L. Zhang, G. Chen, Anal. Chem. 84 (2012) 171−178.

[184] X. Yao, X. Xu, P. Yang, G. Chen, Electrophoresis 27 (2006) 3233−3242.

[185] M. Moreno, A. Sanchez, E. Bermejo, A. Zapardiel, M. Chicharro, Electrophoresis 32 (2011) 877−883.

[186] Z. Chen, L. Zhang, G. Chen, Electrophoresis 30 (2009) 3419−3426.

[187] S. Sheng, L. Zhang, G. Chen, Food Chem. 145 (2014) 555−561.

[188] A. Crevillen, M. Avila, M. Pumera, M. Gonzalez, A. Escarpa, Anal. Chem. 79 (2007) 7408−7415.

[189] R.S. Chouhan, A.C. Vinayaka, M.S. Thakur, Anal. Bioanal. Chem. 397 (2010) 1467−1475.

[190] T. Taton, C. Mirkin, R. Letsinger, Science 289 (2000) 1757−1760.

[191] F. Zamborini, L. Bao, R. Dasari, Anal. Chem. 84 (2012) 541−576.

[192] S. Gao, N. Koshizaki, H. Tokuhisa, E. Koyama, T. Sasaki, J.-K. Kim, J. Ryu, D.-S. Kim, Y. Shimizu, Adv. Funct. Mater. 20 (2010) 78−86.

[193] S. Kim, S. Lee, H. Lee, Talanta 81 (2010) 1755−1759.

[194] G. Pelossof, R. Tel-Vered, X.-Q. Liu, Chem. Eur. J. 17 (2011) 8904−8912.

[195] J. Gao, D. Liu, Z. Wang, Anal. Chem. 82 (2010) 9240−9247.

[196] D.G. Babar, S.K. Sonkar, K.M. Tripathi, S. Sarkar, J. Nanosci. Nanotechnol. 14 (2014) 2334−2342.

[197] A. Pearson, A.P. O'Mullane, S.K. Bhargav, V. Bansal, Electrochem. Commun. 25 (2012) 87−90.

[198] H. Dai, Y. Chi, X. Wu, Y. Wang, M. Wei, G. Chen, Biosens. Bioelectron. 25 (2010) 1414−1419.

[199] Y. Zhang, S. Deng, J. Lei, Q. Xu, H. Ju, Talanta 85 (2011) 2154−2158.

[200] M. Hu, J. Tian, H.-T. Lu, L.-X. Weng, L.-H. Wang, Talanta 82 (2010) 997−1002.

[201] J.G. Bruno, T. Phillips, M.P. Carillo, R. Crowell, J. Fluores. 19 (2009) 427−435.

[202] Z. Wang, H. Xu, J. Wu, J. Ye, Z. Yang, Food Chem. 125 (2011) 779−802.

[203] C. Carrillo-Carrión, B. Simonet, M. Valcárcel, Anal. Chim. Acta 692 (2011) 103−108.

[204] G.M. Durán, A.M. Contento, A. Ríos, Talanta 131 (2015) 286−291.

[205] L. Trapiella-Alfonso, J.M. Costa-Fernandez, R. Pereiro, A. Sanz-Medel, Talanta 106 (2013) 243−248.

[206] M. Zhang, X. Cao, H. Li, F. Guan, J. Guo, F. Shen, Y. Luo, Ch Sun, L. Zhang, Food Chem. 135 (2012) 1894−1900.

[207] Z. Chen, X. Ren, X. Meng, D. Chen, Ch Yan, J. Ren, Y. Yuan, F. Tang, Biosens. Bioelectron. 28 (2011) 50−55.

[208] J. Godoy-Navajas, P. Aguilar Caballos, A. Gómez-Hens, Anal. Chim. Acta 701 (2011) 194−199.

[209] X. Gao, G. Tang, Z. Su, Biosens. Bioelectron. 36 (2012) 75−80.

[210] A. Sarreshtehdar, E. Mohammad, N. Danesh, P. Lavaee, M. Ramezani, K. Abnous, S.M. Taghdisi, Food Chem. 190 (2016) 115−121.

[211] J. Pollet, F. Delport, K.P.F. Janssen, D.T. Tran, J. Wouters, T. Verbiest, J. Lammertyn, Talanta 83 (2011) 1436−1441.

[212] L. Zamfir, I. Geana, S. Bourigua, L. Rotariu, C. Bala, A. Errachid, N. Jaffrezic-Renault, Sens. Actuator B 159 (2011) 178.

[213] V.A. Zamolo, G. Valenti, E. Venturelli, O. Chaloin, M. Marcaccio, S. Boscolo, V. Castagnola, S. Sosa, F. Berti, G. Fontanive, M. Poli, A. Tubaro, A. Bianco, F. Paolucci, M. Prato, ACS Nano 6 (2012) 7989−7997.

[214] L. He, T. Rodda, C.L. Haynes, T. Deschaines, T. Strother, F. Diez-Gonzalez, Anal. Chem. 83 (2011) 1510−1513.

[215] G. Naja, P. Bouvrette, J. Champagne, R. Brousseau, J.H.T. Luong, Appl. Biochem. Biotechnol. 162 (2010) 460−475.

[216] Z. Zhang, Q. Yu, H. Li, A. Mustapha, M. Lin, J. Food Sci. 80 (2015) N450−N458.

Chapter 8

Biocompatible Integration of Electronics Into Food Sensors

L.M. Dumitru,[1,*] M. Irimia-Vladu[2] and N.S. Sariciftci[1]

[1]*Johannes Kepler University, Linz, Austria;* [2]*Institute for Surface Technologies and Photonics, Weiz, Austria*

Corresponding author: E-mail: liviu_mihai.dumitru@jku.at

Chapter Outline

1. INTRODUCTION

We live in a society dominated by technology and electronic devices. We communicate or book our holidays using smart phones; read the news on our ultralight touch screen tablets or order gifts for our loved ones with a simple click of a mouse. Technology is part of our daily routine and has positively impacted all economical sectors. It has also changed the way we communicate and socialize.

But 100 years ago, all these would have seemed drawn from a science fiction novel. When Julius Edgar Lilienfeld predicted and described the field-effect transistor (FET) [1], more than 90 years ago, he did not foresee that this laboratory curiosity would radically change and reshape our society. Other 34 years had to pass till Dawon Kahng and Martin Atalla realized the first metal-oxide-semiconductor field-effect transistor (MOSFET) [2]. Few years after, the device was commercially available. Nowadays almost every electronic device has a MOSFET integrated. The most common semiconductor

Comprehensive Analytical Chemistry, Vol. 74. http://dx.doi.org/10.1016/bs.coac.2016.04.009

used is highly doped crystalline silicone, which is also the substrate of the circuit. Without a doubt the main beneficiary of this advancement is the IT industry. In the last decade, computers and laptops have become more affordable, lighter and high performing mainly due to the impressive development made in this field, allowing millions of transistors to be integrated and miniaturized on the same chip as predicted by Moore's Law [3].

Thin-film transistors (TFTs) are electronic devices based on the FET principle, where the semiconductor is deposited as a thin film on an insulating substrate such as glass, metallic oxides or plastic substrates. Traditional inorganic TFTs are based on amorphous silicon (a-Si:H) or polysilicon. The first organic transistor was fabricated and tested in the late 1980s, having an electrolyte as gating medium. This transistor was not a pure FET but rather an electrochemical transistor (ECT). In 1986, the first device showing clear transistor behaviour was made and tested successfully by Tsumura et al. [4]. The story behind organic thin-film transistors (OTFTs) starts in the late 1970s when Heeger, Shirakawa and MacDiarmid obtained a novel conductive polymer by doping polyacetylene with arsenic iodine, chlorine and bromine [5]. This was considered the starting point of a new era — conductive polymers. For this crucial discovery, in 2000, they were awarded with the Nobel Prize for Chemistry. Since then, organic semiconductor (OSC) materials have been intensively studied and upgraded (in terms of electrical performance) thanks to the enormous research done in the field of organic chemistry, leading to the synthesis of new solution-processed molecules such as oligo and polythiophenes [6]. These efforts were the origin for a completely new technological field, based on the properties of conductive organic materials. 'Plastics' that can exhibit electrical properties comparable to those of silicon-based semiconductors have opened up the possibility of making a large range of electronic devices that can be thin, flexible and eventually completely disposable such as organic light-emitting diodes (OLEDs), OTFTs and solar cells.

In addition, OSCs have proved to be among the best candidates to be interfaced to many biological molecules [7,8] due to many other properties such as synthesis, tunability, processing, softness and self-assembling ability. Their better mechanical compatibility with tissue than traditional 'hard' electronic materials and improved biocompatibility [9,10] with mechanically flexible substrates well suits the nonplanar form factors often required for medical implants [11].

2. BIOELECTRONICS

Without a doubt, electronics and microelectronics have completely reshaped research areas such as medicine and biology. Thanks to this 'symbiosis', between biology and electronics that gave birth to 'bioelectronics', scientists have now a whole new range of analytical tools currently used in medical diagnosis, novel drugs discovery, food safety, environmental monitoring and national security.

In biology, for example, characterizing biological systems at a molecular and cellular level is possible thanks to the existing available techniques: optical and fluorescence microscopy, DNA and protein analysis or Patch clamps [12].

But also sensors, or more specifically biosensors, can be designed when surface chemistry is used as a bridge between biological molecules and electronics transducing devices. The biomolecule in contact with its complementary molecule (analyte) will produce a specific effect measured by the transducer, which converts the information into a measurable effect, eg, electrical signal as depicted in Fig. 1.

The first enzymatic biosensor [13] was proposed by Clark and Lyons more than 50 years ago. Since then many types of biosensors have been developed interfaced with all kind of bio-receptors: antibody/antigen interactions, DNA and ARN interactions, enzymes and cellular interactions and even interactions due to synthetic bio-receptors. Depending on the physicochemical properties of the interest analyte, optical, electrochemical or mass-sensitive transducers can be used.

Antibody-based biosensors are the most specific and reliable sensing devices exploiting the unique lock and key fit property of antibodies, meaning that the specific geometrical configuration of the antibody (unique key) enables it to open a specific lock (antigene).

By using radioactive-marked antibodies (radioimmunoassay), scientists have been able to detect specific analytes where other analytical techniques have been failed. Pharmacology, clinical chemistry, forensics, environmental monitoring and food safety are just few of the fields where this analytical method can be applied to. However, the use of highly unstable, potential dangerous radioactive materials plus the high overall costs are limiting factors for using this method. Research groups all around the world have dedicated

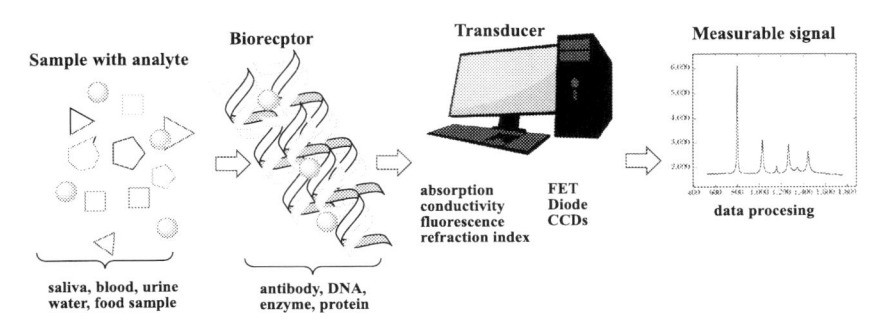

FIGURE 1 Schematic representation of a biosensor. *FET*, Field-effect transistor.

their efforts to develop new, simpler and practical immunochemical instrumentation techniques comparable to the high sensitivity and selectivity of radioimmunoassay. Advances in device miniaturization, biotechnology and nanotechnology have allowed fabrication of antibody-based biosensors capable of detecting analytes of interest inside a single cell [14]. Such nanoscale devices were used for optical detection of benzo[a]pyrene metabolites when a specific monoclonal antibody was used as a bio-receptor. This method permits cellular probing for both analyte detection [15] and cell signal monitoring.

Enzymes also can be used as bio-receptors and have both specific binding and catalytic capabilities. In this case the bio-detection is amplified by a reaction catalysed by the biocatalyst. The catalytic activity of enzymes allows reaching lower limits of detection as compared to the common binding techniques. However, this extra boost in method sensitivity is strictly related to the integrity of the native protein conformation. The enzyme will become catalytic inactive if it is denatured or broken down into subunits. Both the enzymes and the bioaffinity biosensors were originally found and used mainly in labs or hospitals. But now, thanks to the advances made in the field of easy-to-use bioelectronics, portable devices for non-specialists (eg, glucose meters or pregnancy tests) are available in pharmacies or even in supermarkets.

Another type of biosensors uses nucleic acids as bio-receptors allowing the detection of DNA or RNA in human samples. A labeled complementary sequence is added to the samples after the unwinding of the DNA double helix into single strands. If the sequence of bases of the target analyte (eg, virus or bacteria) is known, the labeled probe will hybridize to its complementary sequence, giving rise to a fluorescent signal that is then transmitted to a signal generator.

Microorganisms can also be employed as bio-receptors interfaced with electronic devices. Bacteria, fungi or even viruses are extremely sensitive and can be successfully used as indictors of toxicity for various chemical substances. These biosensors usually offer extremely low detection limits due to signal amplification. This is mainly due to the complex biochemical structure of the microorganism and their catalytic properties.

But also man-made or artificial receptors can be used to fabricate biosensors, thanks to the advances made in the field of genetic engineering and molecular imprinting. By using these techniques, scientists are able to alter clones and genes of an organism to produce new biological substances such as membranes, proteins or hormones. Molecular polymer stamps can be designed when the analyte molecules are mixed with monomers and cross-linkers. After polymerization, the analyte molecules are removed from the polymer network. The final polymer structure will have molecular binding sites complementary to the interest analyte. These molecular imprinted polymers can withstand harsh environments such as temperature or chemicals that would normally denature a protein, thus making them extremely attractive. However, these biomimetic receptors exhibit lower selectivity as compared with their natural receptors.

The next section will discuss how sensors and biosensors can be used in quality control for food safety applications.

3. SENSORS AND BIOSENSORS FOR FOOD INDUSTRY

Within the last decades the food industry has experienced an impressive growth both in terms of the amount of foodstuff produced and sold and in terms of product quality. Food products have been created to meet our daily lifestyles and new eating habits. The perishable nature of food products along with a stiff competition has forced the food industry to rapidly adapt and comply with the increasingly stringent worldwide food regulations. In the same time the food companies have realized the financial benefits deriving from a better quality control and quality assurance concerning foodstuff production. Safe, high quality and easy traceable products translate into lower costs, higher market share and customer satisfaction.

New lines for intelligent manufacturing, packaging and food quality control have been developed to increase the shelf-life while maintaining the product quality. New equipments and technologies for processing and packaging technologies incorporating sensing platforms, identification tags such radio frequency identification, automated sampling and imaging systems were specially designed for the food industry.

Released gasses, flavored compounds, volatile molecules or chemical-specific species associated with food freshness or ripening stages can be qualitatively and quantitatively detected using many chromatographic techniques. Near-infrared spectroscopy such as FTIR (Fourier transform spectroscopy) is extremely attractive for surface analysis allowing simultaneous identification of different chemical compounds.

Even though these traditional analytical techniques are extremely robust, reliable and offer good reproducibility, the main drawback is that they cannot be integrated in the production or packaging flow, which in many cases is critical. Other handicaps include expensive instrumentation, time-consuming analysis or well-trained personnel. For these reasons, (bio)sensing online platforms can become an extremely attractive alternative to traditional techniques. If properly designed, these platforms could be fully integrated either in the manufacturing flow or directly in/onto the foodstuff package providing a good traceability of the product as well as guaranteeing the product quality.

Various types of sensors (physical, chemical, biological) have been developed that allow food quality monitoring. Food freshness [16], pathogens detection [17], carbon dioxide [18] and oxygen [19] concentration, pH [20] and temperature [21] can be monitored using these devices.

Smart or intelligent packaging is the new trend in food industry that allows incorporation of such sensors into the food packaging technology. This technology is able to prolong shelf time and protect food from environmental contaminants and allows online quality control information and traceability of the product. In other words, this packing technology can sense the environment inside or outside the package and provide information to the producer, consumer or retailer about product parameters. Generally speaking, this term can also refer to product identity, authenticity and traceability.

Smart labels and sticker [22] are already used to provide consumers with simple and reliable information regarding product ingredients and how the product should be used. But smart labels include also time-temperature indicators [23] that can reveal changes in the colour or shape of the product.

For example, Omenetto et al. developed different silk-based solutions that can be mixed with enzymes, antibiotics, growth factors or antibodies. These functional inks [24] were then used and printed on different substrates. As a proof-of-principles, the authors printed a pattern on a surgical glove, Fig. 2, that changes colour when exposed to bacteria. This technique could be used for large-scale fabrication of smart food packaging foils patterned with functional inks, for different contaminant's detection.

Many companies, universities, and research institutes are directly involved in this field by developing biosensor devices or novel biosensing techniques that are/could be used in smart food packaging.

Enzymatic-based biosensors for pesticide [25] detection or label-free immunochemical sensors for antibiotics screening in milk [26] were developed in Higson's group. An electrochemical sensor designed for the detection of polyphenols in olive oil [27] or a disposable electrochemical DNA array for hazelnut allergens detection [28] were also successfully developed in

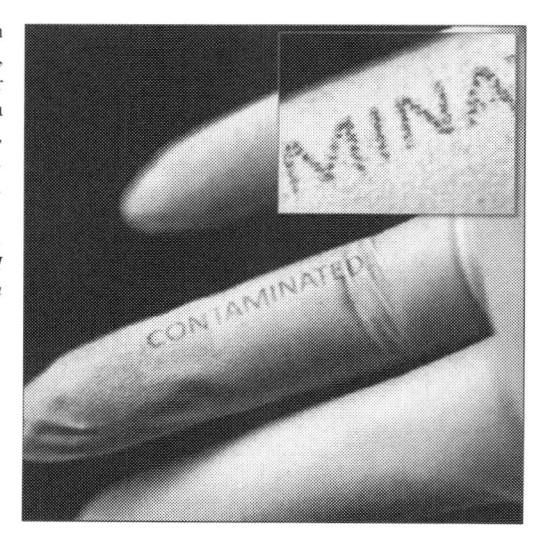

FIGURE 2 Surgical gloves with printed patterns turning to red, showing 'CONTAMINATED', after exposure to *Escherichia coli* at a concentration of $\approx 10^4$ CFU/mL, indicating contamination. *From H. Tao, B. Marelli, M. Yang, B. An, M.S. Onses, J.A. Rogers, et al. Adv. Mater. 27 (2015) 4273–4279. Copyright Wiley-VCH Verlag GmbH & Co. KGaA. Reproduced with permission.*

Mascini's group. Electrochemical biosensors for the detection of mycotoxins such as Aflatoxin M1 in milk [29] or Ocharatoxin A [30] were proposed by Marty's group. The latter group is also involved in developing enzyme-based sensors for the detection of organophosphorus pesticides [31].

Other biosensors development by universities for food safety analysis includes antibiotics [26,32] detection (in food and milk), bacteria (eg, *Escherichia coli* or *Salmonella*) [33,34], herbicides and pesticides (in milk or fruit juice) [35,36], ethanol [37], aldehydes [38] and catechol [39] in alcoholic beverages, amines in fish [40], fruits and vegetables [41], oxalates [42] and sulphites [43].

Unfortunately, not all of the biosensor prototypes developed by academia are commercially available. R&D companies such as Neogen Corporation (www.neogen.com) or Roche Diagnostics AG (www.roche-diagnostics.com) provide clients in the food industry, with rapid-screening solution for food quality control applications.

The Food Safety Division of Neogen offers kits and ready-to-use screening tools for foodborne bacteria, spoilage organisms, mycotoxins, food allergens, genetic modifications and drug residues. The company also commercializes an extremely simple dipstick test that detects dairy antibiotics in the beta-lactam group, requiring only minimal training and equipment. The Listeria Genus Detection Kit from Roche detects rapidly Listeria DNA that is isolated from cultures inoculated with food sample material. The presence of the pathogenic species Listeria monocytogenes in positive samples might be further confirmed by using other detection kits. This detection kit is rapid, highly sensitive with specific response and also minimizes the risk of sample contamination and false-positive as well as false-negative results.

Other commercially available biosensors for food industry include glucose, ethanol and methanol sensors (Biometra Biomedizinische Analytik GmbH, Germany), glucose and lactate (Solea-Tacussel, France), fish freshness (Oriental Electric, Japan) and bacteria (Sweedish BIACORE AB/Sweden, Malthus Instruments/United Kingdom, Biosensori SpA/Italy, Biotrace/United Kingdom).

Ideally, these sensors should be fully disposable and have a minimum environmental impact. However, finding biocompatible or biodegradable materials for interfacing electronics with enzymes, antibodies or proteins is still a big challenge. For this reason, in the next section we will present a highlight of the state-of-the-art reports for organic electronics applications involving biocompatible, biodegradable, natural-like and easy tunable materials.

4. MATERIALS

Biological materials or bio-inspired materials are usually regarded as niche materials used in specific areas, such as medicine, medical engineering, biotechnology, or for pharmaceutical applications. Biological materials are present in living organisms or produced by them and do not contain any synthetic substances and are biodegradable. Biomaterial or a bio-inspired material is a mixture of synthetic and natural substances that can be easily interfaced with living cells, tissues or organs. If this interfacing does not produce any immune reaction, the material is regarded as biocompatible.

To be biocompatible, biological materials must have low or no toxicity or cytotoxicity (cell killing ability).

4.1 Substrates for Bioelectronics

Edible polymers can be used for food wrapping [44], packaging [45] or as physical support for biosensing devices [46]. Human or animals can safely consume these materials without any harmful effects to the health. They degrade more rapidly than their synthetic counterparts, making them viable solutions to the huge amount of non-biodegradable plastic wastes generated currently by the food packaging industry.

Edible polymers are usually used as coating layers to protect food products against the loss of nutrients. But can be used also as implantable bioresorbable scaffolds for bone surgery [47] or polymer—anticancer drug conjugates [48].

There are four categories of edible polymers:

A. Hydrocolloids: cellulose derivatives, collagen, chitosan, gelatine, starch, soy protein, lentil protein, whey protein, peanut protein, Mung Bean;
B. Polypeptide: zein, casein and whey proteins, gluten and soy proteins;
C. Lipids: carnauba wax, beeswax, candelilla wax and shellac;
D. Synthetic and composite edible polymers: lipids and hydroxypropyl methylcellulose, methylcellulose and lipid, methylcellulose and fatty acids, corn zein, methylcellulose and fatty acid, gelatin and fatty acid, soy protein isolates and gelatin and soy protein isolate and polylactic acid, polyvinyl acetate, polymethyl methacrylate (PMMA).

But also paper [49—52], leather, silk [24,46] or other polymers [54] can be used as biodegradable substrates for bioelectronics [53].

Paper is a versatile material and produced by mixing cellulose fibres together with additives, clays and dies. Flexible paper sheets are obtained after pressing and drying the cellulose-based mixture. Paper is used for writing, printing, packaging, cleaning and also as substrate for fabrication of electronic circuits (Fig. 3) due to its low cost, versatility and biodegradable properties. Recent reports have demonstrated that even solar cell devices can be integrated on paper substrates [52]. By incorporating state-of-the-art polymers and electrode materials, the authors reported 4% efficiency for these paper-based solar cells.

Pentacene-based OTFTs and ring oscillators were developed by Eder and his group, on commercially available paper [49]. The transistors exhibit carrier

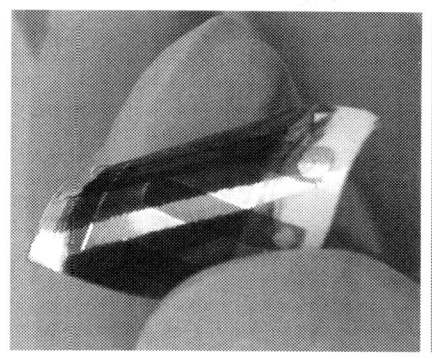 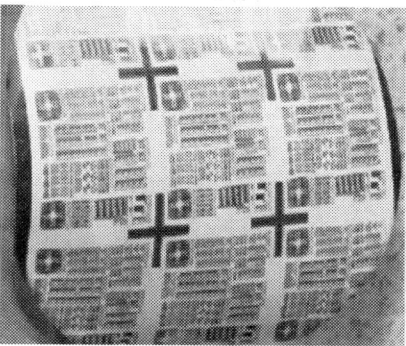

FIGURE 3 (Left) 4% efficient organic solar cells fabricated on paper substrates with a zinc coating acting as the back contact and evaporated MoO3/Ag/MoO3 semitransparent top electrodes. *(From L. Leonat, M.S. White, E.D. Głowacki, M.C. Scharber, T. Zillger, J. Rühling, et al. J. Phys. Chem. C. 118 (2014) 16813–16817. Copyright American Chemical Society Reproduced with permission.)* (Right) Organic field-effect transistors (OFETs) and integrated circuits on paper substrates. *(From F. Eder, H. Klauk, M. Halik, U. Zschieschang, G. Schmid, C. Dehm, Appl. Phys. Lett. 84 (2004) 2673. Copyright American Institute of Physics. Reproduced with permission. Colour image courtesy of Hagen Klauk.)*

mobility of 0.3 cm^2/V and a current on/off ratio of 10^6, similar to the results reported for pentacene TFTs on PEN films.

Polysaccharide circuit boards (PCBs) were developed and characterized by using laminate composite nanocellulose thin films on water-soluble sugar substrates as a bilayer [55]. Conductive pads and tracks were afterwards defined onto the PCB by means of conventional evaporation and printing techniques. A simple light-emitted circuit (light-emitting diode, LED) was developed on a PCB, as showed in Fig. 4C and transferred on a petal flower. Even though the luminescence was lower after the transfer, no cracking or mechanical wear was observed during operation.

Organic field-effect transistors (OFETs) were successfully designed and tested by Irimia-Vladu et al. using hard gelatine [10] as a biocompatible and biodegradable substrate (Fig. 4, right). Gelatine is mass-produced from collagen by processing animal skin, tissues and bones. The food and pharmaceutical industries use important quantities of gelatine for the production of jams, gummy bears, as drug excipient or hard gelatine capsules for drugs.

Edible food sensors [46] based on wireless passive antenna technology were developed and integrated on silk substrates in Omenetto's group. The sensors, operated at different frequency regions, were conformably attached to fruits or cheese, and the spoilage process was monitored. As a proof-of-principle, the resonant frequency response of these sensors attached to foodstuff was measured, and ripening of bananas and spoilage of cheese could be monitored, as depicted in Fig. 5.

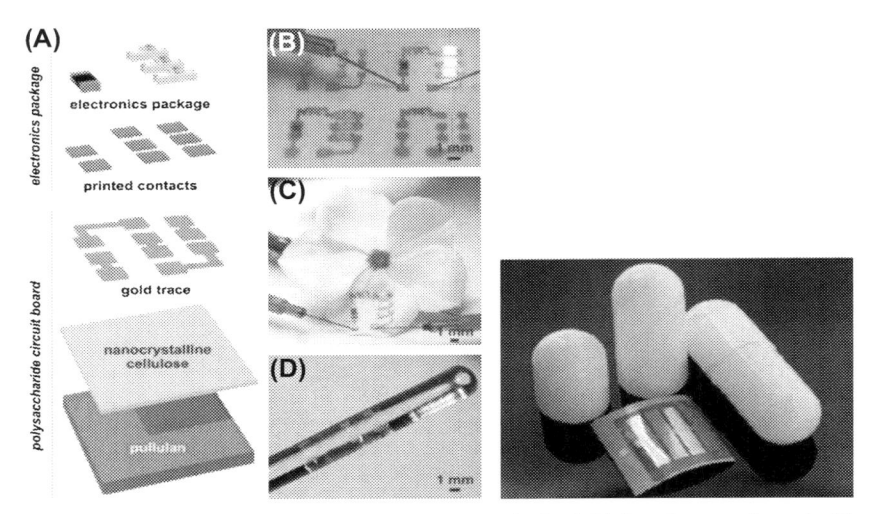

FIGURE 4 (Left) (A) Schematic of a typical electronic decal fabricated on a polysaccharide circuit board, in which the dissolution of the water-soluble pullulan substrate results in a conformal electronic device with nanofibril cellulose substrate. (B) Operation of the PCB-LED circuits on a full-sheet. (C) The PCB-LED circuit was excised and transferred to a biological surface. (D) A polysaccharide circuit board laminated around a glass rod to illustrate flexibility and transparency. *(From M.A. Daniele, A.J. Knight, S.A. Roberts, K. Radom, J.S. Erickson, Adv. Mater. (2015) 1−7. Copyright Wiley-VCH Verlag GmbH & Co. KGaA.)* (Right) Organic field-effect transistors fabricated on edible hard gelatin capsule. *(From M. Irimia-Vladu, P.A. Troshin, M. Reisinger, L. Shmygleva, Y. Kanbur, G. Schwabegger, et al. Adv. Funct. Mater. 20 (2010) 4069−4076. Copyright Wiley-VCH Verlag GmbH & Co. KGaA. Reproduced with permission.)*

These are just few examples of how unconventional materials can be successfully used for developing new sustainable sensors and devices.

4.2 Semiconducting Materials

The field of 'solution-processed semiconductors' is huge, and hundreds of companies and universities are developing every day new organic and inorganic molecules with semiconducting properties. These materials are not fully compatible with bio-interfaces or with bio-molecules, have a high cost and degrade rapidly when exposed to air. There are, however, some exceptions since naturally or naturally inspired compounds can be used as semiconductors for bioelectronics fabrication.

OFETs, OLEDs or organic solar cells can be fabricated using natural semiconductors, used since ancient times, such as indigo, a pigment extracted from plants [56,57] or tyrian purple which was originally extracted from sea snails [58,59]. These two materials as well as other 'nature-inspired' indigoids exhibit good ambipolar properties. Molecular structure of indigo, epindolidione and quinacridone, as well as the colour of these compounds in solution, is presented in Fig. 6.

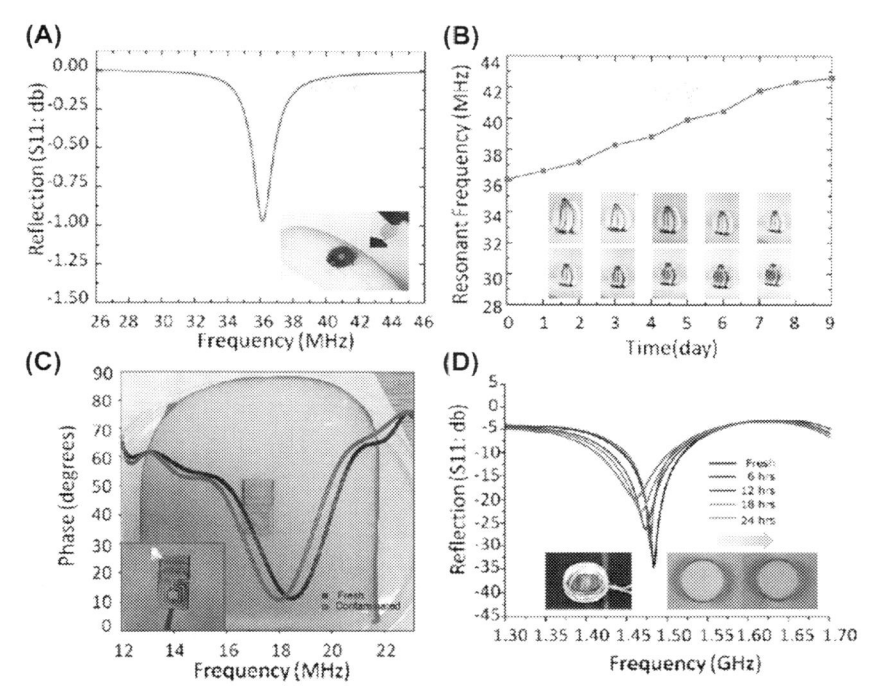

FIGURE 5 (A) Reflection spectra of a silk radio frequency identification-like antenna attached to a banana. (B) Measured time-dependent resonant frequencies of the silk antenna measuring a ripping banana over a 9-day period. (C) Measured frequency-dependent impedance phase angle of a silk sensor applied to a slice of cheese (*blue curve* (dark grey in print versions)) to detect bacterial contamination (*red curve* (dark grey in print versions)). (D) Measured frequency responses of a silk sensor for a milk spoilage detection. *From H. Tao, M.A. Brenckle, M. Yang, J. Zhang, M. Liu, S.M. Siebert, et al. Adv. Mater. 24 (2012) 1067–1072. Copyright Wiley-VCH Verlag GmbH & Co. KGaA. Reproduced with permission.*

Good electronic performances were reported for epindolidone-based low-voltage FETs (Fig. 7). These devices exhibited hole mobilities of $0.05-1$ cm^2 V^{-1} s^{-1} and electron mobilities up to 0.1 cm^2 V^{-1} s^{-1} [60]. Also, these materials are extremely stable in a wide range of electrolyte solutions and show no observable degradation after hundreds of cycles of operation. These findings demonstrate the potential of using such highly stable pigments as versatile organic semiconducting materials. But even oligo-olefins such as carotenoids can be used as natural semiconductors, although the mobilities values $(1 \times 10^{-4}$ V^{-1} s$^{-1})$ obtained for these materials are much lower compared to indigoids or epindolidiones [60].

4.3 Dielectric Materials

Nature provides also an impressive number of biodegradable and biocompatible materials with dielectric or insulating properties. DNA, for example, is

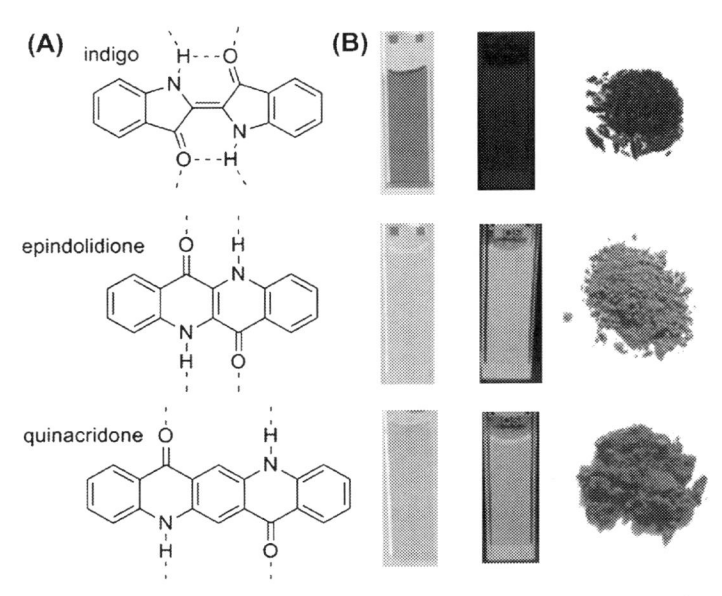

FIGURE 6 (A) Molecular structures of indigo, epindolidione and quinacridone. (B) Solutions of each material in dimethyl sulphoxide (DMSO), with photoluminescence excited at 365 nm. *E.D. Głowacki, G. Romanazzi, C. Yumusak, H. Coskun, U. Monkowius, G. Voss, et al. Adv. Funct. Mater. (2014) 1−12. Copyright Wiley-VCH Verlag GmbH & Co. KGaA. Reproduced with permission.*

mainly known for its encoding properties, but this versatile molecule was also used and successfully employed as functional layer in organic electronics [7,61−63].

Electron blocking layers for OLEDs were formed by spin-casing-modified DNA polar solutions. The luminance and luminous performance of these DNA-based LEDs were 10 times higher than those of LEDs using common polymers [63].

Gate dielectrics based on DNA were also successfully used for fabricating OFETs [7]. However, high hysteresis is a major drawback of these devices, most probably due to mobile ionic contaminants present in the dielectric layer. This problem can be solved by incorporating additional blocking layers to the FET structure (such as metal oxides) or by using cross-linkers. Irimia-Vladu et al. demonstrated that also adenine, guanine, thymine and cytosine (DNA sequence bases) could be used as gate dielectric for OFETs [9]. These vacuum-processed nucleobase-based OFETs exhibited good electrical performances and showed minimal or no hysteresis in the transfer characteristics.

Investigations revealed that comparable dielectric constants [64], to that of polyvinyl alcohol (PVA), can be achieved when using solution-processed sucrose, glucose and lactose-based dielectric layers.

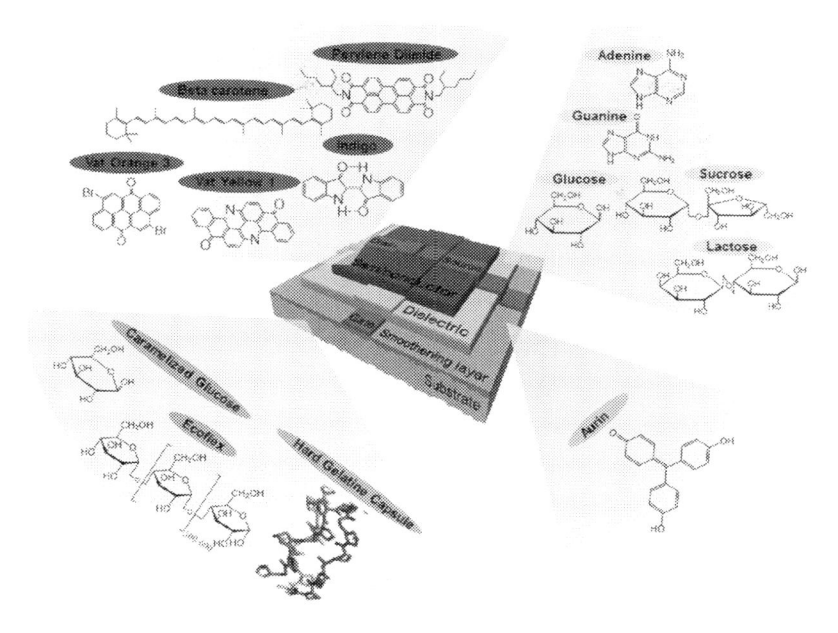

FIGURE 7 An organic field-effect transistor constructed using only natural materials or materials inspired by nature. Hard gelatine or starch-based foils such as Ecofex were used as substrates; while natural aurin was used as a smoothing layer. Natural 'sweet' dielectrics such as glucose, fructose and sucrose can be used as insulating materials. Both natural semiconductor such as indigo and β-carotene can be used as semiconductors inspired by nature. *From M. Irimia-Vladu, P.A. Troshin, M. Reisinger, L. Shmygleva, Y. Kanbur, G. Schwabegger, et al. Adv. Funct. Mater. 20 (2010) 4069–4076. Copyright Wiley-VCH Verlag GmbH & Co. KGaA. Reproduced with permission.*

4.4 Field-Effect Transistors for Food Control Applications

From flexible displays [65] to disposable sensors [66], cyborg roses [67], TFTs and FETs have attracted a lot of attention in the recent decades, mainly because of the diverse areas in which these devices find application.

This last section aims to provide readers with an overview of some high-profile scientific articles and findings reporting sensing applications in which TFTs and FETs were used as transducers.

TFTs are three-terminal devices consisting of a thin semiconductor layer, a gating medium (insulating/dielectric layer) and three conductive terminals, called source, drain and gate electrodes. Source and drain are in direct contact with the semiconductor layer (organic or inorganic) and separated by a short distance, thus forming a semiconducting channel. The third contact, the gate, is separated from the semiconductor by a gate medium. Upon the application of a source-drain voltage (V_{DS}), a current (I_{DS}) may flow in the semiconducting channel. The magnitude of the I_{DS} is modulated by a voltage applied to the gate electrode (V_{GS}), which generates an electric field perpendicular to the channel and thus may attract mobile charges carriers into it. In addition to the manifold applications of OFETs in electronics, these devices have attracted interest as transducers for sensing applications. Thanks to gating, OFETs may display much higher currents (I_{DS}) than simple (ungated) organic chemiresistors, where low current may lead to poor signal/noise ratio. Also, multiple independent parameters can be extracted from the single characteristics of a single transistor, thus providing rich sensor information. Interest in OFET sensors has multiplied through the recent introduction of electrolytes, instead of dielectrics, as gate media. These were initially introduced for their high capacitance, leading to low operational voltages. However, since electrolytes are ubiquitous in life- and environmental sciences, the 'EGOFET' (electrolyte-gated organic field-effect transistor) opens up manifold possibilities for novel bio- and environmental sensors. In this section we will focus only on the OFETs and EGOFETs as transducers for gas and liquid sensing applications.

In the case of OFETs, charges can be induced at the insulator—semiconductor interface, when a gate voltage is applied. Electrons or negative charges will be induced in the semiconductor layer upon applying a positive gate voltage, while a negative voltage will induce holes or positive charge holes. These charges (holes or electrons) will increase the conductivity of the semiconductor, and a conducting channel will be formed between the source and drain electrodes. A positively charged channel is called p-channel, and a negatively charged channel is called n-channel. By varying the gate voltage, the channel conductance can be modulated.

EGOFET device mechanism is based on the formation of nanometric scale electrical double layers (EDLs) at the gate—electrolyte interface and electrolyte—semiconductor interface, respectively. The high capacitance of electrolytes ($\approx \mu F/cm^2$ range) allows EGOFETs operation at a sub-volt regime.

The FETs are excellent candidates for sensing applications mainly because these devices are capable of current amplification, and changes in the transport properties can be measured upon exposure to a gas, for example. These devices can be used as multi-parameter sensors by monitoring simultaneously the variation of mobility (μ), current *on/off* ratio, threshold voltage (V_T) and bulk conductivity in the semiconductor layer. Torsi et al. proposed this model for the first time in the year 2000, for an OFET sensor exposed to a gaseous analyte [68].

The same group demonstrated that high sensitivity and selectivity can be achieved when using an alkoxyphenilene—thiophene bilayer semiconductor for making an OFET-based sensor, Fig. 8 [69]. The proposed gas sensor was able of chiral discrimination between (S)-β-citronellol and (R)-β-citronellol as well as to 'sense' the racemic mixture.

Four years after these findings, Angione et al. proposed an OFET gas sensor incorporating a functional bio-receptor, the so-called FBI configuration

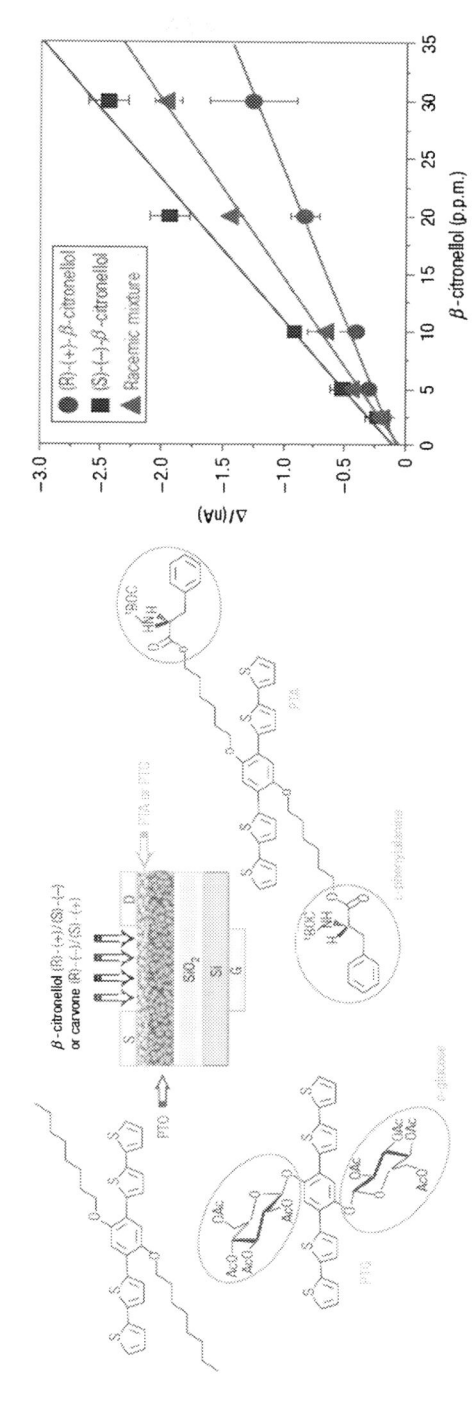

FIGURE 8 (Left) Structure of a bilayer organic field-effect transistor chiral sensor. (Right) Calibration curves of the chiral bilayer organic thin-film transistors exposed to (S)-(−)-β-citronellol, (R)-(+)-β-citronellol and to the racemic mixture. *Adapted from L. Torsi, G.M. Farinola, F. Marinella, F. Marinelli, M.C. Tanese, O.H. Omar, L. Valli, et al. Nat. Mater. 7 (2008) 412−417. Copyrights Nature Publishing Group. Reproduced with permission.*

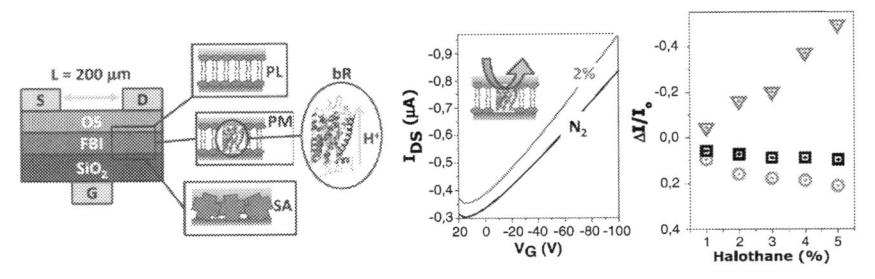

FIGURE 9 (Left) Structure of an FBI-OFET bearing a phospholipid bilayer (PL), *purple membrane* (dark grey in print versions) (PM) and streptavidin (SA). (Centre) FBI-PM OFET response to 2% halothane vapours and (Right) calibration curve for the FBI-PM OFET when exposed to different concentrations of halothane. *FBI*, Functional BioInterlayer; *OFET*, organic field-effect transistor. *Adapted from M.D. Angione, S. Cotrone, M. Magliulo, A. Mallardi, D. Altamura, C. Giannini, et al. Proc. Natl. Acad. Sci. 109 (2012) 6429–6434. Reproduced with permission.*

(Functional BioInterlayer, Fig. 9) [70]. As a proof-of-principle the OFET was interfaced with a purple membrane bio-receptor, and its selective response to an anaesthetic, halothane vapours, was investigated and reported.

The sensing mechanism behind OFET gas sensors can be pictured as a disruption at the semiconductor/dielectric interface when the semiconductor is exposed to the gas analyte. This will cause changes in the FET electrical performance due to trapping/detrapping of charges and increase/decrease of potential barrier between grains. The gas molecules can be absorbed on, be trapped between grain boundaries or even permeate the semiconductor layer.

A comparative study [71] of gas sensing behaviour (Figs 10 and 11) for P3HT- and PBTTT-based OFETs was published in 2014. The authors

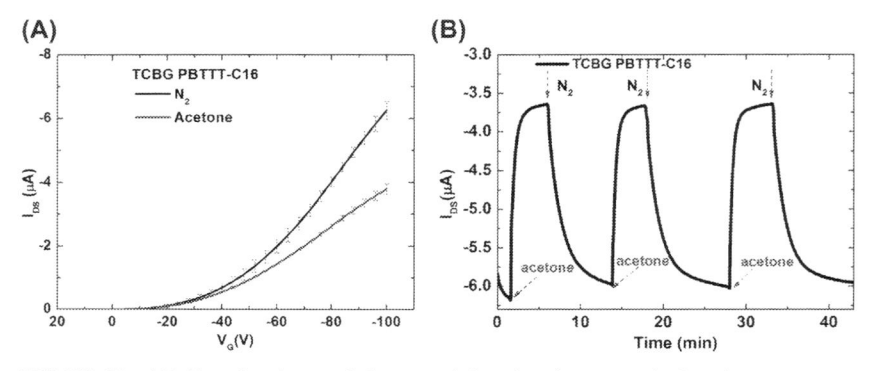

FIGURE 10 (A) Transfer characteristics recorded under nitrogen and after the top-contacts bottom-gate (TCBG) PBTTT-C16 sensor was exposed to acetone. (B) Transient drain current (I_{DS}) measurements when the sensor is exposed three times to acetone vapours (as $= 0.5$ at $-6.6°C$). *From K. Manoli, L.M. Dumitru, M.Y. Mulla, M. Magliulo, C. Di Franco, M.V. Santacroce, et al. Sensors (Basel) 14 (2014) 16869–16880. Reproduced with permission.*

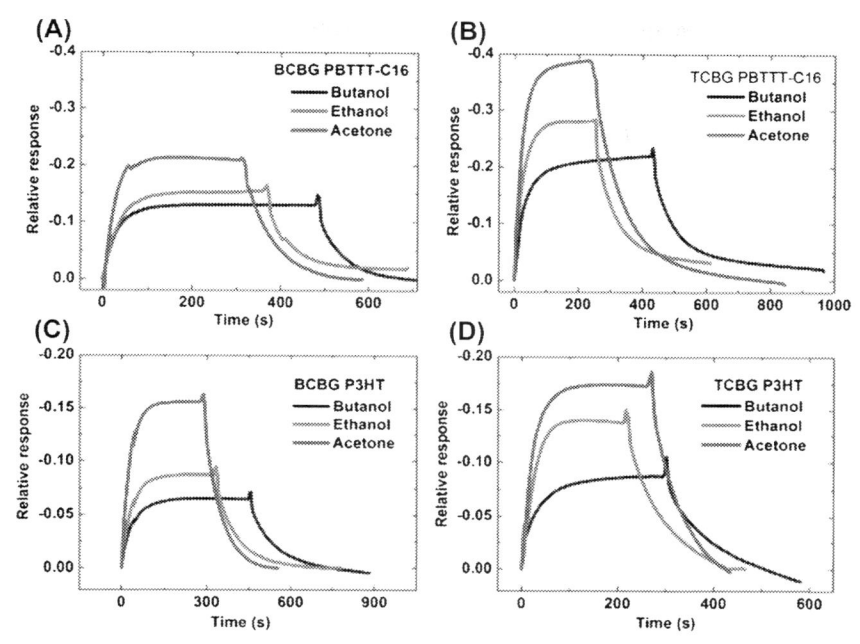

FIGURE 11 Transient drain current (I_{DS}) measurements of the OTFT-based sensors exposed to the three analytes at the same activity (as = 0.5 at −6.6°C). (A) Top-contacts bottom-gate (TCBG) PBTTT-C16. (B) Bottom-contacts bottom-gate (BCBG) PBTTT-C16. (C) TCBG P3HT. (D) BCBG P3HT. *From K. Manoli, L.M. Dumitru, M.Y. Mulla, M. Magliulo, C. Di Franco, M.V. Santacroce, et al. Sensors (Basel). 14 (2014) 16869−16880. Reproduced with permission.*

investigated the response of two OFET sensor architectures [top-contacts bottom-gate (TCBG) and bottom-contacts bottom-gate (BCBG)] when exposed to butanol, ethanol and acetone. A decrease of the drain current was observed for all sensors when exposed to polar analytes.

The results showed a higher drop in the current with the increase of the dipole moment for each analyte. Also the morphological structure of the two OSCs used in this study contributed to the different sensing response of the devices when exposed to the same analyte, as reported in Fig. 11.

The PBTTT-based sensors exhibited higher responses for the same analyte as compared to the P3HT-based sensors. Remarkably an increased sensitivity in the response for the TCBG sensor configuration was observed, as compared to the BCBG sensor. The authors ascribe this to bulk conductivity and a possible adsorption of analyte vapours on the gold electrodes.

Like in the example discussed above, analyte nature (dipole moment, electron affinity and molecular size) and the morphology of the semiconductor layer can play a crucial role in the sensor's response. By introducing functional groups to the side chain of the semiconductor, an increase in sensitivity can be achieved [72]. It is very difficult to achieve a reasonably good gas sensing discrimination, especially when it comes to complex gas samples. This inconveniency can be somehow overcome by using bilayers in which one of the semiconductors is more sensitive to the analyte than to another one, or by using sensor arrays.

In some cases, the analyte can produce a positive or negative shift in the threshold voltage V_T. This behaviour was observed when OFET sensors were used to detect ammonia or nitrogen oxides. The threshold voltage can be very sensitive to any extra charge induced to or 'taken' from the system. Oxidant gas molecule can dope p-type OSCs, shifting the threshold to more positive values and increasing the drain-source current. Reducing agents (eg, NH_3) cause an opposite response, shifting the threshold to more negative values, and a decrease in the drain-source current can be observed.

All these examples demonstrated that these OFETs are excellent multi-parameter gas-sensing platforms, reaching limits of detection in the range of parts-per-million due to signal amplification. These devices are cheap, give a fast response and are compatible with mass production techniques and can be easily integrated into smart packaging. However, even though remarkable new materials are being synthetized every day, selectivity of these devices remains an open issue.

OFET-based sensors can be produced also by interfacing the OSC with pure water [73], an electrolyte solution [74] or a solid electrolyte layer [75]. [76] In this case the devices are referred to as EGOFETs and are suitable candidates for liquid-sensing applications [77]. Sub-volt operation is possible due to the high capacitance of the two electric double layers formed when operating the device. Such a device is depicted in Fig. 12, along with the schematic representation of the operating mechanism.

The gating is achieved through the formation of EDL or Debye-Helmholtz layer at the gate—electrolyte and electrolyte—OSC interfaces. Both of these two interfaces can be used to integrate bio-receptors. Biscarini's group was one of the pioneers to prove that by functionalizing the top gold electrode of an EGOFET with boronic acid, picomolar concentrations of dopamine can be detected in a phosphate buffer solution [78].

A remarkably sensitive EGOFET, showed in Fig. 13, able to differentiate odorant chiral molecules was proposed by Mulla et al. in 2015. A mutant odorant-binding protein (OBP) was immobilized on the top gold electrode by means of a self-assembled monolayer (SAM) technique [79].

The OBP-functionalized EGOFET was able to detect and discriminate (S)- and (R)-carvone enantiomers down to picomolar concentration. A limit of detection of 50 pM was estimated for this detection method, which is almost six order of magnitude better than the limit reported in previous studies where carvone was detected using FET-based sensors [69] (Fig. 14).

Also the OSC—electrolyte interface can be used for sensing applications. Biotin can be immobilized on a phospholipid bilayer (PL), anchored on a

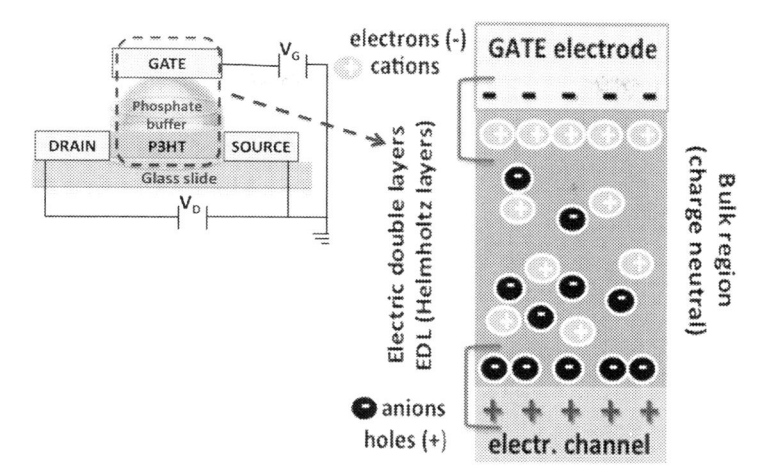

FIGURE 12 (Left) Schematic cross-section of an electrolyte-gated organic field-effect transistor gated with a droplet of phosphate buffer. (Right) Operation mechanism and formation of electric double layer at the two interfaces, when the gate is polarized. *Adapted from S.H. Kim, K. Hong, W. Xie, K.H. Lee, S. Zhang, T.P. Lodge, et al. Adv. Mater. 25 (2013) 1822−1846. Copyright Wiley-VCH Verlag GmbH & Co. KGaA. Reproduced with permission.*

P3HT layer. The PL is used both for immobilization of biotin, but it can also act as a barrier preventing chemical doping of the OSC [74]. This phenomenon can sometimes occur during EGOFET operation as small free ions, present in the electrolyte, might penetrate into the semiconductor's bulk and dope it. The proposed sensor was afterwards used for streptavidin detection and an increase in the drain-source current could be observed. It is speculated that this increase

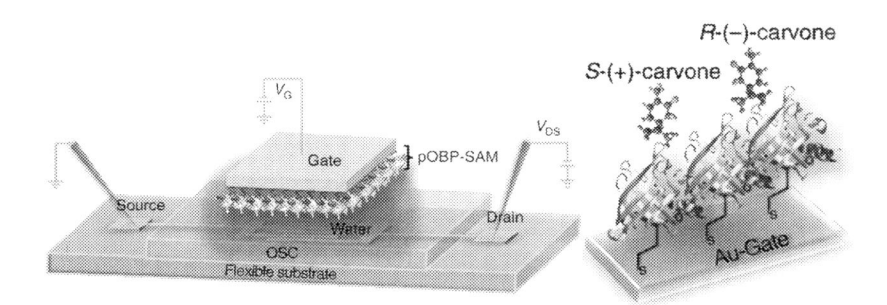

FIGURE 13 (Left) Schematic representation of a water-gated organic field-effect transistor for chiral discrimination. (Right) Top gold electrode functionalized with the odorant-binding protein (OBP). *OSC*, Organic semiconductor; *SAM*, self-assembled monolayer. *Adapted from M.Y. Mulla, E. Tuccori, M. Magliulo, G. Lattanzi, G. Palazzo, K. Persaud, et al. Nat. Commun. 6 (2015) 6010. Copyrights Macmillan Publishers Limited − Nature Communications. Reproduced with permission.*

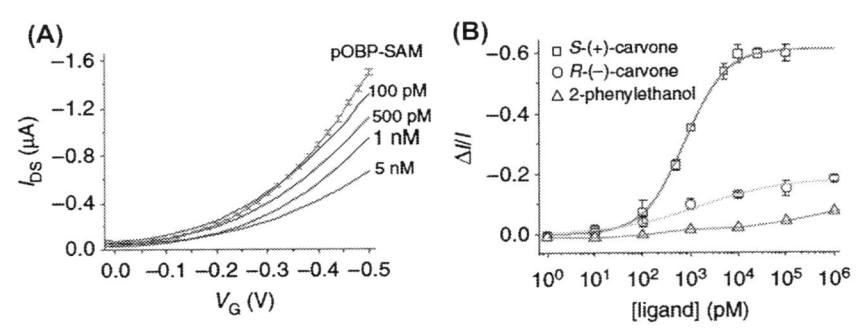

FIGURE 14 (A) Transfer characteristics are for a pristine pOBP-SAM gate (*red curve* (light grey in print versions)) and for a gate exposed to concentrations of (S)-(β)-carvone in the 100pM−5 nM range (*black*). (B) Transfer characteristics normalized source-drain current changes ($\Delta I/I_0$) when the devices were exposed to the (R)-(β)- and (S)-(β)-carvone well as the 2-phenylethanol ligands. *SAM*, Self-assembled monolayer; *OBP*, odorant-binding protein. *Adapted from M.Y. Mulla, E. Tuccori, M. Magliulo, G. Lattanzi, G. Palazzo, K. Persaud, et al. Nat. Commun. 6 (2015) 6010. Copyrights Macmillan Publishers Limited − Nature Communications. Reproduced with permission.*

in current is due to the extra negatively charges brought by the streptavidin (negatively charged at that pH), after binding to biotin.

Other strategies for EGOFET biosensors include the use of cross-linked alginic acid capsules (Fig. 15) as ion-permeable polymeric gating electrolyte [80]. This versatile, biodegradable and biocompatible polymer was used both as electrolyte medium and as matrix for incorporating bio-receptors.

The strategy of using such a material is truly revolutionary, and this polymeric membrane can not only prevent liquid from evaporating during device operation but also serve as reactor for bio-enzymatic reactions. To prove this, the authors immobilized streptavidin inside the capsule, and the capsule was afterwards incubated with a solution containing fluorescence-labelled biotin. By means of fluorescent microscopy, the streptavidin−biotin

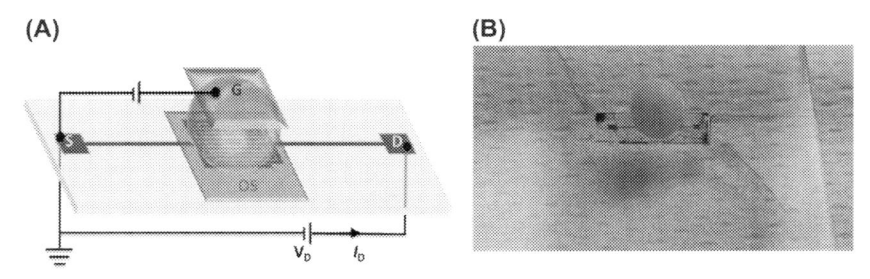

FIGURE 15 (A) Schematic representation of an electrolyte-gated organic field-effect transistor gated using an alginic acid capsule illustration. (B) Picture of an actual device bearing the alginic acid capsule. *From L.M. Dumitru, K. Manoli, M. Magliulo, T. Ligonzo, G. Palazzo, L. Torsi, APL Mater. 3 (2015) 014904. Copyright AIP Publishing. Reproduced with permission.*

binding could be observed. Also the well-known colorimetric assay for glucose detection was reproduced inside the alginic acid capsules. For this purpose, glucose oxidase (GOx) and horseradish peroxidase (HRP) were previously mixed with the biopolymer. When the prepared capsules were immersed in different concentration of glucose, a specific blue-green colour could be observed, as showed in Fig. 16.

In a proof-of-principle, the authors demonstrated that an EGOFET could monitor the reactions taking place inside the capsules. Even though this glucose biosensor is far from being a state-of-the-art glucose sensor, these preliminary results are extremely promising for low-cost, low-power, flexible bio-analytical applications for food industry.

4.5 Conclusions

The purpose of this chapter was to highlight the recent advances in the field of bio-organic electronics concerning analytical applications for the food industry. Nature is a massive reservoir of inspiration and offers endless possibilities in terms of biodegradable, biofunctional and biocompatible materials that can be used for the fabrication of smart sensing devices. So far many naturally occurring small molecules and polymers were demonstrated to function with good performance in organic electronic devices. Examples comprise DNA and nucleobases [9,10,55,81−83], various sugars [10], the biopolymer shellac [84,85], indigo and tyrian purple [84,86], cellulose [87,88], silk [89−91], waxes [92], protein, peptides and aminoacids [93,94] as well as biodegradable polymers in various forms [95,96]. Although such materials have been so far only part of small-scale laboratory demonstrations, the fact that silk, for example, could be used as robust platform for fabrication of high

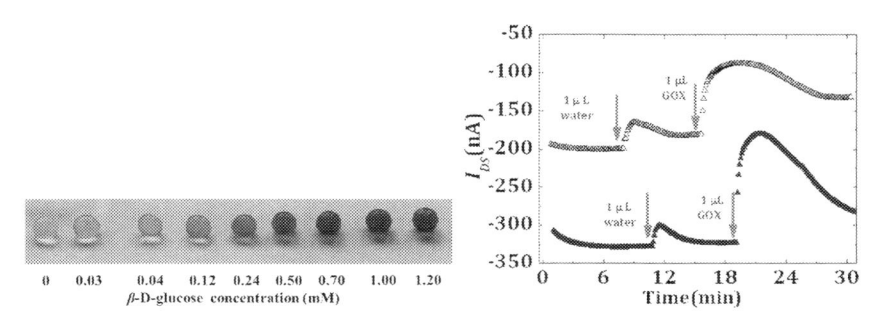

FIGURE 16   (Left) Alginate GOx/HRP-based capsules colorimetric assay for different glucose concentrations. (Right) Transient electronic detection of an enzymatic assay using an alginate capsule gating the EGOFET sensor. *HRP*, Horseradish peroxidase; *GOx*, glucose oxidase; *EGO-FET*, electrolyte-gated organic field-effect transistor. *Adapted from L.M. Dumitru, K. Manoli, M. Magliulo, T. Ligonzo, G. Palazzo, L. Torsi, APL Mater. 3 (2015) 014904. Copyright AIP Publishing. Reproduced with permission.*

performance, transient, integrated electronics [97] demonstrates the immense versatility of natural materials that can be tailored to virtually any specific application involving biocompatibility. There is still a long way till such devices will become reliable and cheap enough to replace the traditional analytical techniques; however, we can only imagine the possibilities.

REFERENCES

[1] J.E. Lilienfeld, Method and apparatus for controlling electric currents, US Patent No. 1745175 A (1930).

[2] K. Dawon, Electric Field Controlled Semiconductor Device, 1963. US Patent No. 3102230 A.

[3] R. Schaller, Spectr. IEEE. 34 (1997) 52−59.

[4] A.T. Tsumura, H. Koezuka, Appl. Phys. Lett. 49 (1986) 1210−1212.

[5] H. Shirakawa, E.J. Louis, A.G. MacDiarmid, C.K. Chiang, A.J. Heeger, J. Chem. Soc. Chem. Commun. 16 (1977) 578−580.

[6] D. Fichou, Handbook of Oligo-and Polythiophenes, John Wiley & Sons, 2008.

[7] P. Stadler, K. Oppelt, T.B. Singh, J.G. Grote, R. Schwödiauer, S. Bauer, H. Piglmayer-Brezina, D. Bauerle, N.S. Sariciftci, Org. Electron. 8 (2007) 648−654.

[8] P. Lin, F. Yan, Adv. Mater. 24 (2012) 34−51.

[9] E.D. Glowacki, L.N. Leonat, G. Voss, M. Irimia-Vladu, S. Bauer, N.S. Sariciftci, Proc. SPIE-Organic Semicond. Sens. Bioelectron. IV. 8118 (2011) 81180−81190.

[10] M. Irimia-Vladu, P.A. Troshin, M. Reisinger, L. Shmygleva, Y. Kanbur, G. Schwabegger, M. Bodea, R. Schwodiauer, A. Mumyatov, J.W. Fergus, V.F. Razumov, H. Sitter, N.S. Sariciftci, Adv. Funct. Mater. 20 (2010) 4069−4076.

[11] D. Feili, M. Schuettler, T. Doerge, S. Kammer, T. Stieglitz, Sens. Actuators A Phys. 120 (2005) 101−109.

[12] H. Kitano, Science 295 (2002) 1662−1664.

[13] L.C. Clark, C. Lyons, Ann. N. Y. Acad. Sci. 102 (1962) 29−45.

[14] V.-D. Tuan, K. Paul, Anal. Bioanal. Chem. 382 (2005) 918−925.

[15] R. Koncki, G.J. Mohr, O.S. Wolfbeis, Biosens. Bioelectron. 10 (1995) 653−659.

[16] M.S. Maynor, T.L. Nelson, C.O. Sullivan, J.J. Lavigne, Org. Lett. 9 (2007) 3217−3220.

[17] P. Leonard, S. Hearty, J. Brennan, L. Dunne, J. Quinn, T. Chakraborty, R. O'Kennedy, Enzym. Microb. Technol. 32 (2003) 3−13.

[18] P. Puligundla, J. Jung, S. Ko, Food Control 25 (2012) 328−333.

[19] A.K. Mcevoy, C. Mcdonagh, D. Brian, I. Klimant, C. Krause, O.S. Wolfbeis, Analyst 127 (2002) 1478−1483.

[20] C. Bohnke, H. Duroy, J. Fourquet, Sens. Actuators B. 89 (2003) 240−247.

[21] N. Wang, N. Zhang, M. Wang, Comput. Electron. Agric. 50 (2006) 1−14.

[22] K.L. Yam, P.T. Takhistov, J. Miltz, J. Food Sci. 70 (2005) 1−10.

[23] H. Vaikousi, C.G. Biliaderis, K.P. Koutsoumanis, Appl. Environ. Microbiol. 74 (2008) 3242−3250.

[24] H. Tao, B. Marelli, M. Yang, B. An, M.S. Onses, J.A. Rogers, D.L. Kaplan, F.G. Omenetto, Adv. Mater. 27 (2015) 4273−4279.

[25] K.A. Law, S.P.J. Higson, Biosens. Bioelectron. 20 (2005) 1914−1924.

[26] F. Davis, S.P.J. Higson, Pediatr. Res. 67 (2010) 476−480.

[27] C. Capannesi, I. Palchetti, M. Mascini, A. Parenti, Food Chem. 71 (2000) 553−562.

[28] F. Bettazzi, F. Lucarelli, I. Palchetti, F. Berti, G. Marrazza, M. Mascini, Anal. Chim. Acta 614 (2008) 93—102.

[29] N. Paniel, A. Radoi, J.L. Marty, Sensors 10 (2010) 9439—9448.

[30] L. Barthelmebs, A. Hayat, A.W. Limiadi, J.L. Marty, T. Noguer, Sens. Actuators B Chem. 156 (2011) 932—937.

[31] G. Jeanty, J.L. Marty, Biosens. Bioelectron. 13 (1998) 213—218.

[32] W. Haasnoot, R.O.N. Verheijen, 0105: 131—134, 2001.

[33] Y.H. Che, Y. Li, M. Slavik, D. Paul, J. Food Prot. 63 (2000) 1043—1048.

[34] D.R. DeMarco, E.W. Saaski, D.A. McCrae, D. V Lim, J. Food Prot. 62 (1999) 711—716.

[35] N.F. Starodub, B.B. Dzantiev, V.M. Starodub, A. V Zherdev, Anal. Chim. Acta 424 (2000) 37—43.

[36] I. Palchetti, A. Cagnini, M. Del Carlo, C. Coppi, M. Mascini, A.P.F. Turner, Anal. Chim. Acta 337 (1997) 315—321.

[37] B. Leca, J.L. Marty, Anal. Chim. Acta 340 (1997) 143—148.

[38] T. Noguer, J.L. Marty, Anal. Lett. 30 (1997) 1069—1080.

[39] B.R. Eggins, C. Hickey, S.A. Toft, Z.D. Min, Anal. Chim. Acta 347 (1997) 281—288.

[40] P. Bouvrette, K.B. Male, J.H.T. Luong, B.F. Gibbs, Enzym. Microb. Technol. 20 (1997) 32—38.

[41] M. Esti, G. Volpe, L. Massignan, D. Compagnone, E. La Notte, G. Palleschi, J. Agric, Food Chem. 46 (1998) 4233—4237.

[42] S. Milardović, Z. Grabarić, V. Rumenjak, M. Jukić, Electroanalysis 12 (2000) 1051—1058.

[43] M. Situmorang, J.J. Gooding, D.B. Hibbert, Anal. Chim. Acta 394 (1999) 211—223.

[44] F. Debeauforta, J.-A. Quezada-Gallo, A. Voilley, Crit. Rev. Food Sci. Nutr. 38 (1998) 299—313.

[45] V. Coma, A. Deschamps, A. Martial-Gros, J. Food Sci. 68 (2003) 2788—2792.

[46] H. Tao, M.A. Brenckle, M. Yang, J. Zhang, M. Liu, S.M. Siebert, R.D. Averitt, M.S. Mannoor, M.C. McAlpine, J.A. Rogers, D.L. Kaplan, F.G. Omenetto, Adv. Mater. 24 (2012) 1067—1072.

[47] S. Gogolewski, Injury 31 (2000) 28—32.

[48] N.R.I. Duncan, M.J. Vicent, F. Greco, Endocr. Relat. Cancer 12 (2005) 189—199.

[49] F. Eder, H. Klauk, M. Halik, U. Zschieschang, G. Schmid, C. Dehm, Appl. Phys. Lett. 84 (2004) 2673.

[50] B. Lamprecht, R. Thunauer, M. Ostermann, G. Jakopic, G. Leising, Phys. Status Solid. A 202 (2005) 50—52.

[51] A.C. Siegel, S.T. Phillips, B.J. Wiley, G.M. Whitesides, Lab Chip. 9 (2009) 2775—2781.

[52] L. Leonat, M.S. White, E.D. Głowacki, M.C. Scharber, T. Zillger, J. Rühling, et al., J. Phys. Chem. C 118 (2014) 16813—16817.

[53] M. Irimia-Vladu, Chem.Soc.Rev. 43 (2014) 588—610.

[54] C.J. Bettinger, Z. Bao, Adv. Mater. 22 (2010) 651—655.

[55] M.A. Daniele, A.J. Knight, S.A. Roberts, K. Radom, J.S. Erickson, Adv. Mater. 27 (2015) 1600—1606.

[56] A. Baeyer, V. Drewsen, Berichte Der Dtsch. Chem. Gesellschaft. 15 (1882) 2856—2864.

[57] E. Steingruber, Indigo and Indigo Colorants. Ullmann's Encyclopedia of Industrial Chemistry, 2004.

[58] E.D. Glowacki, G. Voss, L. Leonat, M. Irimia-Vladu, S. Bauer, N.S. Sariciftci, Isr. J. Chem. 52 (2012) 540—551.

[59] C.J. Cooksey, Molecules 6 (2001) 736—769.

[60] E.D. Głowacki, G. Romanazzi, C. Yumusak, H. Coskun, U. Monkowius, G. Voss, M. Burian, R. Lechner, N. Demitri, G. Redhammer, N. Sunger, G. Suranna, N.S. Sariciftci, Adv. Funct. Mater. 25 (2015) 776–787.

[61] Y.-W. Kwon, C.H. Lee, D.-H. Choi, J.-I. Jin, J. Mater. Chem. 19 (2009) 1353–1380.

[62] P. Lin, X. Luo, I.M. Hsing, F. Yan, Adv. Mater. 23 (2011) 4035–4040.

[63] J.A. Hagen, W. Li, A.J. Steckl, J.G. Grote, Appl. Phys. Lett. 88 (2006) 171109–171111.

[64] M. Irimia-Vladu, P.A. Troshin, M. Reisinger, G. Schwabegger, M. Ullah, R. Schwoediauer, A. Mumyatov, M. Bodea, J.W. Fergus, V.F. Razumov, H. Sitter, S. Bauer, N.S. Sariciftci, Org. Electron. 11 (2010) 1974–1990.

[65] G.H. Gelinck, H.E.A. Huitema, E. van Veenendaal, E. Cantatore, L. Schrijnemakers, J.B.P.H. van der Putten, T.C.T. Geuns, M. Beenhakkers, J.B. Giesbers, B.-H. Huisman, E.J. Meijer, E.M. Benito, F.J. Touwslager, A.W. Marsman, B.J.E. Van Rens, D.M. De Leeuw, Nat. Mater. 3 (2004) 106–110.

[66] Q. Zhang, V. Subramanian, Biosens. Bioelectron. 22 (2007) 3182–3187.

[67] E. Stavrinidou, R. Gabrielsson, E. Gomez, X. Crispin, O. Nilsson, D.T. Simon, et al., Sci. Adv. 1 (2015) 1–8.

[68] L. Torsi, A. Dodabalapur, L. Sabbatini, P.G. Zambonin, Sens. Actuators B. 67 (2000) 312–316.

[69] L. Torsi, G.M. Farinola, F. Marinelli, M.C. Tanese, O.H. Omar, L. Valli, F. Babudri, F. Palmisano, P.G. Zambonin, F. Naso, Nat. Mater. 7 (2008) 412–417.

[70] M.D. Angione, S. Cotrone, M. Magliulo, A. Mallardi, D. Altamura, C. Giannini, N. Cioffi, L. Sabbatini, E. Fratini, P. Baglioni, G. Scamarcio, G. Palazzo, L. Torsi, Proc. Natl. Acad. Sci. 109 (2012) 6429–6434.

[71] K. Manoli, L.M. Dumitru, M.Y. Mulla, M. Magliulo, C. Di Franco, M.V. Santacroce, G. Scamarcio, L. Torsi, Sensors 14 (2014) 16869–16880.

[72] L. Torsi, M.C. Tanese, N. Cioffi, M.C. Gallazzi, L. Sabbatini, P.G. Zambonin, G. Raos, S.V. Meille, M.M. Giangregorio, J. Phys. Chem. B 107 (2003) 7589–7594.

[73] K. Loig, L. Herlogsson, D. Braga, B. Piro, M.-C. Pham, X. Crispin, M. Berggren, G. Horowitz, Adv. Mater. 22 (2010) 2565–2569.

[74] M. Magliulo, A. Mallardi, M.Y. Mulla, S. Cotrone, B.R. Pistillo, P. Favia, et al., Adv. Mater. 25 (2013) 2090–2094.

[75] L.M. Dumitru, K. Manoli, M. Magliulo, G. Palazzo, L. Torsi, Microelectron. J. 45 (2014) 1679–1683.

[76] L.M. Dumitru, K. Manoli, M. Magliulo, L. Sabbatini, G. Palazzo, L. Torsi, Appl. Mater. Interfaces. 5 (2013) 10819–10823.

[77] S.H. Kim, K. Hong, W. Xie, K.H. Lee, S. Zhang, T.P. Lodge, C.D. Frisbie, Adv. Mater. 25 (2013) 1822–1846.

[78] S. Casalini, F. Leonardi, T. Cramer, F. Biscarini, Org. Electron. Phys. Mater. Appl. 14 (2013) 156–163.

[79] M.Y. Mulla, E. Tuccori, M. Magliulo, G. Lattanzi, G. Palazzo, K. Persaud, L. Torsi, Nat. Commun. 6 (2015) 6010.

[80] L.M. Dumitru, K. Manoli, M. Magliulo, T. Ligonzo, G. Palazzo, L. Torsi, APL Mater. 3 (2015) 014904–014914.

[81] Y. Zhang, P. Zalar, C. Kim, S. Collins, G.C. Bazan, Adv. Mater. 24 (2012) 4255–4260.

[82] C. Yumusak, T.B. Singh, N.S. Sariciftci, J.G. Grote, Appl. Phys. Lett. 95 (2009) 263304–263306.

[83] E.F. Gomez, V. Venkatraman, J.G. Grote, A.J. Steckl, Adv. Mater. 27 (2015) 7552–7562.

[84] M. Irimia-Vladu, E.D. Głowacki, P.A. Troshin, G. Schwabegger, L. Leonat, D.K. Susarova, O. Krystal, M. Ullah, Y. Kanbur, M. Bodea, V. Razumov, H. Sitter, S. Bauer, N.S. Sariciftci, Adv. Mater. 24 (2012) 375−380.

[85] M. Irimia-Vladu, L. Leonat, H.Z. Akpinar, H. Sitter, S. Bauer, N.S. Sariciftci, Green Chem. 15 (2013) 1473−1476.

[86] E.D. Głowacki, L. Leonat, G. Voss, M.-A. Bodea, Z. Bozkurt, A.M. Ramil, M. Irimia-Vladu, S. Bauer, N.S. Sariciftci, AIP Adv. 1 (2011) 042132−042137.

[87] A. Petritz, A. Wolfberger, A. Fian, A. Haase, H. Gold, T. Rothländer, T. Griesser, B. Stadlober, Appl. Phys. Lett. 103 (2013) 153303−153307.

[88] A. Petritz, A. Wolfberger, A. Fian, T. Griesser, M. Irimia-Vladu, B. Stadlober, Adv. Mater. 27 (2015) 7645−7656.

[89] R. Capelli, J.J. Amsden, G. Generali, S. Toffanin, V. Benfenati, M. Muccini, D.L. Kaplan, F.G. Omenetto, R. Zamboni, Org. Electron. 12 (2011) 1146−1151.

[90] C. Wang, C. Hsieh, J. Hwang, Adv. Mater. 23 (2011) 1630−1634.

[91] M.K. Hota, M.K. Bera, B. Kundu, S.C. Kundu, C.K. Maiti, Adv. Funct. Mater. 22 (2012) 4493−4499.

[92] B. Stadlober, E. Karner, A. Petritz, A. Fian, M. Irimia-Vladu, IEEE J. Solid-State Circ. 41 (2015) 10−17.

[93] J. Chang, C. Wang, C. Huang, T. Tsai, T. Guo, Adv. Mater. 23 (2011) 4077−4081.

[94] A. Dezieck, O. Acton, K. Leong, E.E. Oren, H. Ma, C. Tamerler, M. Sarikaya, A.K.-Y. Jen, Appl. Phys. Lett. 97 (2010) 13305−13307.

[95] H. Cheng, P.S. Hill, D.J. Siegwart, N. Vacanti, A.K.R. Lytton-Jean, S.-W. Cho, A. Ye, R. Langer, D.G. Anderson, Adv. Mater. 23 (2011) 95−100.

[96] C. Yang, W. Lin, Z. Li, R. Zhang, H. Wen, B. Gao, G. Chen, P. Gao, M.M.F. Yuen, C.P. Wong, Adv. Funct. Mater. 21 (2011) 4582−4588.

[97] H. Tao, D.L. Kaplan, F.G. Omenetto, Adv. Mater. 24 (2012) 2824−2837.

Chapter 9

Food Microfluidics Biosensors

M.A. López, M. Moreno-Guzman, B. Jurado and A. Escarpa*

Universidad de Alcalá, Alcalá de Henares, Madrid, Spain

*Corresponding author: E-mail: alberto.escarpa@uah.es

Chapter Outline

List of Abbreviations	**BoNT** Botulinum neurotoxin
	CCD Contactless detection
AC Alternate current	**CFU** Colony formation units
BGS Background solution	**CL** Chemiluminescence

CNT Carbon nanotube
CT Cholera toxin
DC Direct current
ED Electrochemical detection
EFGF Electric field gradient focusing
EIS Electrochemical impedance spectroscopy
ELISA Enzyme-linked immunosorbent assay
EOF Electroosmotic-driven flow
FAS Field amplified stacking
FRET Föster resonance energy transfer
GCE Glassy carbon electrode
HAB Harmful algal blooming
HRP Horseradish peroxidase
LED Light-emitting diode
LIF Laser-induced fluorescence
LOC Lab-on-a-chip
LOD Limit of detection
LTCC Low-temperature co-fired ceramic

MC Microfluidic chip
MCE Microchip capillary electrophoresis
MEMS Microelectromechanical system
Micro-TAS Micro total analysis systems
MIP Molecular imprinted polymers
OA Okadaic acid
PCR Polymerase chain reaction
PDMS Poly(dimethylsiloxane)
PMMA Poly(methyl methacrylate)
PMT Photomultiplier detector
RI Refractive index
S Sample
SAM Self-assembled monolayer
SEB Staphylococcal enterotoxin B
SPR Surface plasmon resonance
TFG Temperature gradient focusing
TIR Total internal reflection
UV Ultraviolet visible radiation

1. INTRODUCTION

During the last decades, (bio)-sensors in combination with microfluidic platforms have received significant research interest. The notable and rapid progress in micro- and nanotechnology as well as in biotechnology has enabled the integration of a variety of analytical functions in a single platform. In this sense, the possibility to carry out all sample handling and analysis steps on a small scale using miniaturized devices is very appealing. From the analytical point of view, one of the main characteristics of miniaturized systems is the large surface-to-volume ratios, which enhance molecular diffusion and heat transfer. This small scale reduces the required time to analyse a product, as greater control of molecular interactions is achieved at the microscale level; makes easier automation and simplification and provides additional advantages such as the speed of result generation (high throughput), the amount of information (simultaneous or multiparametric), the autonomy allowing field test (portability), reduction of reagent cost or the amount of chemical waste.

Microfluidics refers to the science and technology of systems that moves volumes below the microscale (10^{-9} to 10^{-18} L) through microsized channels on microsystems [1]. This term is closely related to the concept of micro total analysis systems (μTAS), which represents the miniaturization and integration of all analytical process steps onto a single monolithic device [2,3].

Miniaturization is more than simply the scaling down of well-known systems since the relative importance of forces and processes changes with scale. One of the most relevant characteristics of analytical Microsystems is the omnipresence of laminar flow (Reynold's number are typically very low) in which viscous forces dominate over inertia. This means that turbulence is often unattainable and that molecule transportation only occurs through diffusion, which has direct consequences on the designs of this type of micro system.

Microfluidic platforms on a microchip format (microfluidic chips) are made up of a network of channels and reservoirs to perform multiple operations such as sampling, filtration, preconcentration, separation and detection. These platforms involve at least one injector (where a sample plug is loaded) and pumping microchannels (where the transport of analytes and reagents is performed) that are suitably interfaced to reservoirs (where different solution/samples are deposited). Microchannels and reservoirs are fabricated on microchips using mainly photolithography or micromoulding on planar substrates. In the following sections a general overview of the different aspects related to microfluidic analytical systems will be presented.

Biosensors comprise a biological sensing material (enzymes, antibodies, DNA, aptamers, tissues, organism), a biologically derived material or a biomimic, coupled to a chemical or physical transduction system, which converts a biospecific reaction into a quantifiable and processable signal [4]. The high selectivity and sensitivity of the biorecognition molecule to the analyte, which allow to measure it with minimal preparation steps in a complex sample matrix, the low-cost production and the potential for miniaturization are some of the main advantages of biosensors. Stimulated by the current trend of µTAS and lab-on-a-chip (LOC) technologies, they are currently incorporated into microfluidic systems for the realization of highly sophisticated systems. Actually, microfluidic chips are an ideal platform for performing microscale flow-injection analysis. This concept includes fluid motion (mainly, electrokinetic or hydrodynamic), injection and analyte detection, which allows accurate, ultrasmall sample volumes to be introduced and has accurate fluid control and manipulation capabilities. Those microchip sensing strategies demonstrated distinguishing merits of integration, automation and parallelization, allowing improved sensitivity, lower cost and shorter time-to-result. Owning to its miniaturization, it can be used as a homemade portable microinstrument, especially for on-site detection [5].

As in the case of large-scale biosensors, different classifications according to multiple criteria can be used. The most common way to group sensors relies on considering their transduction mechanism (electrical, optical, mass, thermal, piezoelectric, etc.), their biorecognition principle [use of enzymes, DNA, antibodies, receptors, molecular imprinted polymers (MIP)] or their applications (environmental, food, medical diagnosis).

Although microfluidic analytical biosensor systems have been mainly used in biomedical applications, food analysis field is receiving great interest for the development of high-performance analytical microdevices. Recently, Escarpa [6] has presented the lights and shadows of one emerging and exciting topic introduced as *Food Microfluidics*. It is defined as microfluidic technology for food analysis and diagnosis in important areas such as food safety and food quality. Although food microfluidics is still in its *adolescence*, the integration of micro-, nano- and biotechnologies constitutes an extraordinary chance for future implementation of applications of microfluidic biosensors in the field of food analysis as illustrated in Fig. 1.

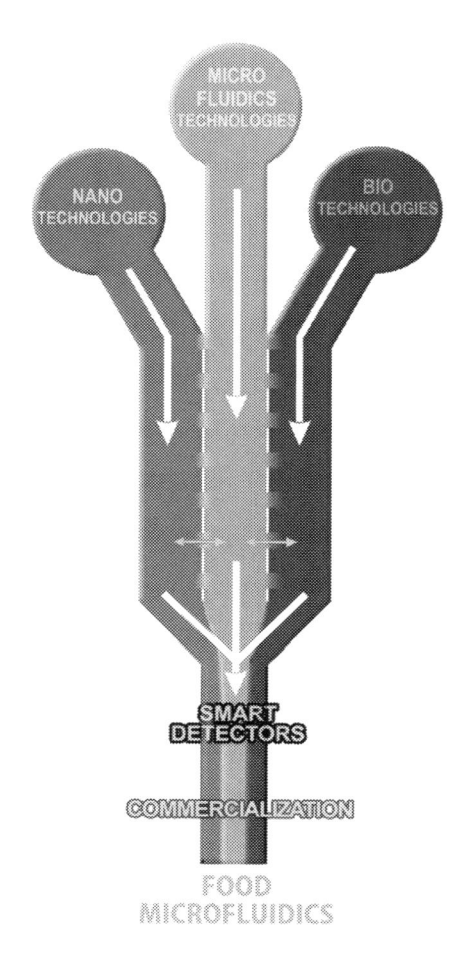

FIGURE 1 Analogy of the operation mode in a microfluidic chip with the technologies available and challenges in the future development of microfluidic biosensors in food analysis. Conceptual and realistic solutions departing from key-technology reservoir flowing towards *Food Microfluidics. Reproduced with permission from A. Escarpa, Lights and shadows on food microfluidics, Lab Chip 14 (2014) 3213−3224.*

Development and innovation of food industry rely on the concepts of food safety and food quality. As the food chain is increasingly complex, there is a high demand for the development of efficient traceability systems which are able to guarantee the firmness of the whole chain [7]. These systems should possess high sensitivity, ability to be implemented rapidly, low cost, portable and permit automatic screening. In this sense, microfluidics is a technology that allows constructing small, fast and cheap microfluidic analytical systems.

Food quality can be understood as a set of factors which are able to differentiate food products according to their organoleptic characteristics, composition and functional properties. This term relates to appearance, taste, smell, nutritional value content, functional ingredients, freshness, flavour, texture and chemicals. An increased regulatory action together with a more and more consumer demand for information have led to the extensive labelling of major and minor constituents of the foodstuffs. The scientific evaluation of the food freshness is another important task concerning food quality assessment. Moreover, continuous monitoring of food industrial processes allows real-time detection of possible errors in the chain production as well as taking decisions to rectify such errors in an immediate manner [7].

The assessment of food safety is the other key axe for the modern food industry [8]. This concept entails the production and the commercialization of food which do not represent a risk to the consumer, so it must be free from allergens, pesticides, fertilizers, heavy metals, organic compounds, pathogens and toxins. Food contamination usually deals with harmful substances or microorganisms that are not intentionally added to the food. Contaminants may enter the food chain during growth, cultivation or preparation; accumulate in food during storage; form in the food through the interaction of chemical components or may be concentrated from the natural components of the food [9]. On the other hand, chemicals are also added during food processing in the form of additives. Pathogen microorganisms, pesticides, animal-drug residues and antimicrobial drug resistance are the main concerns for food safety at present. Food regulatory agencies have established control programmes, such as the Hazard Analysis Critical Control Point (HACCP) programme, to avoid the entering of these substances into the food chain [10].

2. MICROFLUIDIC ANALYTICAL SYSTEMS: COMPONENTS AND OPERATIONS

2.1 Materials and Microfabrication

Without any doubt, a relevant key of the miniaturized analytical systems is the necessity of the development of microfabrication techniques. Microfluidics traces its history from the microelectronics industry and silicon-based micromachining processes using photolithography, etching and bonding techniques. Although a deep description of the microfabrication techniques is obviously outside of the scope of this chapter (readers can find more information in excellent books [11]), a brief description of materials properties and the most important aspects involved in both, glass and polymer microfabrication technologies for microfluidic platforms, mainly in the microchip format, will be described.

It is clear that material substrate used for microchip fabrication is a key factor since the performance of the microfluidic platform will be strongly dependent of this material. Glass, polymers and in lower extension silicon are the three main types of materials used for microfluidic device fabrication. However, it is important to take into account that one of the limitations for microfluidic device development is the high fabrications cost due to the need for specific fabrication facilities (ie, clean room facility), which is also strictly dependent on the materials employed.

Silicon was the first material used due to the well-established microfabrication processes developed for semiconductor microelectronics industry. The chemical and physical properties of silicon are well characterized. It possesses good thermal conductivity, it is resistant to high temperatures and structurally strong, and can be produced virtually in any geometrical microstructure with high precision, and its surface chemistry has been studied extensively giving the capability to attach molecules easily. However, it has some important limitations [12]: (1) it is not optically transparent in the wavelength range for optical detection; (2) it is electrically conductive, so it is unsuitable for electrochemical detection or electrokinetic flow transport or reagents; (3) proteins and other biomolecules tend to bind to silicon surface groups, reducing the sensitivity; (4) the microchip construction is relatively expensive, time-consuming, requires of clean room facility and (5) the wet-etching process typically used in fabrication produces a low aspect ratio. Although silicon has been a significant material in microfluidic devices, these drawbacks have made it replaced or combined with glass or polymer substrates for most analytical applications.

Besides, the use of glass as material substrate can overcome some of the problems associated to silicon. Glass is less expensive, it presents good chemical resistance and it is optically transparent through the visible spectrum allowing laser-induced fluorescence (LIF) detection. Furthermore, thanks to be none electrically conductive, glass is compatible with electrokinetic fluid transport. However, glass also has some drawbacks. As with silicon substrates, the wet-etching method used for fabrication does not produce a high aspect ratio, toxic chemicals are involved, and this material is more fragile. Structures on glass substrates are usually generated using standard photolithographic technologies [13–15].

In the last years, polymers have emerged as an interesting alternative to glass and silicon in the microfluidic platforms fabrication. Polymers offer the advantage of being low cost and easily produced since microfabrication techniques (eg, injection moulding, compression moulding, extrusion, hot embossing, soft lithography, laser photoablation, screen printing and replication) are highly reproducible, while expensive patterning procedures are only required for constructing the mould. In addition, polymers have good chemical resistance, optical transparency, low autofluorescence, nontoxic and do not absorb UV. Moreover a variety of surface modification methods are availably improving the efficiency of these devices. There are a variety of materials [eg, plexiglas, cyclic olefin copolymers, polyester, polycarbonate, poly(methyl methacrylate) (PMMA)], each with specific properties offering the possibility of tailoring a material to a specific application.

Without any doubt poly(dimethylsiloxane) (PDMS) has emerged as the most used polymer material. PDMS is flexible, optically transparent, impermeable to water and permeable to gases, and very compatible with biological studies. Nevertheless due to its hydrophobic surface, this material presents some limitations as poor wettability, forming bubbles in aqueous solutions, lower durability and the tendency to adsorb proteins and other molecules. However, due to the elastomeric nature of PDMS that allows an easy seal with smooth and flat surfaces, hybrid devices using mainly PDMS and other different type of material as glass, polyester have been reported.

Related to polymer microfabrication microchips, the two major ways are replication from a master (moulding methods) and direct machining [16,17]. Replication methods often produce a microstructure by allowing a polymer workpiece to form an inverse copy of a mould. These methods are hot embossing, injection moulding and casting. The formation of microchannels and other structures using moulding methods generally involves two primary steps: (1) moulder (also known as master) fabrication and (2) channel pattern transfer from that moulders to polymer substrates. These methods have in common that a very soft or even liquid form of polymer is poured or pressed into a mould, after which the material is hardened and removed from the mould. On the other hand, direct machining methods remove small amounts of polymer in places where the microstructures (microchannels, reservoirs) should be located. This type of micromachining is laser ablation.

Hot embossing is the process of pressing a mould into heated, softened thermoplastic, followed by cooling, thereby producing an inverted replicate of the mould. The first step in hot embossing consists of heating the mould and the polymer to glass transition temperature. Once the polymer begins to soften it takes the form and the shape of the mould. The mould and the polymer are then cooled below the glass transition temperature to harden the polymer, and afterwards the polymer is removed from the mould. Embossing can take several minutes per device and can be a useful tool in rapid prototyping devices.

In the injection moulding technique the molten polymer material is injected into the mould. For microinjection moulding, the mould is heated to the softening temperature of the polymer to prevent the injected polymer material from hardening too early. After injection of molten polymer into the mould, this is slowly cooled, so that the polymer hardens and then, the polymer microstructure can be removed from the mould. Injection moulding allows very high-throughput production with low production costs (moulded devices are released every 5−10 s).

Casting uses a chemical process to harden the polymer [11]. Two components, a base and a hardener or cure, are mixed just prior to use. Immediately upon mixing, the chemical curing process starts. After some time, this process results in hardening of the polymer (at atmospheric pressure and temperature). The liquid mixture is poured in the mould and as the polymer sets it takes the shape of the mould. The polymer structure can then be removed from the mould. This technique is very popular with elastomers such as PDMS because they form hermetic, reversible seals and smooth surfaces such as glass or silicon by adhesion. Casting is the simplest among these three moulding processes, but requires contact with the mould for minutes or hours.

On the other hand, laser micromachining (such as photoablation) is a direct machining method based on the removal of polymer material by using intense UV or infrared radiation provided by a laser. The photoablation process involves absorption of a short-wavelength laser pulse to break covalent bonds in long-chain polymer molecules with production of a shock wave that ejects decomposed polymer fragments [11]. Many commercially available polymers can be photoablated, including polycarbonate, PMMA, polystyrene, nitrocellulose and poly(tetrafluoroethylene). The laser energy can be specially patterned using a mask with the subsequent generation of microcavities and channels in various geometries or controlling position of the laser with x-y stages. The resulting structures are generally characterized as having little thermal damage, straight vertical walls and well-defined depth. However, laser ablation does not lend itself well to mass production.

Apart from these well-established materials, a low-temperature co-fired ceramic (LTCC) has also been used for constructing electrochemical microfluidic devices. Using this material, highly intricate three-dimensional structures can be designed, integrating fluid networks, embedded passives and electronic circuits by means of unfired flexible sheets. After firing, the LTCC becomes rigid with very good thermal, mechanical and electrical properties [18].

2.2 Fluid Handling Modalities

In microfluidic platforms, the movement of reagents and samples is required to perform the different analytical stages, including delivery, mixing, reaction, analysis and detection. A precise control of the flow rate is essential to increase the accuracy and reliability of analytical results [19]. Liquid transport can be performed in a variety of ways mainly categorized such as pressure, electric and passive fluid handling forces. The difference lies mainly in the nature of the driving agent behind the transport. When flow is generated mechanically by a pump, it is called pressure-driven flow; when flow is driven by a voltage, it is called electroosmotic flow (EOF) and when not external power sources are required to drive liquids, it is called passive flow.

2.2.1 Pressure-Driven (or Hydrodynamic) Microfluidic Systems

In this modality, a pump is used to set fluids into motion. A large variety of principles have been developed and wider information can be found in interesting reviews [20–23]. Within these microsystems, the pumping mechanism is based on the use of mechanical displacement actuators. They can be defined as those that exert oscillatory or rotational pressure forces on the working fluid through a moving solid–fluid (vibrating diaphragm, peristaltic, rotary pumps) or fluid–fluid (ferrofluid, phase change, gas permeation pumps) boundary. There are a large combination of actuators for pumping and valves (electrostatic, piezoelectric, thermopneumatic, electromagnetic, etc.) which determine the overall performance, in terms of generated pressure, stroke displacement, response time and reliability. However, one of the most used pumping methods includes the use of syringe or piston micropump, with capillary tubing attached with interconnects to the microchip, although it is increasingly difficult to use in micro/nanochannels due to the hydrodynamic resistance increases with the reduction in dimension and are not amenable for miniaturization. Furthermore, pressure-driven flow can be used for both aqueous and nonaqueous liquids.

2.2.2 Electroosmotic Flow

The liquid flow is originated in the presence of an electric field when an ionic solution is in contact with a charged solid phase. Oppositely charged ions in the fluid shield the surface charge and can be manipulated with a DC or AC electric field. DC electroosmotic pumping is normally used. As a well-known example, in silica material in contact with an aqueous electrolytic solution, the solid surface has an excess of negative charge due to their ionization of the surface's silanol groups. A high number of counterions of these anions are found on the interphase between the capillary wall and the solution originating the electric double layer, which is formed by a stagnant layer adjacent to the capillary wall (Stern layer) and a mobile diffuse layer (Gouy-Chapman layer). The cationic counterions in the diffuse layer migrate towards the cathode. As they are solvated, they drag solvent with them, originating the EOF. The linear velocity of the EOF depends on the potential originated across the double layer and the zeta potential. The flow is simply created by filling a microchannel with a buffer solution and applying a suitable voltage at the channel ends. Some advantages with respect to hydrodynamic flow systems are the absence of mobile parts such as valves, actuators and also the particular flow profile. Contrary to hydrodynamic flows, where one finds a parabolic distribution of the flow velocities which the largest velocity at the centre of the channel and zero velocity at the walls, in electrokinetics, EOF is generated close to the wall and, therefore, it produces a flat profile with a very uniform velocity distribution across the entire cross section of the channel. Furthermore, EOF is pulse free and there is no back pressure (as there is with mechanical pumps). The main problem arises from the strong dependence on the chemical factors involved in the microsystems as the pH values. This dependency makes EOF hard to control, since every change in pH, dielectric constant or concentration, due to reactions or the mixing process, has an immediate effect on the magnitude of the EOF. On the other hand, this strong dependency also means that there are many parameters that can be exploited to control and manipulate the EOF. Furthermore, other major limitations are the high voltage and the electrically conductive solution required.

2.2.3 Electrowetting

Discrete liquid droplets are manipulated in the presence of programmed voltage sequences applied to an electrode array. By applying sequences of AC or DC electric potentials between ground and actuation electrodes, droplets of reagents or samples can be driven to move, to merge, to divide and to dispense from reservoirs by combining electrostatic and dielectrophoresis forces. In electrowetting, the fluid is transported using surface tension (an interfacial force which dominates at microscale). Voltage is applied on the dielectric layer, decreasing the interfacial energy of the solid and liquid surface, which results in fluid flow. Thus, when applying the electric potential along the interface of a liquid metal droplet in an electrolyte, charge redistribution occurs, resulting in a gradient in surface tension at the interface, which causes movement of the droplet to regions of lower surface tension. Switching the direction of the applied potential also changes the direction motion [24].

2.2.4 Passive Forces

Microfluidic systems that do not require external power sources to drive liquids have been also reported for microfluidic devices. In these cases, only gravity and capillary forces are used to drive fluidics within the microchannels [25]. Capillary flow, due to its simplicity and efficiency, is a quite used mechanism for fluid transport in well-known microfluidic devices such as lateral-flow test. This movement can be produced either by modifying the solid–liquid surface tension (including flows due to electrowetting, surface gradients and reactive flows), or by inducing a gradient in the liquid–gas surface tension including thermocapillary, electrocapillary and solutocapillary motion [19].

2.3 Sample Preparation

Although the characteristic high selectivity of biosensors habitually allows sample analysis with minimal sample preparation process, the extraordinary complexity of food samples and the low abundance of contaminants make preparative cleanup and preconcentration steps to be required. While sample preparation techniques are widely developed in conventional formats and off-line, it is important to point out that from the different steps related to the analytical process, sample preparation is notably the least developed in miniaturized devices. However, very exciting unique opportunities could be expected because of (1) the intrinsic possibilities of microfabrication technology to create sophisticated designs and microstructures in connection with sample preparation requirements; (2) the natural possibilities offered by microfluidics: the presence of laminar flow and diffusion and (3) the ease of using electrokinetic phenomena to move fluidics into the network channels with accuracy (focusing on flow techniques). The integration of sample pretreatment in microfluidic systems can eliminate sample handling loss or contamination problems arising from off-line sample manipulation.

Isolation/cleanup and sample preconcentration will be briefly discussed in the following paragraphs. For a deeper study of sample pretreatment in microfluidic devices, excellent literature is proposed [20,26−29].

2.3.1 Isolation/Cleanup From the Sample Matrix

A challenging task faced by the analytical chemist when dealing with raw samples is extracting/isolating the analyte of interest from the sample matrix. It is fair to say that the majority of nonideal samples arrive in a format incompatible with most analytical instrumentation and requires some degree of cleanup. Even well-defined samples (eg, aqueous solutions) require basic filtration prior to analysis. Other desirable techniques may include liquid/liquid extraction and solid-phase extraction.

2.3.1.1 Filtration Approaches

Possibly the most essential step when performing analysis in microfluidic systems is the filtration of sample prior to processing. Due to the typical small dimensions in microstructures, particulates can cause serious operational problems, providing sites for nucleation or blockage. The simplest solution is to filter all reagents and sample on-chip prior to analysis.

Related to filtration, two approaches have been employed: structurally based filters (filtering and retention by integrated flow restrictions and controlled by manufacturing process) and diffusion-based filtration (filtering by diffusion in laminar flows). Structurally based or microfabricated filters have been proposed in popular architectures such as frits, pillar structures or flow restrictions within fluidic channels to mimic conventional filters [30]. The effectiveness and application of these filter structures is determined by the resolution limits of the manufacturing process. Consequently, filtering of submicron-sized particulates using physical structures puts stringent demands on device fabrication.

On the other hand, filtration can be induced by allowing the analytes of interest to migrate across a laminar boundary between two adjacent layers (sample and solvent stream) where material transport only occurs by diffusion due to the predominance of laminar flow in microfluidic systems [31,32]. An interesting quality of laminar flow is that material transport between two adjacent streamlines only takes place by diffusion and not by turbulent mixing as it is done in a test tube. Since species of low molecular mass have greater mobility (larger diffusion coefficients) than large molecular species, the process can be further controlled by altering the time in which the two fluids come into contact. The approach is highly configurable since the time allowed for diffusional transfer between streams is directly controlled by fluid velocity and the length of the channel. Importantly, this approach addresses many of the problems associated with structurally based filters, since its operation is reliant on the control of molecular diffusion rather that the resolution of the manufacturing process.

2.3.1.2 Extraction Approaches

Liquid/liquid extraction, technique widely used in conventional sample pretreatment methodologies, describes the physical process by which a compound (or a mixture of compounds) is transferred from one liquid phase to another and could play a predominant role in miniaturized systems. The high surface-to-volume ratio and the short diffusion distances, typically within microfluidic environments, combined with laminar flow conditions, offer the possibility of performing liquid–liquid extraction within microchannels without shaking. Although a potential drawback could be the low unit throughput (normally between 1 and 100 µL/min), this problem can be obviated by operating arrays of parallel channels concurrently.

Solid-phase extraction is a broadly used technique in which a target molecule is retained by a chromatographic stationary phase material and subsequently eluted in an appropriate (and selective) solvent. Solid-phase extraction functions either as a sample cleanup method and a preconcentration method. This fact is not only due to that the target analyte is retained within the stationary phase and the unwanted components of the sample matrix flow to waste but also after elution in the desired volume entails preconcentration of the analyte. Several and creative approaches for performing solid-phase extraction in microfluidics have been proposed, including to coat channel walls with a high-affinity stationary phase, packing the microchannels with stationary phase using beads or monolithic microchannels [33,34].

2.3.2 Preconcentration: Electrokinetic Approaches

In addition to the complexity of food samples, analytes may be present at extremely low concentrations. This combined with the ultrasmall detection volumes encountered in microfluidic systems (pL–nL) makes often desirable to incorporate sample preconcentration prior to detection within microfluidic systems.

Electrokinetic flow-driven systems have a relevant role in microfluidic devices due to their inherent and easy miniaturization, simplicity of fabrication and high-controlled fluidic manipulation as previously discussed. Besides, this principle has shown to be very useful to analyte preconcentration in microchips. These techniques have been studied for several decades and are commonly used in microfluidic capillary electrophoresis as a single-step preconcentration method for achieving high sensitivity. There are two different strategies. The first one is based on the velocity change of the analytes between the sample and separation solution zones (stacking). The second technique (focusing), the analytes are focused at the points where the migrating velocities become zero.

Briefly, in field amplified stacking (FAS), a background solution (BGS) with a high conductivity and a sample solution (S) with a low conductivity are prepared. The S is introduced into the capillary filled with the BGS, and then an appropriate voltage is applied to both ends of a capillary. The local electric field in the sample zone is higher than that in the BGS as the electric current in the capillary is constant. Therefore, the electrophoretic velocity of the analytes in the sample zone is faster than that in the BGS. This difference in the electrophoretic velocities between the S and BGS zones generates the 'stacking' effect at the S/BGS boundary, so that the analytes are concentrated around the boundary [35]. A major problem of this technique is that neutral analytes cannot be concentrated. This drawback has been solved using electrokinetic chromatography or using ionic liquids inside the microchannel [36,37].

Related to 'focusing' techniques, electric field gradient focusing (EFGF) and temperature gradient focusing (TGF) will be discussed.

In EFGF, sample molecules migrate across the channel due to the applied electric field gradient. In the microchannel, a hydrodynamic counterflow is also introduced with a constant velocity. When the electric field at the inlet is higher than that at the outlet and the counterflow is introduced from the outlet, a faster electrophoretic migration of the sample overcomes the hydrodynamic flow. Thus, the sample moves towards the outlet. With decrease in the electric field along the microchannel, the electrophoretic velocity of the sample is gradually decreased and finally inversed due to the counterflow, so that the sample is focused and separated in the order of their electrophoretic mobilities [38].

In TGF, the electrophoretic velocity of the sample is gradually changed by the variation of the ionic strength of the BGS which is dependent on the temperature. TGF is more advantageous than EFGF in respect of easier operation and the applicability to a wide range of analytes [39].

In addition to chemical preconcentration methods, some relevant applications make use of biological reagents. In this sense, antibody-modified surfaces such as microchannel walls or micro/nanoparticles allow the specific enrichment and determination of selected analytes. The use of antibodies in microfluidic platforms for food analysis will be more deeply discussed later.

2.4 Detection

From the beginning, detection has been one of the main challenges for analytical microsystems, since very sensitive techniques are needed as a consequence of the ultrasmall sample volumes used in micron-sized environments. In principle, a wide variety of detection alternatives can be used in microfluidic systems, which can be categorized into optical, electrochemical and mechanical detection schemes.

2.4.1 Optical Detection

Optical detectors are commonly used due to their accessibility in laboratories and the simplicity of microfluidics—detector interfaces. Among them, LIF has been the most used detection technique because of its inherent sensitivity. However, other conventional optical detection methods, including absorbance, chemiluminescence (CL) and surface plasmon resonance (SPR), have all been applied in microfluidic biosensors. Although microscopes, lasers, spectrophotometers, charge-coupled devices and photomultiplier tubes can be precisely coupled to microfluidic systems, they are difficult to miniaturize into low-cost and portable detection devices [40]. Alternatively, optoelectronic technology (waveguides, photodiodes) has been successfully integrated in microfluidic systems.

2.4.1.1 Laser-Induced Fluorescence

The remarkable success of LIF as detection method in microfluidic systems comes from that it can be easy adapted to low-concentration/low-volume systems [41].

Fluorescence is a physical phenomenon that englobes a three-stage process (excitation, excited-state lifetime and fluorescence emission) and occurs in a series of molecules called fluorophores. Sometimes, the analyte does not present this native fluorescence, but can be measured through adequate labelling with fluorescent dyes (small molecules, proteins, quantum dots). The availability of highly sensitive, highly selective fluorescent labelling techniques makes fluorescence a widely used optical method for molecular sensing in microfluidic systems.

The instruments consist of the following basic parts: a laser excitation source, an optical system to focus and collect the light, and a photosensitive detector. Advances in laser technology have produced stable light sources that cover a rapidly increasing range of wavelengths from the UV to the IR region of the electromagnetic spectrum, that are relatively inexpensive and that can easily be focused onto micron-sized detection areas. The use of a pinhole at the focus point along the optical path (confocal LIF) even allows for 3D focusing and further reduction of the background signal (scattered light, autofluorescence of microchip material). These very low levels of background signal combined with very sensitive photon detection techniques (photo-multiplier tube, photon counting systems) result in the lowest limits of detection (LODs) of all microchip detection systems.

However, the high cost and the large size of the instrumental setup for LIF are sometimes incompatible with the concept of μTAS. In addition, tedious derivatization schemes are needed to use LIF with nonfluorescent compounds.

2.4.1.2 Lamp-Based Fluorescence

After LIF detection, lamp-based approaches form the second largest group of optical detection techniques used for microchip devices.

Lamp-based excitation is the most common method used with fluorescence microscopy to image biological samples, and it can be easily applied to detection in microfluidic systems through commercially available microscope setup. The common epifluorescence microscope uses condensing optics to collimate and parallelize the light generated by the lamp. After reflection by a dichroic mirror, the light is focused with an objective on to the microscope stage. Fluorescence emission is collected by the same objective and is usually detected by a photomultiplier detector (PMT) after encountering a dichroic mirror and emission filter [20].

2.4.1.3 Fluorescence Excited by Light-Emitting Diodes

Light-emitting diodes (LEDs) constitute an inexpensive and powerful alternative light source for fluorescence detection. LEDs exhibit very high output power covering the whole visible and UV wavelength range light, being highly effective light sources and requiring a low power driving currents. These advantages, combined with their very compact dimensions, mean that they are perfectly suited for integration into miniaturized devices [20].

However, the half-bandwidth of the light emitted from LEDs often exceeds 20 or 30 nm, which implies the need for additional filters and frequently generates a higher level of background signal. Since the photons in high-output LED are usually generated over an area of a few square millimetres, LEDs are not considered to be point light sources. This fact, combined with the divergence of the emitted light, means that sophisticated collimation and focusing optics are needed to make use of the whole radiation power of the LED.

2.4.1.4 Absorbance

This universal detection technique involves the measurement of absorbance/attenuation of a specific wavelength of incident light by the analyte.

Translation to the microfluidic systems has proven to be challenging, meanly because of the short optical path length (channel depth) in miniaturized systems and the difficulties in coupling the light into and out of these channels. Besides, absorbance detection in the UV range implies additional requirement of UV-transparent microdevices (quartz or fused silica) which significantly increases the cost of the devices.

2.4.1.5 Chemiluminescence

It is a very sensitive and selective optical method for analyte detection in which the light is generated by a chemical reaction. In this sense, no excitation light source is required and a filter system to reduce the background is not necessary. This simplified setup is very attractive when attempting to integrate CL detection into microfluidic devices.

2.4.1.6 Surface Plasmon Resonance

SPR is a physical process that can occur when photons of plane-polarized light strike a reflecting surface under total internal reflection (TIR) conditions. The reflecting surface is typically a thin film of metal, such as gold, which allows the electrical field of the photons to extend about half a wavelength beyond the metal surface. When the energy of the photon electrical field is optimized, it can interact with the free electron constellations in the gold surface. The photons from the incident light are absorbed and converted into the surface plasmons used for detection. SPR is a powerful technique for measuring biomolecular interactions in real-time and in a label-free environment. When reactants in solution (eg, high-affinity ligand or antigen) are exposed to a surface where a sensor molecule (eg, receptor or antibody) is immobilized, binding results in a change in the refractive index (RI) at the sensor surface, which is measured as a change in resonance angle or resonance wavelength, and normally converted to arbitrary units and displayed in a plot termed as sensorgram.

2.4.2 Electrochemical

Electrochemical detection is very important because it presents the inherent ability for miniaturization without loss of performance and its high compatibility with microfabrication techniques. Likewise, it possesses high sensitivity, its responses are not dependent on the optical path length or sample turbidity, and it has low-power supply requirements, which are additional advantages [42]. Different electrochemical techniques have been applied to microfluidic devices.

2.4.2.1 Amperometry

This is the simplest voltammetric technique, and it has been the most widely used in electrochemical sensors. The data are provided as a current generated by an electrode reaction. A three-electrode system consisting of a working electrode, a reference and an auxiliary electrode is the basic configuration. By applying the desired potential to the working electrode with respect to the reference electrode, the current is prompted to flow between the working electrode and the auxiliary to be measured. The conditions are fixed to be the diffusion of the analyte to the surface of the working electrode, which is the limiting factor. This way, the current changes linearly with respect to analyte concentration. The miniaturization and integration of this type of sensors is easily achieved by creating a three-electrode system in the form of thin- or thick-film patterns [43,44]. As the electrode diameter is decreased into the low micrometre range, a shift to nonplanar diffusion occurs (compared to planar diffusion in conventional macroelectrodes) causing an increase in the collection efficiency of the electroactive species at the surface. The practical result is an increase in the signal-to-noise ratio which generally translates into a lower detection limit. The disadvantages are that the decreased current that accompanies miniaturization can require improved electronics/shielding for it to be measured accurately, and that a stable reference electrode is necessary [45]. The combination of electrodes often seen in these microsensors is platinum used for the working electrode, Ag/AgCl for the reference electrode, and a platinum auxiliary electrode. Depending on the purpose, other materials, such as gold, carbon, iridium or the use of nanomaterials such as graphene, carbon nanotubes (CNTs) or nanowires can be used for the working electrode. The shape, size and position of the electrodes are precisely defined and this system can be formed in a network of microchannels. This technique presents good detection limits but is restricted to electroactive analytes. Also, selectivity can be tuned through a judicious choice of the detection potential.

The coupling of electrochemical sensors onto microfluidic platforms can be adapted from those previously reported for microchip capillary electrophoresis (MCE). In this sense, and according to the relative position between both working electrode/separation channel, the configurations can be classified as *end-channel*, *in-channel* and *off-channel* detection [46]. This configuration is especially important in case of MCE since minimal band broadening and electrical isolation (decoupling) from the high separation voltage must be ensured. In *end-channel* detection, the electrode is placed just outside the separation channel. For *in-channel* detection, the electrode is placed directly into the separation channel, and *off-channel* detection involves grounding the driving voltage before it reaches the detector by means of a decoupler.

2.4.2.2 Conductometry

In this case, a less sensible but universal detection technique has been applied as a detection mode in capillary electrophoresis-microchips and microfluidic platforms in general. Conductometry can be either in the galvanic (a pair of electrodes is placed in the separation channel for liquid impedance measurement) or the contactless mode (no contact between the pair of electrodes and separation channel solution). Contactless detection (CCD) is preferred for different reasons: (1) in case of MCE the electronic circuit is decoupled from the high voltage applied for separation (no direct DC coupling between the electronics and the liquid in the channel), (2) the formation of glass bubbles at the metal electrodes is prevented and (3) electrochemical modification or degradation of the electrode surface is prevented; (4) can be used with a wide range of background electrolytes and can take place at any location along a channel.

2.4.3 Mechanical

Nanometre scale mechanical detection systems as cantilevers have been widely used for sensor applications. Cantilever-based devices generally operate in two different modes upon analyte binding: (1) static deflection, where binding on one side of a cantilever causes unbalanced surface stress resulting in a measurable deflection; (2) dynamic, resonant mode, where binding on a cantilever causes variations of its mass and consequently shifts the resonant frequency. Various physical methods can be employed for actuating or sensing cantilever motion, including mechanical, optical, electrostatic, and electromagnetic methods. Cantilever-based devices can be realized with different shapes and sizes using conventional microelectromechanical system (MEMS) photolithography processes and bulk or surface micromachining. The flexibility of device design indicates the possibility of incorporation in microfluidic systems and miniaturized LOCs. With appropriate chemical functionalization the devices can specifically detect different chemical and biological entities. Mechanical-based detection may require no labelling of biomolecules. Often, labels make the detection method more complicated, time-consuming and costly and could interfere with the function of antigens or antibodies. Other characteristic of cantilever technology is the potential to fabricate large arrays of sensors for multimolecular sensing [40].

3. MICROFLUIDIC BIOSENSING FOR FOOD ANALYSIS

After description of the main components and operations of microfluidic analytical systems, Fig. 2 illustrates and identifies the main strengths (left panel) as well as the main weaknesses, which constitute the challenges (right panel) in the field of Food Microfluidics.

As previously stated, microfluidic platforms have gained a great attention as a network of integrated channels and reservoirs to perform the multiple operations associated to the sensing methodology. This approach is clearly advantageous since it allows ultrasmall sample volumes to be introduced and it has accurate fluid control and manipulation capabilities. Although detection continues being one of the main challenges for microfluidic platforms due to the small volumes used on micron-sized environment, mainly both optical and electrochemical detection techniques have proven to be well-suited detection

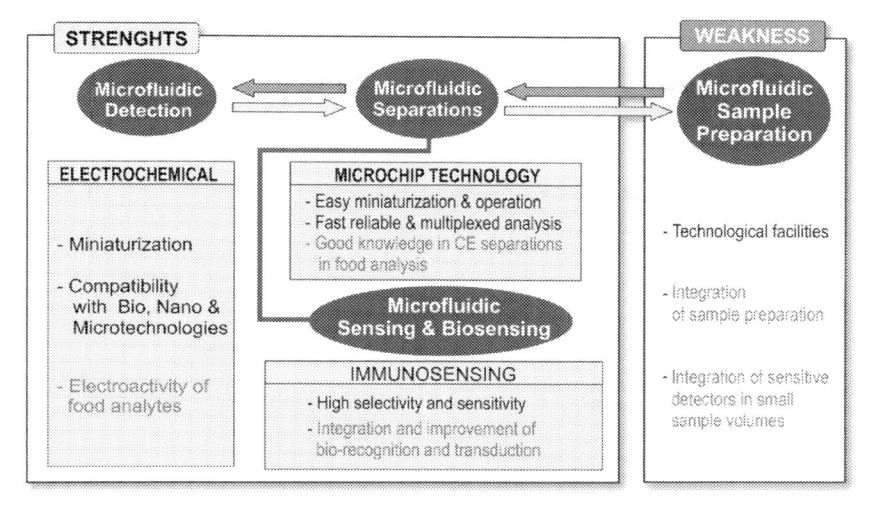

FIGURE 2 Strengths and weaknesses of *Food Microfluidics*. General strengths and weaknesses are pointed out in black, while the ones specifically related to Food Microfluidics are in red (grey in print versions). *Reproduced with permission from A. Escarpa, Lights and shadows on food microfluidics, Lab Chip 14 (2014) 3213—3224.*

principles. However, because of the complexity of food samples, the complete integration of sample preparation remains as one of the most important challenges in Food Microfluidics, since it requires sophisticated micro-fabrication facilities to develop microstructures for filtering, preconcentration, cleanup and even derivatization to make the analytes compatible with the very sensitive detectors. Stimulated by the current trend of µTAS and LOC technologies, biosensor strategies have currently been incorporated in microfluidic systems for the realization of highly sophisticated systems.

Fundamental groups of food concern analytes which have been analysed combining the exceptional characteristics of both biosensors and microfluidic platforms includes foodborne pathogens, toxins and biochemicals. Taking into account the nature of the biological molecule, which is responsible for the inherent selectivity of biosensors, two main approaches have been reported in the literature for food analysis in microfluidic platforms: antibodies (microfluidic immunosensors) and enzymes (microfluidic enzyme biosensors). Although DNA-based microfluidic devices have also been reported in the literature, they usually relay in the polymerase chain reaction (PCR) assay or array platforms and then will not be covered in this chapter.

3.1 Microfluidic Immunosensors in Food Analysis

Combination of immunoassay and microfluidic platforms offers advantages related to the selectivity and sensitivity, associated to the antigen—antibody interaction, with the remarkable features stated earlier for microfluidic platforms. Microfluidic immunoassays make use of a network of micro-channels and/or immunoreactor chambers usually built in a monolithic platform (microchip) from different materials as silicon, glass or polymers with part of all the necessary components of an immunoassay procedure integrated. Furthermore, these microfluidic platforms are especially suitable for immunoassays from different points of view:

1. The long-time associated to incubation stages in conventional immuno-assay can be attributed to the inefficient mass transport for the immunoreagents to move them from the bulk solution to the wall surface where interaction takes place. In microchannels, the surface area to volume ratio is higher which makes the diffusional distances dramatically reduced and produce lower analysis times.

2. Miniaturization reduces drastically the consumption of the precious samples and expensive reagents.

3. Fluid movement can be easily controlled, especially with electrokinetic fluidic motivation, through the adequate control of applied electric fields, or in a more intricate way by using of pumps, valves, mixers, etc.

The advantages previously mentioned have converted the microfluidic immunoassay platforms into an emergent and powerful alternative in the field of food and agricultural analysis in which a complicated sample preparation step is usually highly required.

3.1.1 Toxins Detection

Today, one of the most important fields of food analysis is the accurate determination of extremely low levels of toxins. Toxins are a heterogeneous group of compounds able to interfere with biochemical processes, such as membrane function, ion transport, transmitter release and macromolecule synthesis [47]. Toxins can be produced by animals, plants and microorganism, and their toxic response causes diseases or even the death. Toxin contamina-tions can be widely spread in the whole work, threatening the health or life of humans and livestock, and affecting the international trade [5].

Some of the most common toxins affecting food and agricultural products are mycotoxins. They are a natural and structurally diverse group of secondary metabolites produced by fungi on agricultural commodities in the field and during storage under a wide range of climatic conditions. Several hundred different mycotoxins have been discovered and they can be found as contaminants in cereals, related products used for feed, beverages as fruits juices and wines, foodstuffs and their products worldwide. Mycotoxins are potent toxins and have a wide range of actions on animals and humans (eg, cyto-, nephron- and neurotoxic, carcinogenic, mutagenic, immunosuppressive and oestrogenic effects) causing important health and economic problems. Thus, it has made necessary to establish maximum permitted levels of these mycotoxins in different foods and feedstuffs and develop fast, low-cost, sensitive and selective analytical methods.

Related to electrochemical detection in microfluidic immunosensors, a microfluidic cell of 7.50 µL has been used for performing a heterogeneous competitive electrochemical immunosensor for citrinin determination in rice samples [48]. The cell consists of two bodies: the top body, made of stainless, contains the counter electrode, internal holes that allow entry and exit of solution flow, as well as another channel that communicates with the reference electrode; the bottom body comprises the glassy carbon electrodes (GCEs), one of them used to immobilize the citrinin-ovalbumin conjugate (CIT-OVA) and the other used as detection unit. Different solutions flow in a stop and go mode for performing the heterogenous competitive immunoassay. A limited amount of mouse anti-CIT IgG antibody is added to CIT from rice samples, competing immunologically with CIT-labelled ovalbumin bonded at the GCE. This electrode surface was previously modified by electrodeposition of gold and subsequent modification with a cysteamine self-assembled monolayer (SAM) to bind CIT-OVA. Formation of an immune-complex with a secondary antibody rabbit anti-mouse labelled with the enzyme horseradish peroxidase (HRP) allowed the determination of the analyte. The HRP catalyses the oxidation of catechol to benzoquinone in presence of hydrogen peroxide which back electrochemical reduction to catechol can be amperometrically detected on the GCE surface at -0.15 V versus Ag/AgCl. The immunosensor presents a range of work between 0.5 and 50 ng/mL; an LOD of 0.1 ng/mL and a total assay time of 45 min, avoiding any sample pretreatment.

The same group has developed similar microfluidic electrochemical immunosensors for determination of ochratoxin A, *Botrytis Cinerea* and zearalenone in fruits and feedstuffs, respectively [49−51]. The main body of these devices was made of polymeric materials as Plexiglas or PMMA, where a central channel (100 µm internal diameter) is designed as flow-through chamber containing the microfluidic immunosensor and the detector system. Other accessory channels are used to inject the different solutions for performing the immunoassay. These microfluidic immunosensors make use of magnetic particles confined at the central channel as platform for specific antibody or analyte-derivatized immobilization. Following the principles of heterogeneous competitive immunoassay, the analyte from the sample and an HRP-analyte conjugated compete for binding to the specific antibody. After washing steps and using a working gold layer electrode deposited at central channel by sputtering technique, the enzymatic product is amperometrically measured and related to the analyte concentration in the sample.

The use of beads as support for antibody immobilization in microfluidic devices has been broadly reported. Compared to microtitre wells used in ordinary immunoassays, the use of beads (mainly magnetic beads) present relevant advantages such as an increased surface-to-volume ratio, even compared to microchannels, reducing the diffusion distances. This fact outcomes assays with higher sensitivity due to high efficiency of interactions between samples and reagents. Easy manipulation and transport of the beads in a fluidic system, a variety of surface modifications available for an easy biomolecules attachment are other relevant benefits.

Another interesting example of mycotoxin analysis has been developed using a total integrated electrokinetic magnetic bead-based electrochemical immunoassay for control of permitted levels of zearalenone in infant foods [52,53]. This work makes use of electrokinetic EOF flow as pumping force to delivery sample and reagents into the microfluidic device. Electrokinetic EOF-based fluidic motivation benefits these analytical platforms since the flow in multiple channels on a microchip can easily be controlled with a few electrodes. Electrokinetics is especially valuable in immunosensing devices, since it allows to modify the injected amounts of reagents and samples in the different analytical steps by simply changing the electric field and time applied (acting as a variable well-controlled loop), as well as manipulating the fluidics in different directions including reversed and stopped flows with extreme simplicity and automation. In this work, following the mentioned principles, a strategy which implies the creative definition and use of simple double-T microchip layout to integrate all required immunoassay steps has been proposed as it is depicted in Fig. 3. Different reagents and the sample are automatically moved following a predefined potential and time programme by electrokinetic injection and pumping. This strategy is based on physical separation of both zones (immunological complex formation and enzymatic reaction) in different channels of the microchip, avoiding one of the greatest problems associated with microfluidic immunoassays as nonspecific adsorption to the channel walls surfaces. Immunoassay is carried out using a competitive scheme where the mycotoxin zearalenone and an HRP-labelled derivative compete for the binding sites of the specific antibody oriented immobilized onto magnetic beads through the use of protein G. Detection made use of HRP enzyme, which in presence of hydrogen peroxide and hydroquinone as electrochemical mediator, produces an electrical current proportional to the activity of the enzyme and, consequently, to the amount of zearalenone mycotoxin in the sample. Using this methodology, either solid or liquid food samples were analysed, and those contaminated above the permitted levels of the mycotoxin could be easily discriminated. A suitable LOD of 0.4 micrograms per litre (mg/L) well below the legislative requirements, and an extremely low systematic error of 2% from the analysis of a maize certified reference material, revealing additionally an excellent accuracy, have been obtained using this integrated microfluidic approach.

Other interesting group of toxins is those related to bacterial toxins. They are by-products of microorganisms, which can negative affect health and life of humans and animals. Included in this category, staphylococcal enterotoxin B (SEB), cholera toxin (CT) and botulinum toxin are the most investigated.

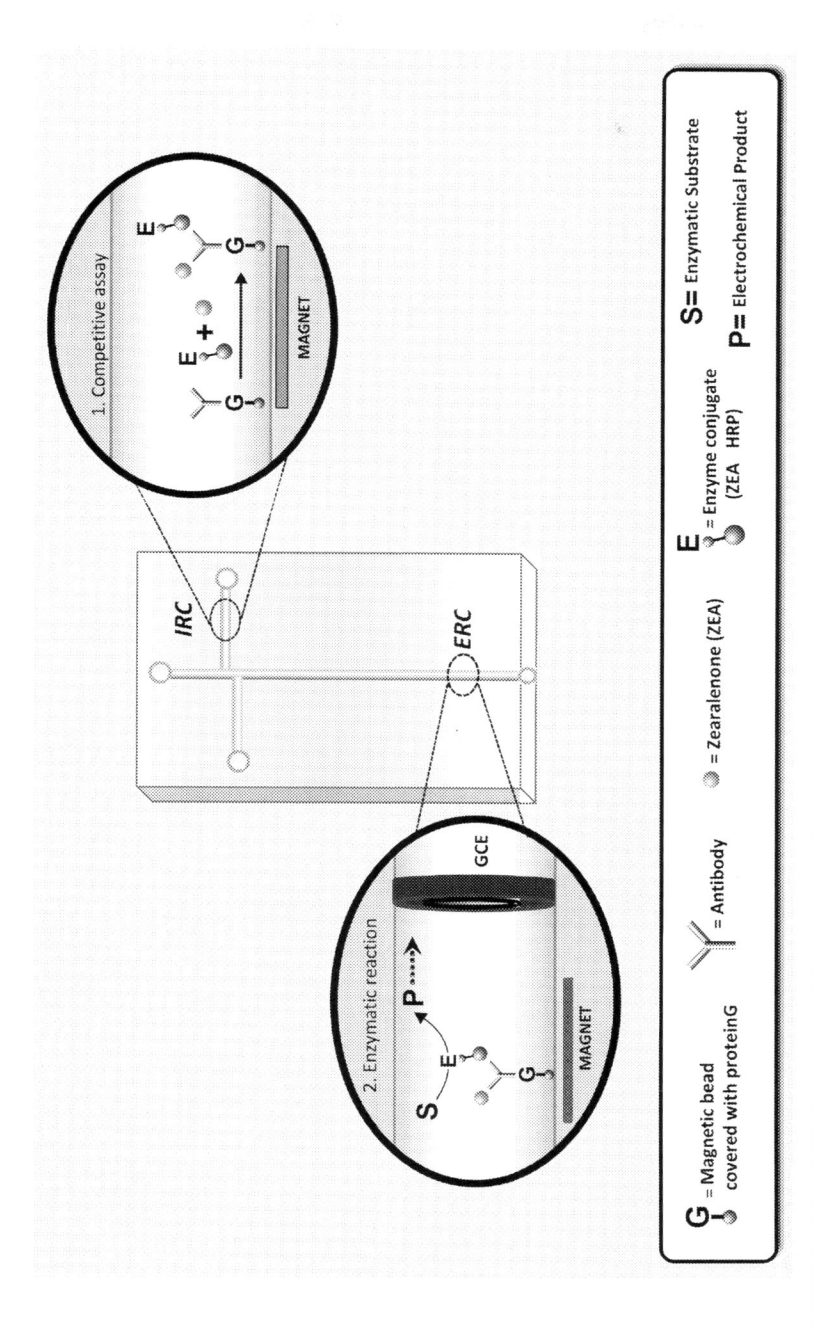

FIGURE 3 Microfluidic chip layout and immunoassay principle (*IRC*, immunological reaction chamber; *ERC*, enzymatic reaction chamber). *Reproduced with permission from M. Hervás, M.A. López, A. Escarpa, Integrated electrokinetic magnetic bead-based electrochemical immunoassay on microfluidic chips for reliable control of permitted levels of zearalenone in infant foods, Analyst 136 (2011) 2131–2138.*

SEB is part of the *Staphylococcus aureus* enterotoxin family that causes huge food poisoning cases worldwide with symptoms of vomiting and severe diarrhea. It is a potent intoxicator at levels as low as 1 ng/mL, with a human LD_{50} of 1 mg. A PDMS microfluidic system using reinforced supported bilayer membranes and fluorescent detection for immunosensing of SEB in milk samples have been reported [54]. In this approach, supported bilayer membranes were generated by fusing vesicles of phosphoethanolamine and phosphatidylcholine in oxidized PDMS microchannels. This membranes minimize the nonspecific adsorption of biomolecules to the PDMS surface, while the addition of a streptavidin layer make them air-stable and allow for subsequent display of biotin-functionalized antibodies. Afterwards, a sandwich format immunoassay was performed by flow injection of the specific biotinylated anti-SEB antibody, the sample and a second antibody labelled with a fluorescent die. An LOD of 0.5 ng/mL was obtained both in Tris buffer and centrifuged milk samples.

A different strategy for microfluidic immunoassay for SEB detection in soymilk samples has also been developed [55]. In this case, an eight-channel LOC fabricated by polymer lamination technology allows the multiplex determination even of eight analytes, thanks to a modular design with interchangeable recognition elements, or parallelized eight sample analyses. Anti-SEB antibodies bounded to CNTs were immobilized onto a polycarbonate strip, which serves as an interchangeable ligand surface bonded onto the LOC. SEB samples are loaded into the device and detected by an ELISA assay using HRP conjugated to a secondary antibody anti-SEB, and detected by enhanced CL on a portable cooled CCD detector. The system presents an LOD of 0.1 ng/mL, where CNT immobilization of the antibody increased the sensitivity of detection sixfold. A syringe used to move fluids and a simple interchangeable immunological surface allows a versatile portable point-of-care for toxin detection.

Cholera toxin, composed of two subunits (A and B) and produced by *Vibrio cholerae*, causes severe diarrhea and can cause death within 3−4 h, if left untreated. The function of B subunit (known as a 'choleragenoid') is to bind specifically to the ganglioside GM1 receptor on the epithelial cell membrane of the small intestine (particularly the jejunum and the duodenum). While the A subunit is the active function protein that causes the diarrhea by activating the production of cAMP, causing the dramatic efflux of ions and water from infected enterocytes, leading to the appearance of the first cholera symptoms: the painless and copious rice-water diarrhea, up to 24 L in 24 h. Development of a microfluidic device using specific antibody anti-CT (B subunit) and GM1 receptor as the recognition system, and liposome-based signal amplification encapsulating a fluorescence dye or electroactive compounds within liposomes was reported [56]. In the electrochemical assay an interdigitated ultramicroelectrode array (IDUA) was integrated as signal transducer into the microfluidic device, while in the fluorescence detection, a fluorescence microscope was used. The microfluidic device was fabricated in PDMS and a glass slide (bearing the IDUA for electrochemical detection) and packaged in a Plexiglas housing that provided leak-proof sealing and a connection to a syringe pump. The assay was carried out in a microcentrifuge tube where anti-CTB antibody conjugated to superparamagnetic beads, the CTB sample and a fluorescent die (or redox molecule) encapsulated GM1 liposomes were incubated. Afterwards, the sample mixture was injected to the microfluidic chip using a syringe pump. After immobilization of the liposome-CTB-beads complexes inside the channel by a magnet, washing of the unbound liposomes and release of the fluorescent die or redox molecules, fluorescence or electrochemical detection was performed. The LODs of both assay formats were found to be 6.6 and 1.0 ng/mL, respectively using a total assay time of 1 h.

Detection of CT using a label-free electrochemical impedance spectroscopy (EIS) immunosensor integrated into a PDMS microfluidic platform has been also reported [57]. The biochip include two reaction chambers/sensing areas, both including four interdigitated electrodes. These gold electrodes were used for immobilization of anti-CT antibodies by deposition of a mixed SAM of 11-mercaptoundecanoic acid and 2-mercaptoethanol, their activation by NHS/EDC and functionalization with protein A. After the flow of the toxin from the sample and the adequate incubation step, EIS measurements were performed. A concentration of 1 ng/mL was detected, where the diagnostic time includes 1 h for the biorecognition event and 5 min for the impedance measurement.

Harmful algal blooming (HAB) has also become a serious problem around the world due to the urbanization and consequential eutrophication in aqueous environment. HAB constitute not only an environmental problem but also produces algal toxin that affect the health of people and foodstuff. Among algal toxins, microcystin, saxitoxin and cylindrospermopsin are of paramount importance and are known as three of the major cyanotoxins. Microcystin is the most widely spread and have been identified as hepatotoxic by inhibiting the protein phosphatases 1 and 2A. This effect causes acute liver failure and haemorrhage, and chronic exposure can exert strong tumour-promoting activity. Saxitoxin, mainly generated by freshwater cyanobacteria and marine dinoflagellates, is responsible for paralytic shellfish poisoning in humans. Contamination in drinking water and bivalve molluscs can cause oral paraesthesia, and soon afterwards, cardiovascular dysfunction and respiratory paralysis in humans. Cylindrospermopsin is a cytotoxic alkaloid generated by multivarious cyanobacteria, and due to its ability to inhibit some protein synthesis mechanisms, this toxin is able to influence many organs, such as the liver, kidney, thymus and lung.

A PDMS microfluidic device for integrated and automatic ELISA sensor for determination of microcystin, saxitoxin and cylindrospermopsin has been presented [58]. The microfluidic analytical chip (30 × 30 mm) includes two layers: the upper patterned fluidic layer and the bottom pneumatic control layer (Fig. 4A). The microchip was fabricated using soft lithography to create the fluidic channels and the control channels for actuating the valves located in the respective layers of the microfluidic chip (Fig. 4A **left side**). Besides, the microfluidic chip is composed of immune-reaction columns and it is capable of executing seven immunoreactions simultaneously. This allows the possibility of analysing different toxins in one single chip. The immunoassay is based on a competitive format where microspheres of protein A-anti-IgGs together with the specific antitoxin antibody were loaded in each immune-reaction column. Thanks to the sample circle and buffer circle (Fig. 4A **right side**) for loading reagents and washing solutions, respectively, and the automatic control of the valves, algal toxins and enzyme-labelled algal toxins

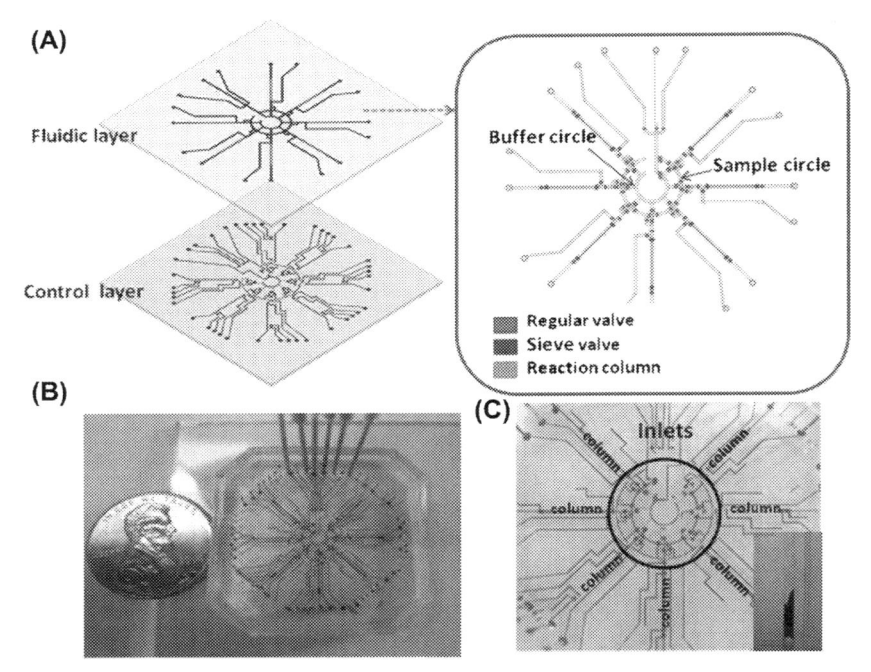

FIGURE 4 (A) Schematic representation of fluidic layers of the immunoreaction chip used in the detection of algal toxins. Valves and columns are clarified by different colours: red (grey in print versions) for regular valves (for isolation), blue (dark grey in print versions) for sieve valves (for trapping protein A beads loaded in the column module) and green (light grey in print versions) indicates the immune columns by loading of microspheres. (B) Optical micrograph of the microfluidic chip. The various channels have been loaded with food dyes to help visualize the different components of the microfluidic chip: control line colours are as in (A), plus green (light grey in print versions) for fluidic channels. A penny coin (diameter 18.9 mm) is shown for size comparison. (C) Optical micrograph of the central area of the chip containing seven immune-reaction columns. Inset: a snapshot of the protein A beads loading process in action. *Reproduced with permission from J. Zhang, S. Liu, P. Yang, et al., Rapid detection of algal toxins by microfluidic immunoassay, Lab Chip 11 (2011) 3516–3522.*

competed for the binding sites of specific antibodies in solution inside the immune column. By sequentially introducing the different assay solution, the immunoassay was completed and detection was performed by colour signals development and recorded by a digital camera. A detailed micrograph of the microfluidic device is shown in Fig. 4B and C. The size of each immuno-reaction column is about 40 nL, the total assay time is lower than 30 min and LODs are 0.020, 0.020 and 0.015 ng/mL, respectively, for saxitosin, micro-cystin and cylindrospermopsin.

In the same way, a surface acoustic wave microfluidic device for okadaic acid (OA) detection as a diarrheic shellfish poisoning toxin was reported [59]. The microfluidics was made in PDMS, while the acoustic wave platform was made in quartz. Movement of fluidics within the device was obtained by a syringe pump. Immunosensitive layer was prepared by immo-bilization of anti-OA antibodies onto a SAM. Following a sandwich assay format, sample solution and second antibody were flowed through the sensor layer. Including immune-sensitive layer coating, sample and antibody injections, with the corresponding incubation steps, the experiment duration time lasted less than 3 h. The method can detect 2 μg of OA in 10 g of shellfish food, 10 times lower than the maximal concentration permitted.

3.1.2 Foodborne Pathogens Detection

From the point of view of food safety, contamination caused by pathogens is a significant public health concern for consumers worldwide. Tolerable levels of these agents are getting more stringent regulations due to the high concern of people for food safety. Some of the most common pathogens involved in food outbreaks include bacteria of the genus *Escherichia, Salmonella, Listeria* or *Staphylococcus*, which requires a rapid, sensitive and reliable detection.

Salmonella is a major foodborne pathogen causing gastroenteritis in humans, and its symptoms include fever, abdominal pain, nausea and vomit-ing, diarrhea, dehydration, weakness and loss of appetite. This pathogen is associated with raw or undercooked eggs, poultry, beef and unwashed fruit. Detection of these bacteria in processed chicken samples was reported using a microfluidic immunosensor [60]. The assay relay in a sandwich format where superparamagnetic particles, bounded with the specific antibody, separate and concentrate the cells from the sample. In a next step, the captured cells are introduced in a PDMS-glass microfluidic chip where antibody-conjugated quantum dots are used for labelling the cells. The movement of different solutions is performed by negative pressure created by a peristaltic pump. The Quantum dots-labelled cells are captured in the detection zone using an external magnetic field and detection is carried out by a portable fluorometer. The effectiveness of the microfluidic immunosensor was tested in complex food samples as chicken extracts, producing positive results at *Salmonella* concentration of 10^3 CFU/mL (LOD) within 30 min.

Escherichia coli O157:H7 which causes patients hemorrhagic diarrhea and kidney failure is considered one of the most dangerous foodborne pathogens.

The potential application of a microfluidic immunoassay for *E. coli* detection, and multiplexed assay for simultaneous detection of different pathogens could be performed in a single microfluidic device by using different-sized glass beads attached to the corresponding specific antibodies [61]. The microfluidic device, fabricated in PDMS and glass slide bounded together, contains three microchambers connected to each other by microchannels of different widths, which allow the different-sized beads to be packed in different microchambers. In this approach, *E. coli* O157:H7 expressing green fluorescence protein was detected using an inverted fluorescence microscope.

A label-free impedance microfluidic immunosensor coupled with magnetic nanoparticles immunoseparation for the direct detection of bacterial cells in food samples was also presented [62]. The microfluidic cell consists in a detection microchamber, and inlet and outlet microchannels fabricated by bonding a gold interdigitated array microelectrode chip to a PDMS microchannel. Magnetic nanoparticle-antibody conjugates were used to separate and concentrate the target bacteria from the food matrix (extract of ground beef samples). The cells attached to nanoparticles were injected into the flow cell with the aid of a syringe pump. Impedance measurements were performed without using redox probe or labelled antibodies. The microfluidic immunosensor was able to detect 1.2×10^3 cells of *E. coli* O157:H7 in 35 min.

A PDMS-based microfluidic immunosensor chip integrated with nanoporous alumina membranes for *E. coli* and *S. aureus* has also been reported [63]. Antibacterial antibodies were immobilized on alumina surface via self-assembled silane monolayer and introduced into a PDMS microfluidic device. Once targeted bacteria were captured by the complementary antibody on the nanoporous membrane, electrolyte current flowing through the nanopores was blocked which led to impedance amplitude increase. Impedance spectrum is used to monitor impedance amplitude change before and after bacteria capturing by complementary antibodies on the nanoporous alumina membrane. The higher the covered area ratio is, the larger the impedance amplitude will be monitored due to the increased blocking efficiency of electrolytic current flow through the nanoporous membrane. Cross-experiments between antibodies and nonspecific bacteria using impedance spectrum and sandwich-type immunoassay confirmed the specificity of this device for bacteria detection. The sensitivity of this technique was around 10^2 CFU/mL which was higher than traditional microelectrodes-based impedance sensor. Similar work was previously reported by the same group [64], where poly(ethylene glycol) hydrogel was fabricated on saline-modified nanoporous alumina surface for bacteria detection through the patterned antibody recognition mechanism. Low-frequency impedance spectrum is used to detect specific bacteria binding at low bacteria concentration of 10^2 CFU/mL.

3.1.3 Biochemicals Detection

Determination of folic acid using a robotized microfluidic platform for automated immunoassay has been reported [65]. The immunochip comprised a linear array of eight independent microchannels with gold electrodes incorporated. As in conventional 96-well microtitre plates, the microchannels have a standard spacing of 9 mm to facilitate automation, being fully compatible with conventional pipetting devices. Fig. 5A shows a drawing of the cartridge containing a housing with a sample deposition reservoir and a microchannel with inlet/outlet in contact with microelectrode tracks. The microchannels have a length of 1 cm and a total volume of 65 nL. Each immunochip was capable of running eight assays in parallel. The chips are inserted into the measuring instrument, which contains an 8-channel peristaltic pump, valves and electrochemical detection electronics (Fig. 5B). The immunoassay developed for folic acid detection in food samples has designed as an indirect competitive assay, where the sample analyte competes with the coated antigen for the free sites of the antibody present in solution. The LOD for folic acid in milk samples was 1.37 ng/mL. This is an interesting example of automated assay for multiplexed analyte detection, where folic acid can be determined within 5 min.

3.2 Microfluidic Enzyme Biosensors in Food Analysis

Enzymes were historically the first molecular recognition elements included in biosensors, and even today they constitute the largest category of biosensors becoming an important tool for detection of chemical and biological components for clinical, environmental and food monitoring.

Enzyme biosensors have several advantages. These include the catalytic amplification of the biosensor response by modulation of the enzyme activity with respect to the target analyte; a very wide range of enzymes available commercially, usually with well-defined and assayed characteristics; and the possibility of modifying the catalytic properties or substrate specificity by means of genetic engineering. Enzyme biosensors have also some limitations such as the limited number of substrates for which enzymes have been evolved, the limited interaction between food contaminants and specific enzymes, stability of enzymes and the capacity to maintain enzymatic activity over a long period of times [66].

Typical examples of enzymes involved in food applications are cholinesterase for organophosphorous and carbamate pesticide analysis; tyrosinase or laccase for analysis of phenols, quinones and related compounds; glucose oxidase for sugar content analysis, carboxyl esterase, alcohol oxidase, carboxypeptidase, L-aspartase, peptidase, aspartate aminotransferase for aspartame, xanthine oxidase to detect the level of freshness in fish, glucose

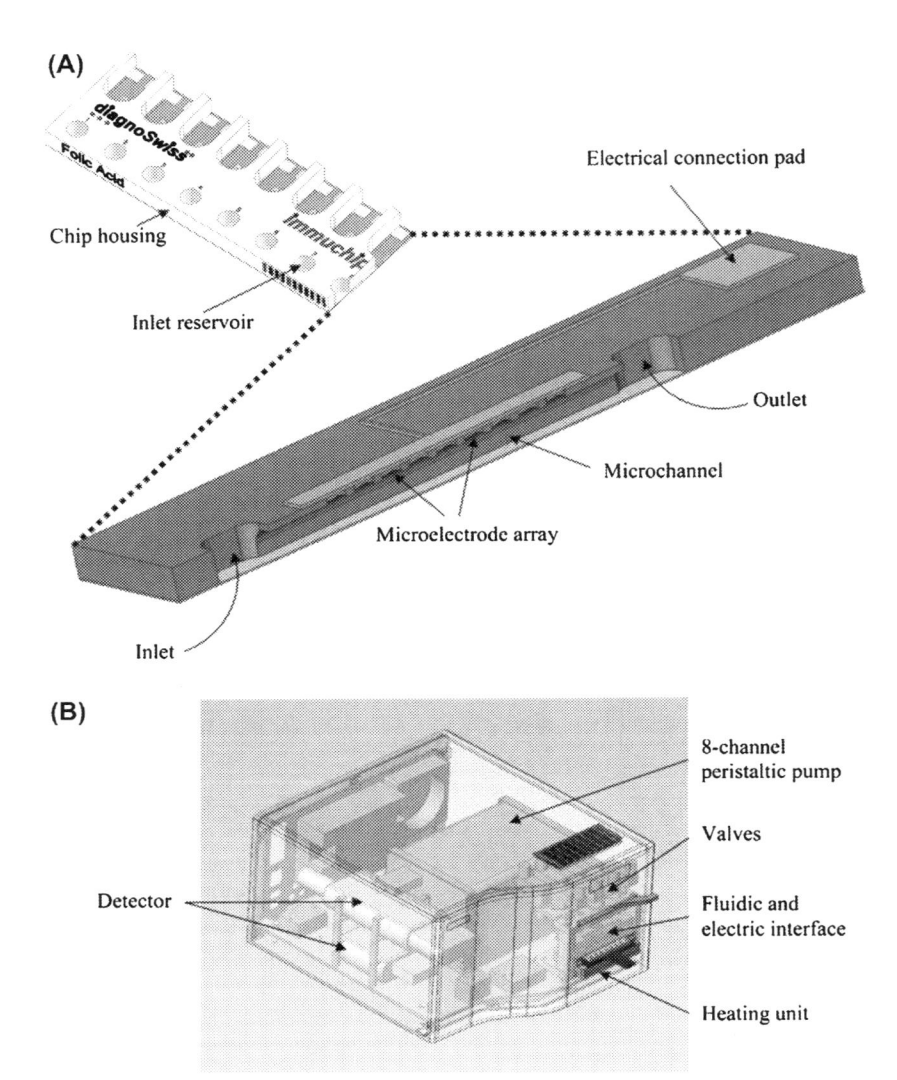

FIGURE 5 (A) Schematic design of the 8-channel immunochip cartridge used for folic acid determination, and detailed view of a microchannel cross section with inlet, outlet and working electrode tracks. (B) Schematic design of the whole automated measuring instrument. *Reproduced with permission from D. Hoegger, P. Morier, C. Vollet, et al., Disposable microfluidic ELISA for the rapid determination of folic acid content in food products, Anal. Bioanal. Chem. 387 (2007) 267–275.*

oxidase, β-galactosidase, fructose dehydrogenase and other dehydrogenase to detect glucose in fruit juice, lactose or lactulose in milk and so on [66].

Related to the operation mode, two main categories can be found. In the so-called direct mode, monitoring of analyte concentration of reactions producing or consuming such specific analyte is performed. However, sometimes, some food contaminant selectively inhibit the activity of certain enzymes, thus the resulting product concentration is affected. The monitoring of inhibitor or activator of enzyme is considered the indirect mode.

Due to the exceptional properties of microfluidic devices considered across this chapter, they also constitute an excellent platform to integrate enzyme biosensors and its application in the food analysis field. In the following paragraphs, some relevant examples from the literature of microfluidic enzyme biosensors will be reported.

3.2.1 Toxin Detection

Botulinum neurotoxin (BoNT), secreted by the bacteria *Clostridium botulinum*, is among the most potent and deadliest known toxin, with an oral lethal dose of 70 μg for an adult human. There are seven serotypes (A—G), each comprising a 100-kDa heavy chain and a 50-kDa light chain. The heavy chain facilitates entry into neurons via cell surface binding and receptor-mediated endocytosis, whereas the light chain is the catalytic component that, once inside the cell, cleaves neuronal proteins necessary for neurotransmitter release and leading to flaccid paralysis associated to botulism [67]. Because of its potency, contamination of food or drink poses a severe threat to human and animal health.

Detection of BoNT type A was accomplished by an enzymatic sensor integrated into arrayed microchannels [68]. The biosensor comprises a SAM for selective conjugation to the C-terminal cysteine of a synthetic BoNT/that mimicked the toxin's in vivo SNAP-25 protein substrate and labelled with a fluorophore (Fig. 6A). These SAMs were formed onto a patterned gold pad, with each pad accommodating a single microchannel (Fig. 6B). Each channel, fabricated using standard soft lithography in PDMS, consisted of an input port connected to a detection port with a total volume of 1.8 μL, and passive forces are used to move liquids inside (Fig. 6C). When exposed to BoNT/A or its catalytic light chain, results in the enzymatic cleavage of the peptide substrate from the surface and releasing of fluorescently labelled fragments into the bulk solution within the microchannel. Then, detection and quantification was performed through a fluorescence microscope. A LOD of 3 pg/mL was obtained in buffer, similar to those found in food samples as vegetable soup with a total assay time of 3 h. In a similar work previously reported by authors, the specific substrate peptide was bound to beads. Although a good LOD of 10 pg/mL was obtained, a great interassay error was produced since output was dependent on the amount of beads present in the sample reservoir [69].

Using a different strategy, a microfluidic biosensor for BoNT/A detection was constructed [70]. The microfluidic device was fabricated using laminated object manufacturing technology where six microchannels and corresponding reservoirs were fabricated in 3D structure with PMMA and polycarbonate layers bonded with adhesive. Samples and reagents were moved by a syringe

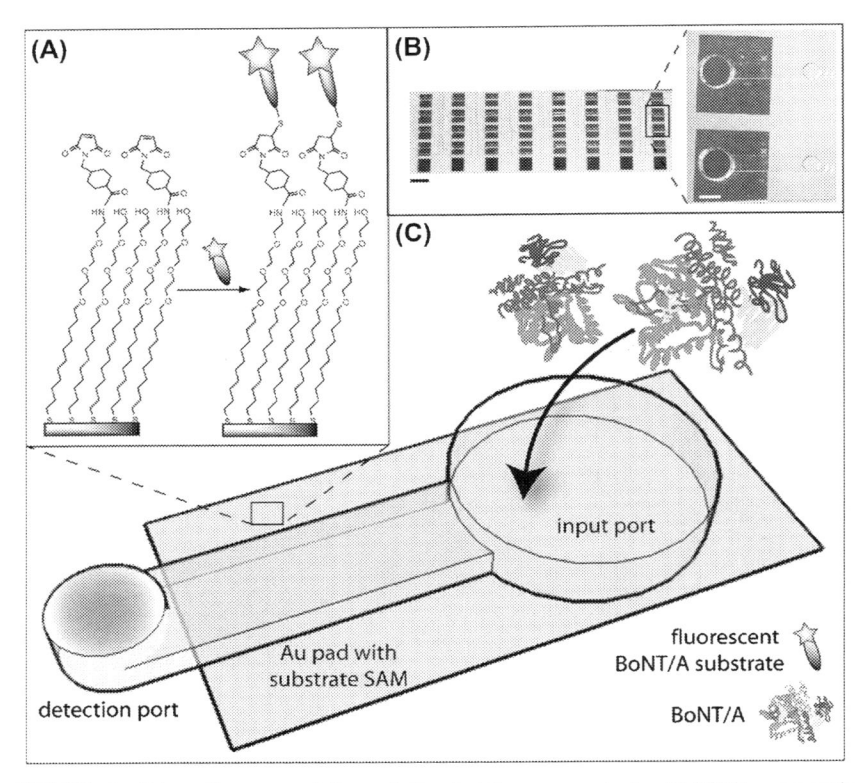

FIGURE 6 Schematic representation of the botulinum neurotoxin (BoNT)/A enzymatic microfluidic sensor. (A) Self-assembled monolayer (SAM) formation onto Au surface presenting the BoNT/A enzymatic substrate. (B) Poly(dimethylsiloxane) microchannels on 40 arrayed Au pads (10.5 mm^2; scale bar 5 mm). Inset image represent two neighbouring channels (scale bar 1 mm). (C) BoNT/A is added at input port and incubated on SAMs, during which time it can cleave the immobilized substrate, releasing fluorescent fragments into solution. Flu-labelled fragments are concentrated at detection port via evaporation. *Reproduced with permission from M.L. Frisk, W.H. Tepp, E.A. Johnson, et al., Self-assembled peptide monolayers as a toxin sensing mechanism within arrayed microchannels, Anal. Chem. 81 (2009) 2760–2767.*

pump. Fluorescence detection is performed with a CCD portable sensor and LED illumination allowing in situ measurements. The BoTN/A assay relays in the specific cleavage activity of the toxin and the Föster resonance energy transfer (FRET) detection. The peptide substrate for BoTN/A (SNAP-25) is labelled with the FITC donor/DABCYL acceptor FRET pair. In this peptide, the close proximity of the DABCYL quenches FITC fluorescence. Interaction of the substrate with the toxin results in cleavage of the peptide sequence, disrupting the FRET and resulting in FITC fluorescence measured by the CCD camera. A total assay time of 2 h with a sample of volume of 10 μL and an LOD of 0.5 nM were obtained.

3.2.2 Biochemicals Detection

As the standard of living has improved around the world and thus the food culture, people have become more aware about the safety and quality of foods. In this sense, there is an increasing demand in analytical methods to check some aspects as freshness of perishable foods at various stages of distribution, which particularly applies to fish.

One of the indexes proposed to measure the fish freshness is that known as the K-value. The basis of such testing is the ratio of components of the postmortem degradation pathway starting from adenosine triphosphate (ATP): A high ATP content indicates a high level of freshness, whereas a high hypoxanthine (Hx) content indicates that spoilage will begin soon. The K-value is defined as follows:

$$\text{K-value}(\%) = \frac{[\text{HxR}] + [\text{Hx}]}{[\text{ATP}] + [\text{ADP}] + [\text{AMP}] + [\text{IMP}] + [\text{HxR}] + [\text{Hx}]}$$

where [ATP], [ADP], [AMP], [IMP], [HxR] and [Hx] are the concentrations of ATP, adenosine diphosphate (ADP), adenosine monophosphate (AMP), inosine monophosphate (IMP), inosine (HxR) and hypoxanthine (Hx), respectively.

Using this principle, a microfluidic device for measurement on-site of the K-value in jack mackerel, yellowtail and sea bream extracts is presented [71]. In this assay, different enzymes were used in different sensing sites of the device. The sum of concentrations of HxR and HX in the numerator of the equation was measured with the enzymes nucleoside phosphorylase and xanthine oxidase (XOD) immobilized on the working electrode of the first sensing site. The sum of the concentration of all compounds in the denominator was measured by converting ATP, ADP, AMP, adenosine and IMP into HxR using enzymes in the liquid phase and detecting HxR and Hx at the second sensing site, as in the first case. In both cases, the amount of hydrogen peroxide produced by the final enzymatic reaction of XOD was detected electrochemically on the platinum working electrodes. The results obtained at different times correlated perfectly with those obtained by HPLC as habitual used technique.

The used device is described in Fig. 7. Structures including flow channels, and injection ports for the sample solution (injection port 1) an enzyme solution (injection port 2) were fabricated in PDMS (Fig. 7A **left side**). In the device, solutions are processed in a long main flow channels. Two sensing sites were formed, one between the two injection ports (sensing site 1) and the other downstream (sensing site 2) (Fig. 7B). To place plugs precisely at the sensing sites, a bypass flow channel structure was formed as shown in Fig. 7C. At each sensing site, a three-electrode system was formed on glass substrate (Fig. 7A **right side**). After injection of the sample and the enzyme solution in the adequate ports, plugs formed spontaneously in the main flow

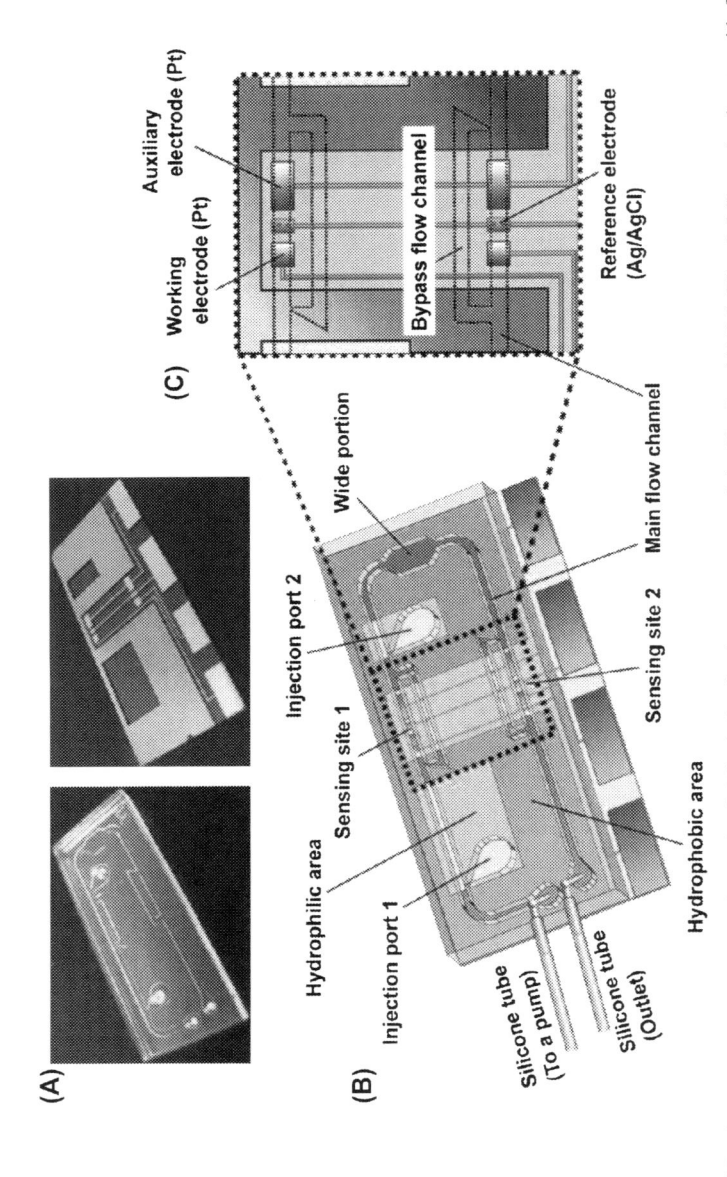

FIGURE 7 Schematic representation of the microfluidic device for the measurement of the K-value. (A) Poly(dimethylsiloxane) substrate with flow channel structures (left) and glass substrate with patterned electrodes (right). (B) Schematic of the completed chip showing the layout of the flow channel network and the electrodes. (C) Magnified view of the sensing site with the bypass flow channels. *Reproduced with permission from D. Itoh, E. Koyachi, M. Yokokawa, et al., Microdevice for on-site fish freshness checking based on K-value measurement, Anal. Chem. 85 (2013) 10962–10968.*

channel through capillary action. To move the plugs, air pressure was applied using a microinjector. Both plugs are merged for enzymatic reaction in the corresponding sensing site and electrochemical measurements were conducted. A similar work was previously reported by the authors, using droplet-base microfluidic sensing system [72].

Related to food safety, L-glutamate is other interesting molecule to analyse. Chinese restaurant syndrome, Parkinson and Alzheimer diseases raise the argument of glutamate safety and, therefore, increased attention for its determination. Two types of microfluidic configurations for detection of the molecule L-glutamate have been reported [73]. The assay is based on two enzymatic cycling reactions, where the enzymes were coimmobilized either on a silicon microchip surface or polystyrene microbeads. In the first case, the microchip contains 32 parallel porous U-shaped channels. In the second, enzymes were coimmobilized on polystyrene beads and packed into a flow-through reactor comprising one microchannel or 7 mm length and 300 μm width. The microfluidic chip containing the immobilized enzymes-beads into a single channel was integrated into a sequential injection analysis flow system for full automation of the system. The immobilized enzymes (L-glutamate dehydrogenase and D-phenylglycine aminotransferase) recycle L-glutamate by oxidation to 2-oxoglutarate followed by the transfer of an amino group from D-4-hydroxyphenylglycine to 2-oxoglutarate. The reaction was accompanied by reduction of nicotinamide adenine dinucleotide (NAD^+) to NADH, which was monitored by fluorescence detection. An LOD of 3 μM and better enzyme stability were obtained when enzymes were coimmobilized onto beads packed in the single microfluidic channel.

By other hand, a PMMA microfluidic device for glucose determination has been developed [74]. Immobilization of glucose enzyme on the wall surface of a single microchannel was performed after functionalization with polyethyleneimine and using glutaraldehyde as cross-linker agent. The hydrogen peroxide generated by the enzymatic reaction was detected in an electrochemical flow cell localized outside of the reactor using a platinum disk as working electrode. The system was applied for determination of glucose content in soft drinks showing an LOD of 1.40×10^{-6} mol/L. A high sampling frequency of 345 samples per hour can be analysed.

A similar biosensing strategy, describes a miniaturized microflow injection biosensing system, containing a microbeads-based enzyme reactor for histamine detection for food analysis [75]. Histamine is a well-known factor to cause an allergy-like food poisoning, and it is produced in fish or fish product as a result of the bacterial decarboxylation of histidine which exists in high amounts in a lean fish of bluefish species like tuna, herring, mackerel, etc. It is important for fishery industries and food-processing companies to manage the freshness of the lean fish. The microfluidic device was fabricated using a build-up technique well known in printed-wiring assembly. An upper Pyrex wafer had an inlet port for the microbeads and solutions, and an outlet port for the waste. The bottom Pyrex wafer has the electrochemical detector where the electrodes were patterned using the photolithography technique. For performing the assay, the reactor contains histamine oxidase immobilized chitosan beads. This enzyme catalyses the oxidation of histamine to imidazole acetaldehyde, ammonia and hydrogen peroxide. The amperometric detector combined with the nanolitre volume reactor by a miniaturized flow injection analysis device measures the current on oxidizing hydrogen peroxide generated from the enzyme reaction. Detection of histamine was carried out with high sensitivity and linearity from 1 μM to 1 mM in phosphate buffer solution and raw tuna samples.

4. CONCLUSIONS AND FUTURE TRENDS

Microfluidic platforms can be considered as an emergent and new approach for high selective and sensitive sensing. Considering the challenge and necessity of development of microfabrication techniques, polymers have shown as very interesting materials offering the possibility of tailoring for specific applications. Electrokinetic and hydrodynamic flows are mainly performed in microfluidic platforms. Up to now, hydrodynamic approach is more commonly used, probably due to its higher reproducibility. However, electrokinetic has a prominent and distinctive role in microfluidic platforms because of flow in multiple channels on a microchip can be easily controlled with a few electrodes. On the other hand, although fluorescence continues to be the most widely used detection technique, electrochemical detection, due to its inherent miniaturization and high compatibility with microfabrication techniques, appears as natural detection principle in these platforms. This detection principle is not expensive and compatible with the disposability of microfluidic platforms.

A growing number of articles in the last decade recognize the utility of microfluidic platforms for food analysis; however, some challenges must be achieved. Even though, a revolution in technology and microfabrication techniques takes places, integration of components into complete and functional systems, small and thus potentially portable, and simple to be used by nonexperts are features required. Moreover, additional efforts have to be made towards the validation of the methods to demonstrate the reliability of microfluidic systems.

Despite these already mentioned challenges, incorporation of biosensors in microfluidic platforms constitutes an excellent and promising tool for the future of food analysis.

ACKNOWLEDGEMENTS

The authors are very grateful for the financial support from the Spanish Ministry of Economy and Competitiveness CTQ2014-58643-R (A.E.) and from the NANOAVANSENS programme from the Community of Madrid (S2013/MIT-3029).

REFERENCES

[1] G.W. Whitesides, The origin and the future of microfluidics, Nature 442 (2006) 368–373.

[2] A. Manz, N. Graber, H.M. Widmer, Miniaturized total chemical analysis system: a novel concept for chemical sensing, Sens. Actuators B 1 (1990) 244–248.

[3] D. Janasek, J. Franzke, A. Manz, Scaling and design of miniaturized chemical-analysis systems, Nature 442 (2006) 374–380.

[4] S. Tombelli, M. Minuni, M. Mascini, Aptamers-based assays for diagnostics, environmental and food analysis, Biomol. Eng. 24 (2007) 191–200.

[5] Z. Zhang, L. Yu, L. Xu, et al., Biotoxin sensing in food and environment via microchip, Electrophoresis 35 (2014) 1547–1559.

[6] A. Escarpa, Lights and shadows on food microfluidics, Lab Chip 14 (2014) 3213–3224.

[7] P. Yáñez-Sedeño, J.M. Pingarrón, Electroanalysis and food analysis, in: A. Escarpa, M.C. González, M.A. López (Eds.), Agricultural and Food Analysis, John Wiley & Sons, 2015, pp. 1–16.

[8] M.A. Hamburg, Advancing regulatory science, Science 331 (2011) 987.

[9] M.I. Pividori, S. Alegret, in: S. Alegret, A. Merkoci (Eds.), Electrochemical Sensor Analysis, Wilson & Wilson's Comprehensive Analytical Chemistry, Elsevier, 2007, p. 467.

[10] R. Rooney, P.G. Wall, Food safety, in: B. Caballero, L. Trugo, P.M. Finglas (Eds.), Encyclopedia of Food Science and Nutrition, Academic Press, 2003, pp. 27–30.

[11] M.J. Madou, Fundamentals of Microfabrication (The Science of Miniaturization), CRC Press, 2000.

[12] M. Hervás, M.A. López, A. Escarpa, Electrochemical immunosensing on board microfluidic chip platforms, Trends Anal. Chem. 31 (2012) 109–128.

[13] J.G. Gardieneers, R.E. Oosterbroek, A. Van der Berg, Silicon and glass micromachining for μ-TAS, in: R.E. Oosterbroek, A. Van der Berg (Eds.), Lab-on-a-Chip: Miniaturized Systems for (Bio) Chemical Analysis and Synthesis, Elsevier, 2003, pp. 37–64.

[14] A.M. Jorgensen, K.B. Mogensen, Silicon and cleanroom processing, in: O. Geschke, H. Klank, P. Telleman (Eds.), Microsystem Engineering of Lab-on-a-Chip Devices, Wiley-VCH, 2004, pp. 117–160.

[15] D. Petersen, K.B. Mogensen, H. Klank, Glass micromachining, in: O. Geschke, H. Klank, P. Telleman (Eds.), Microsystem Engineering of Lab-on-a-Chip Devices, Wiley-VCH, 2004, pp. 161–168.

[16] H. Becker, C. Gärtener, Microreplication technologies for polymer-base μ-TAS application, in: R.E. Oosterbroek, A. van der Berg (Eds.), Lab-on-a-Chip: Miniaturized Systems for (Bio) Chemical Analysis and Synthesis, Elsevier, 2003, pp. 21–35.

[17] G. Binyamin, T.D. Boone, H.S. Lackritz, A.J. Ricco, A.P. Sassi, S.J. Williams, Plastic microfluidic devices: electrokinetic manipulations, life science applications, and production technologies, in: R.E. van der Oosterbroek, A. Berg (Eds.), Lab-on-a-Chip: Miniaturized Systems for (Bio) Chemical Analysis and Synthesis, Elsevier, 2003, pp. 83–112.

[18] E.S. Fakunle, I. Fritsch, Low-temperature co-fired ceramic microchannels with individually addressable screen-printed gold electrodes on four walls for self-contained electrochemical immunoassays, Anal. Bioanal. Chem. 398 (2010) 2605–2615.

[19] Y.T. Atalay, S. Vermeir, D. Witters, et al., Microfluidic analytical systems for food analysis, Trends Food Sci. Technol. 22 (2011) 386–404.

[20] A. Ríos, A. Escarpa, B. Simonet, Miniaturization of Analytical Systems. Principles, Designs and Applications, Wiley, 2009.

[21] P. Woias, Micropumps – past, progress and future prospects, Sens. Actuators B 105 (2005) 28–38.

[22] B.D. Iverson, S.V. Garimella, Recent advances in microscale pumping technologies: a review and evaluation, Microfluid. Nanofluid. 5 (2008) 145−174.

[23] A. Nisar, N. Afzulpurkar, B. Mahaisavariya, A. Tuantranont, MEMS-based micropumps in drug delivery and biomedical applications, Sens. Actuators B 130 (2008) 917−942.

[24] F. Mugele, J.C. Baret, Electrowetting: from basics to applications, J. Phys. Condens. Matter 17 (2005) R705−R774.

[25] X. Li, D.R. Ballerini, W. Shen, A perspective on paper-based microfluidics: current status and future trends, Biomicrofluidics 6 (2012) 011301−011313.

[26] L. Ramos, M.A. López, A. Escarpa, Miniaturization in sample preparation, in: M. Guardia, S. Garrigues (Eds.), Handbook of Green Analytical Chemistry, Wiley, 2012.

[27] A. Rios, A. Escarpa, M.C. Gonzalez, A.G. Crevillén, Challenges of analytical microsystems, Trends Anal. Chem. 25 (2006) 467−479.

[28] A.J. de Mello, N. Beard, Dealing with "real" samples: sample pre-treatment in microfluidic systems, Lab Chip 3 (2003) 11N−19N.

[29] A.G. Crevillén, M. Hervas, M.A. López, M.C. González, A. Escarpa, Real samples analysis on microfluidic devices, Talanta 74 (2007) 342−357.

[30] J. Lichtenberg, N. de Rooij, E. Veerporte, Sample pretreatment on microfabricated devices, Talanta 56 (2002) 233−266.

[31] J.P. Brody, P. Yager, Diffusion-based extraction in a microfabricated device, Sens. Actuators A 58 (1997) 13−18.

[32] B.H. Weigl, P. Yager, Microfluidic diffusion-based separation and detection, Science 283 (1999) 346−347.

[33] Y.H. Tennico, V.T. Remcho, In-line extraction employing functionalized magnetic particles for capillary and microchip electrophoresis, Electrophoresis 31 (2010) 2548−2557.

[34] C. Yu, M.H. Davey, F. Svec, J.M.J. Frechet, Monolithic porous polymer for on-chip solid-phase extraction and preconcentration prepared by photoinitiated in situ polymerization within a microfluidic device, Anal. Chem. 73 (2001) 5088−5096.

[35] H.B. Noh, K.S. Lee, B.S. Lim, et al., Total analysis of endocrine disruptors in a microchip with gold nanoparticles, Electrophoresis 31 (2010) 3053−3060.

[36] J. Palmer, D.S. Burgi, N.J. Munro, et al., Electrokinetic injection for stacking neutral analytes in capillary and microchip electrophoresis, Anal. Chem. 73 (2001) 725−731.

[37] Y. Xu, H. Jiang, E. Wang, Ionic liquid-assisted PDMS microchannel modification for efficiently resolving fluorescent dye and protein adsorption, Electrophoresis 28 (2007) 4597−4605.

[38] P.H. Humble, R.T. Kelly, A.T. Woolley, et al., Electric field gradient focusing of proteins based on shaped ionically conductive acrylic polymer, Anal. Chem. 76 (2004) 5641−5648.

[39] D. Ross, L.E. Locascio, Microfluidic temperature gradient focusing, Anal. Chem. 74 (2002) 2556−2564.

[40] N.M.M. Pires, T. Dong, U. Hanke, et al., Recent developments in optical detection technologies in Lab-on-a-Chip devices for biosensing applications, Sensors 14 (2014) 15458−15479.

[41] M.A. Schwarz, P.C. Hauser, Recent developments in detection methods for microfabricated analytical devices, Lab Chip 1 (2001) 1−6.

[42] M. Pumera, A. Merkoci, S. Alegret, New materials for electrochemical sensing VII. Microfluidic chip platforms, Trends Anal. Chem. 25 (2006) 219−235.

[43] A. De Mello, Miniaturization, Anal. Bioanal. Chem. 372 (2002) 12−13.

[44] H. Suzuki, Advances in the microfabrication of electrochemical sensors and systems, Electroanalysis 12 (2000) 703−715.

[45] A. Bange, H.B. Halsall, W.R. Heineman, Microfluidic immunosensor systems, Biosens. Bioelectron. 20 (2005) 2488–2503.

[46] W.R. Vandaveer, S.A. Pasas-Farmer, D.U. Fischer, et al., Recent developments in electrochemical detection for microchip capillary electrophoresis, Electrophoresis 25 (2004) 3528–3549.

[47] V. Scognamiglio, F. Arduini, G. Palleschi, et al., Biosensing technology for sustainable food safety, Trends Anal. Chem. 62 (2014) 1–10.

[48] F.J. Arévalo, A.M. Granero, H. Fernández, et al., Citrinin determination in rice samples using a micro fluidic electrochemical immunosensor, Talanta 83 (2011) 966–973.

[49] M.A. Fernández-Baldo, G. Messina, M.I. Sanz, et al., Microfluidic immunosensor with micromagnetic beads coupled to carbon-based screen-printed electrodes for determination of *Botrytis cinerea* in tissue of fruits, J. Agric. Food Chem. 58 (2010) 11201–11206.

[50] M.A. Fernández-Baldo, F.A. Gertolino, G. Fernández, et al., Determination of Ochratoxin A in apples contaminated with *Aspergillus ochraceus* by using a microfluidic competitive immunosensor with magnetic nanoparticles, Analyst 136 (2011) 2756–2762.

[51] N.V. Panini, E. Salinas, G.A. Messina, et al., Modified paramagnetic beads ina microfluidic system for the determination of zearalenone in feedstuffs samples, Food Chem. 125 (2011) 791–796.

[52] M. Hervás, M.A. López, A. Escarpa, Electrochemical microfluidic chips couples to magnetic bead-based ELISA to control allowable levels of zearalenone in baby foods using simplified calibration, Analyst 134 (2009) 2405–2411.

[53] M. Hervás, M.A. López, A. Escarpa, Integrated electrokinetic magnetic bead-based electrochemical immunoassay on microfluidic chips for reliable control of permitted levels of zearalenone in infant foods, Analyst 136 (2011) 2131–2138.

[54] Y. Don, S. Phillips, Q. Cheng, Immunosensing of Staphylococcus enterotoxin B (SEB) in milk with PDMS microfluidic systems using reinforced supported bilayer membranes (r-SBMs), Lab Chip 6 (2006) 675–681.

[55] M. Yang, S. Sun, Y. Kostov, Lab-on-a-chip for carbon nanotubes based immunoassay detection of Staphylococcal Enterotoxin B (SEB), Lab Chip 10 (2010) 1011–1017.

[56] N. Bunyakul, K.A. Edwards, C. Promptmas, et al., Cholera toxin subunit B detection in microfluidic devices, Anal. Bioanal. Chem. 393 (2009) 177–186.

[57] M.S. Chiriacò, E. Primiceri, E. D'Amone, et al., EIS microfluidic chips for flow immunoassay and ultrasensitive cholera toxin detection, Lab Chip 11 (2010) 658–663.

[58] J. Zhang, S. Liu, P. Yang, et al., Rapid detection of algal toxins by microfluidic immunoassay, Lab Chip 11 (2011) 3516–3522.

[59] F. Fournel, E. Baco, M. Mamani-Matsuda, et al., Love wave biosensor for real-time detection of okadaic acid as DSP phycotoxin, Sens. Actuators B 170 (2012) 122–128.

[60] G. Kim, J.H. Moon, C.Y. Moh, et al., A microfluidic nano-biosensor for the detection of pathogenic *Salmonella*, Biosens. Bioelectron. 67 (2015) 243–247.

[61] N.Y. Lee, Y. Yang, Y.S. Kim, et al., Microfluidic immunoassay platform using antibody-immobilized glass beads and its application for detection of *Escherichia coli* O157:H7, Bull. Korean Chem. Soc. 27 (2006) 479–483.

[62] M. Varshney, Y. Li, B. Srinivasan, et al., A label-free, microfluidics and interdigitated array microelectrode-based impedance biosensor in combination with nanoparticles immunoseparation for detection of *Escherichia coli* O157:H7 in food samples, Sens. Actuators B 128 (2007) 99–107.

[63] F. Tan, P.H.M. Leung, Z. Liu, et al., A PDMS microfluidic impedance immunosensor for *E. coli* 0157:H7 and *Staphylococcus aureus* detection via antibody-immobilized nanoporous membrane, Sens. Actuators B 159 (2011) 328–335.

[64] J. Yu, Z. Liu, Q. Liu, et al., A polyethylene glycol (PEG) microfluidic chip with nanostructures for bacteria rapid patterning and detection, Sens. Actuators A 154 (2009) 288–294.

[65] D. Hoegger, P. Morier, C. Vollet, et al., Disposable microfluidic ELISA for the rapid determination of folic acid content in food products, Anal. Bioanal. Chem. 387 (2007) 267–275.

[66] I. Palchetti, M. Mascini, Electrochemical enzyme biosensors, in: A. Escarpa, M.C. González, M.A. López (Eds.), Agricultural and Food Analysis, John Wiley & Sons, 2015, pp. 207–220.

[67] G. Schiavo, M. Matteoli, C. Montecucco, Neurotoxins affecting neuroexocytosis, Phys. Rev. 80 (2000) 717–766.

[68] M.L. Frisk, W.H. Tepp, E.A. Johnson, et al., Self-assembled peptide monolayers as a toxin sensing mechanism within arrayed microchannels, Anal. Chem. 81 (2009) 2760–2767.

[69] M. Frisk, W. Tepp, E. Johnson, D. Beebe, Lab Chip 8 (2008) 1793–1800.

[70] S. Sun, M. Ossandon, Y. Kostov, et al., Lab-on-a-chip for botulinum neurotoxin a (BoNT-A) activity analysis, Lab Chip 9 (2009) 3275–3281.

[71] D. Itoh, E. Koyachi, M. Yokokawa, et al., Microdevice for on-site fish freshness checking based on K-value measurement, Anal. Chem. 85 (2013) 10962–10968.

[72] D. Itoh, S. Fumihiro, T. Nishi, et al., Droplet-based microfluidic sensing system for rapid fish freshness determination, Sens. Actuators B 171-172 (2012) 619–626.

[73] W. Laiwattananpaisal, J. Yakovleva, M. Bengtsson, et al., On-chip microfluidic systems for determination of L-glutamate base on enzymatic recycling of substrate, Biomicrofluidics 3 (2009) 014104–014112.

[74] L.M.C. Ferreira, E.T. da Costa, C.L. do Lago, et al., Miniaturized flow system based on enzyme modified PMMA microreactor for amperometric determination of glucose, Bionsens. Bioelectron. 47 (2013) 539–544.

[75] T. Ito, T. Hiroi, T. Amaya, et al., Preliminary study of a microbeads based histamine detection for food analysis using thermostable recombinant histamine oxidase from *Arthrobacter crystallopoietes* KAIT-B-007, Talanta 77 (2009) 1185–1190.

Section IV

Biosensor Trade in Agrifood Sector

Chapter 10

Commercially Available (Bio)sensors in the Agrifood Sector

A. Antonacci,[1] F. Arduini,[2] D. Moscone,[2] G. Palleschi[2] and V. Scognamiglio[1,*]

[1]IC-CNR Istituto di Cristallografia, Rome, Italy; [2]Università di Roma Tor Vergata, Rome, Italy

*Corresponding author: E-mail: viviana.scognamiglio@mlib.ic.cnr.it

Chapter Outline

1. INTRODUCTION

Biosensors were introduced for the first time by Updike and Hicks in 1967 [1], which developed a biosensor based on the immobilization of glucose oxidase enzyme on an oxygen electrode for the measurement of glucose concentration in biological fluids. Then in 1973, Yellow Springs Instruments commercialized the first glucose biosensor [2]. In 1987, Medisense Inc. launched on the market the first pen-sized biosensor for self-monitoring of blood glucose by diabetic patients [3]. Since then, constant research and development efforts have been undertaken in biosensor technology, as demonstrated by the rise in the number of publications and patents (Fig. 1), highlighting the high potential of this technology for in field application in areas as biomedical diagnosis, health care, and biodefence, as well as in agrifood and environmental control. The constant evolution of this technology is leading to the standardisation of

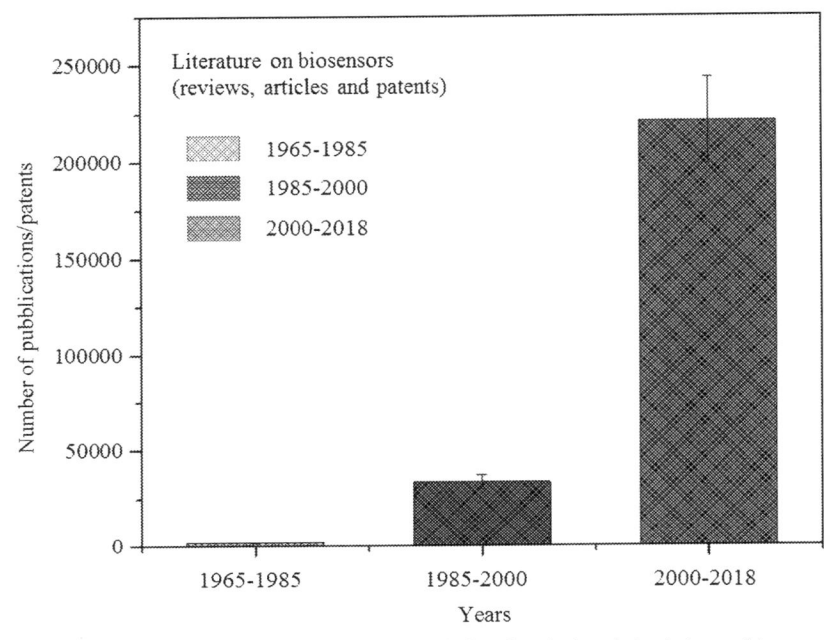

FIGURE 1 Number of publications and patents during the cited period relating to biosensors. *Data taken on March 2016 from www.googlescholar.com using the term 'biosensor'.*

instrumentations and bioelements, as well as to test processes, validation, and scaling-up, resulting in faster and specific applications.

Inline to this trend, the growth for biosensor commercialisation is becoming robust in emerging markets worldwide. The nonmilitary, open market for sensors grew from EUR 81.6 billions in 2006 to EUR 119.4 billions in 2011 and can be expected to grow to EUR 184.1 billions until 2016, according to the new World Report entitled 'Sensors Markets 2016' [4] (Fig. 2). This corresponds to an average annual growth rate of 7.9% between 2006 and 2011, and 9.0% between 2011 and 2016. Among sensors, the global market for biosensors was valued at $12,963.9 millions in 2014, and it is expected to grow at a compound annual growth rate (CAGR) of 9.7% during 2015–20, to reach $22,490 millions by 2020. In particular, applications of sensors for biomedical and life science dominate the market, accounting for 99%, followed by agrifood and environmental monitoring, and remediation applications. Fig. 3 shows income percent from worldwide markets in environmental field, security and biodefense, and home diagnostics from 2006 through 2009, and the forecasts up to 2016 suggest that this growing tendency is constant. Although glucose monitoring dominates the market, biosensor use in agrifood, process industries, environmental monitoring, security and biodefence is growing thanks to novel biosensors developed to satisfy specific detection requirements (Fig. 4).

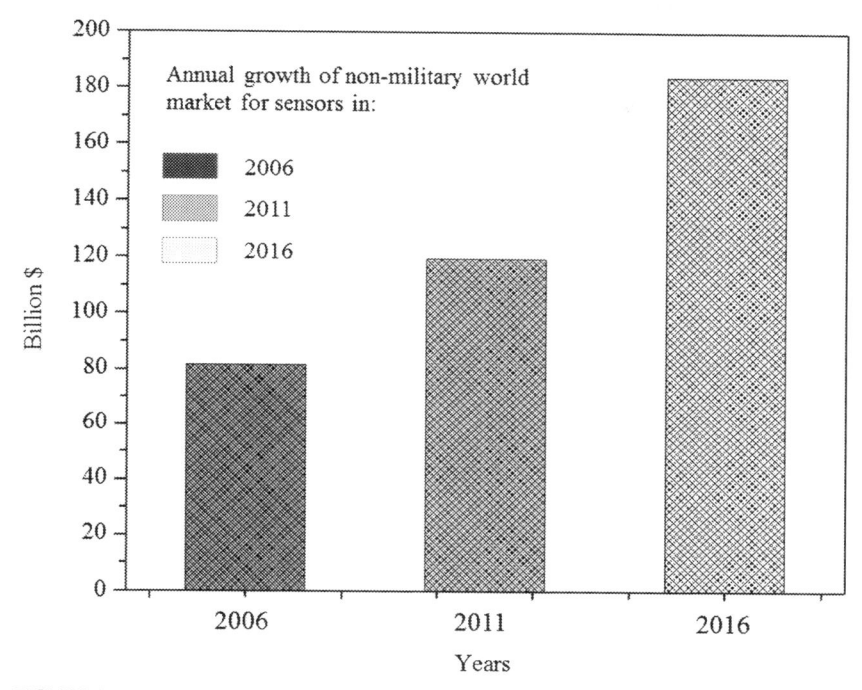

FIGURE 2 Annual growth of nonmilitary world market for sensors.

By a technological point of view, the sensor market can be distributed in electrochemical, optical, thermal and piezoelectric transducers, according to P&S Market Research that releases in 2015 a new research report on 'Global Biosensors Market' during 2015–20 [5]. Among them, electrochemical has a larger segment in the (bio)sensor market, followed by the optical technology.

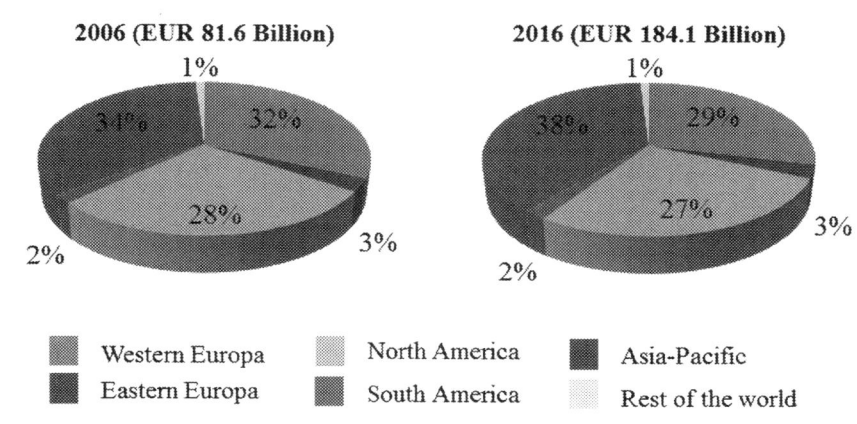

FIGURE 3 Overview of nonmilitary world market for sensing systems, and its growth from 2006 to 2016.

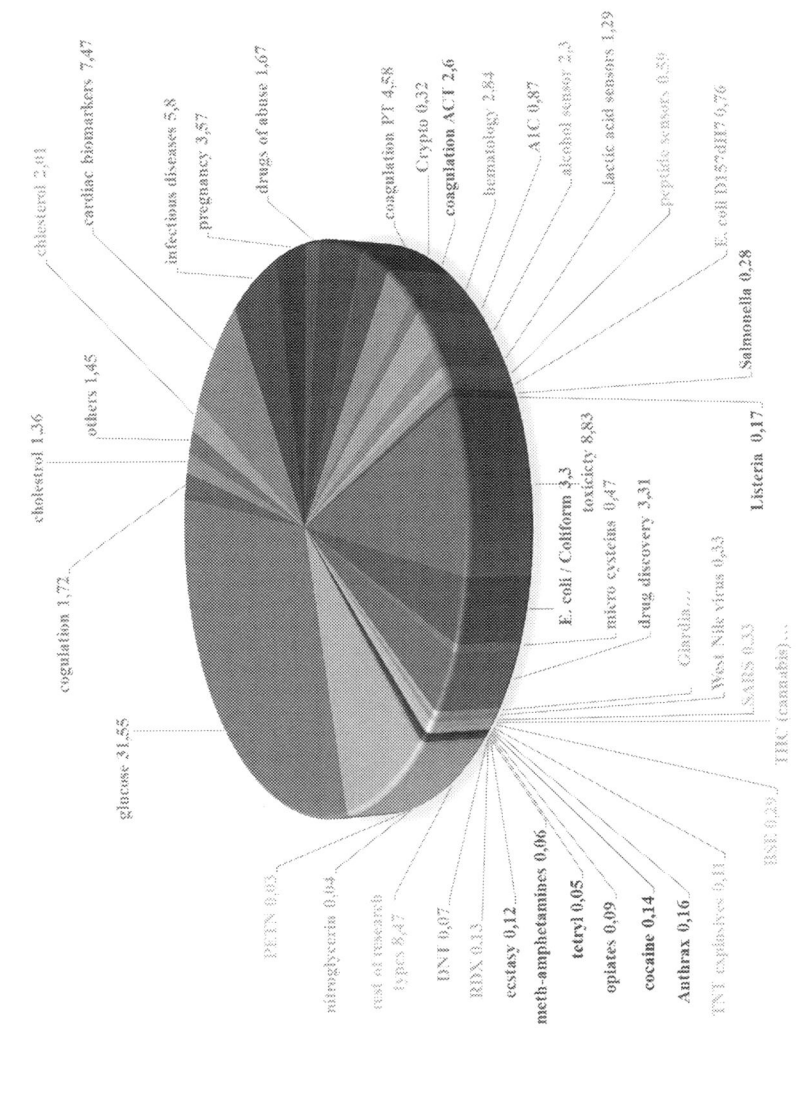

FIGURE 4 The global biosensor market incomes from specific detection application (2009).

These statistical calculations were provided on a list of 225 companies engaged in research and development, equipment manufacturing, supplies and distribution of biosensors [4]. However, there are well over 500 companies worldwide working on biosensors and bioelectronics employed in biosensor fabrication/marketing; design and development of materials, reagents, and instruments; integration of the components and launch into the market [6]. Among all these companies, few multinational companies with huge finance sources for technology acquisition and validation dominate the whole biosensor industry. The main companies involved worldwide in biosensor technology are listed in Table 1, mainly working in biomedical diagnostics, besides to biodefence, food analysis, industrial control, microbiology and proteomics [7].

TABLE 1 Main Companies Involved in Biosensor Technology Worldwide.

Company	Website	Main Expertise
Abbott Laboratories	www.abbottdiagnostics.com	Diagnosis and treatment of diseases/illness; in vitro diagnostic instrument systems and tests for hospitals, reference labs, blood banks, physician offices and clinics
ABTECH Scientific Inc.	www.abtechsci.com	Biosensor systems for the biomedical research community and for point-of-concern biomedical diagnostic applications
Bayer AG	www.bayerdiag.com	Gene chip development for the detection of genetic diseases
Biacore International AB	www.biacore.com/lifesciences/index.html	Health care and medical products industry
Affymetrix	www.affymetrix.com/index.affx	Analytical systems for antibody characterization, immunogenicity, biotherapeutic development/production and proteomics
Cygnus, Inc.	www.cygn.com/homepage.html	Glucose monitoring
DiagnoSwiss	www.diagnoswiss.com	Development and commercialization of biochip technologies for medical diagnostics, microbiology, food, warfare's, pharmaceutical research, proteomics, industrial control
Lifescan	www.lifescan.com	Glucose monitoring

Continued

TABLE 1 Main Companies Involved in Biosensor Technology Worldwide.—cont'd

Company	Website	Main Expertise
Neogen Corporation	www.neogen.com	Products dedicated to food and animal safety for the detection of pathogens in raw ingredients, finished food products, and environmental samples; rapid diagnostic Test kits to detect foodborne bacteria, spoilage organisms, mycotoxins, food allergens, genetic modifications, drug residues, plant diseases, and sanitation concerns
Panbio	www.alere.com	Tests for diagnosis of infectious disease
Roche Diagnostics AG	www.rochediagnostics.com	Biosensors used for analysing blood samples; kits and reagents for genomics and proteomics research, bioreagents for industrial use, products for food safety and pharmaceutical testing
Pelikan Technologies, Inc.	www.pelikantechnologies.com	Hand-held diagnostic and monitoring devices for the testing and management of a variety of medical conditions
Applied Biophysics	www.biophysics.com	Real-time, label-free, impedance-based method to study the activities of cells grown in tissue culture

Among the different application fields, (bio)sensor technology is achieving importance also in the agrifood sector, due to the frequent occurrence of food contamination and the growing apprehension about public safety. Indeed, a huge number of chemical contaminants represents a significant source of foodborne illness. For example, there have been long-standing concerns about chemical safety of food due to a misuse of pesticides during food production and storage, resulting in the occurrence of undesirable residues. More than 2.5 million tons of herbicides valued over US$30 billions are being used in agriculture all over the world increasing the toxic effect on human health. According to the World Health Organization, between 1 million and 25 million people suffer from herbicide poisoning each year. It is estimated that as many as 20,000 people in the United States will develop cancer each year from the residue of herbicides on their food, but the number is much higher when we include the enormous number of people unknowingly suffering from herbicide symptoms [8].

For this reason, the demand for (bio)sensors for food control is estimated to raise in the near future. Undeniably, the total market potential for food diagnostics is gaining momentum due to these pervasiveness of foodborne diseases worldwide, being projected to reach a value of US$16.1 billions by 2020, at a CAGR of 7.4% from 2015. Fig. 5 shows data from the Report 'Food Safety Testing Market by Contaminant, Technology, Food Type and Region — Global Trends & Forecast to 2020' about the food safety testing market revenues in the period from 2015 to 2020, as well as the global size of food safety testing market by region (data have been omitted by the company but provide indications on market volumes) [9].

This food diagnostic industry demands for rapid, accurate and easy-to-use devices, especially detection kits for food safety and quality testing in processed foods, which show essential advantages over the traditional methods that are time consuming and labour intensive, and require laboratory set-up instrumentations. Key features for food analysis include competitive detection in a short time (eg, US Food Safety Inspection Service admits zero-tolerance threshold for *Escherichia coli* O157:H7 contamination in raw meat products; while Environmental Protection Agency infectious admits 10 cells of *E. coli* O157:H7), reliability, integration and portability for in-field analysis, sample

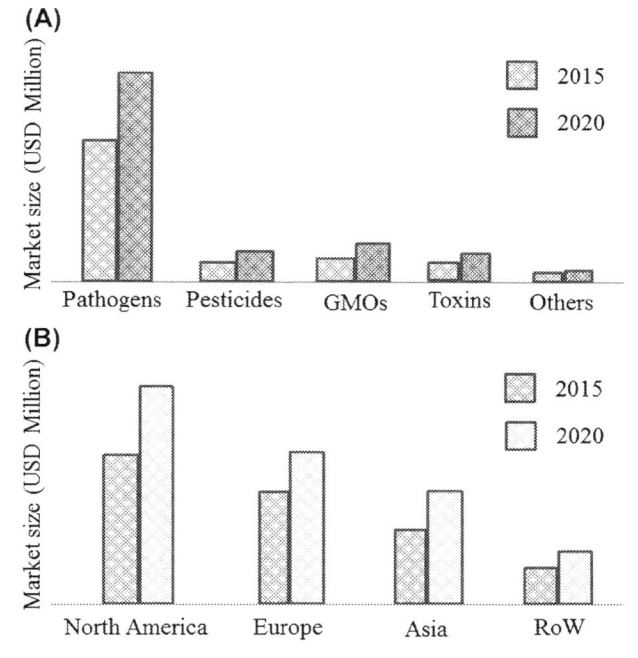

FIGURE 5 (A) Food safety testing market revenues in the period from 2015 to 2020. (B) Global market size of food safety testing market by region.

collection and treatment, real-time and multiplexing detection, low cost, and minimized consumables.

Despite the high number of research studies using different (bio)sensors for the food industry, an open dialogue between academic and industrial professionals still lacks to acquire a reciprocal know-how and to adapt this technology according to the market requirements. This is due to a series of technical challenges that constrain biosensing technology to reach a commercial maturity. For instance, tests of prototypes in real samples have still critical drawbacks owing to immobilization procedures of the bioreceptors, requirement of specific conditions of temperature and pH to preserve the bioreceptor activity and stability, sample preparation, sensor fouling and presence of interfering species. Moreover, very few (bio)sensors are exploited in agrifood especially in online analysis to monitor raw materials, product quality or manufacturing processes, although they can be combined with several available flow injection systems. As stated by Luong et al. [10] 'successful biosensors must be versatile to support interchangeable biorecognition elements, and in addition miniaturization must be feasible to allow automation for parallel sensing with ease of operation at a competitive cost'.

Table 2 reports the main challenges to overcome and that still restrain biosensor awareness and use.

In this context, the last advances accomplished in recent years in biomimetic chemistry, new material design (biohybrids), nanotechnology, micro/nanofluidics and ICT will be addressed in this chapter to highlight the main

TABLE 2 Main Factors Limiting Biosensor Awareness and Use

Key Unmet End Users' Needs
High sensitivity and specificity
Fast readout times
Bioreceptor storage stability
Simple sample pretreatment
Miniaturization
Low costs for mass production
Migration to lab-on-chip biosensors
Availability of wireless connection
Self-configuring biosensor
Multitest detection and monitoring
Integrated biosensing platforms

limitations hindering the transition from the bench to the market. In particular, emerging technologies devoted to improve the stability, sensitivity and reliability of the biosensing/biohybrid elements, as well as miniaturisation and integration, will be discussed in view of their possible commercialization. In addition, an overview of the main commercial biosensors exploited in the agrifood sector, their key limitations and the last trends in biosensing technology will be reported, fostering (bio)sensors to reach a commercial success.

2. COMMERCIAL BIOSENSORS IN THE AGRIFOOD SECTOR

Different kinds of sensing configurations are available on the market for the food industry, including autoanalysers, manual laboratory instruments and hand-held devices [11]. In the following sections, a brief overview of the commercial (bio)sensors for the agrifood sector, which can find main applications in different compartments including food analysis, process control, intelligent packaging and smart agriculture, has been reported.

2.1 Biosensors for Quality and Safety

Innovation and development in the agrifood sector and the recent globalisation of agroindustrial markets are pointed on two fundamental beliefs regarding food safety and quality, which have become a great concern in the last years [12]. Food quality and safety concepts are related to the production of foodstuffs that do not represent a risk for the health of the consumers. The term food quality is related to appearance, taste, nutritional value content, functional ingredient, freshness, flavour, texture and chemical, and to the presence of compounds able to enrich the food. The analysis of food composition allows characterising and attesting if the food contains all the desired constituents such as natural components (sugars, amino acid, alcohol) and additives (vitamins and minerals). The food safety concept entails the production and the commercialisation of safe foods that are free from allergenic or toxic substances, including pesticides, fertilizers, heavy metals, organic compounds, toxins and pathogens.

These requirements of quality and safety can be guaranteed only when the whole food supply chain undergoes to a strict control of a series of links ranging from crop cultivation to production and processing, sales and consuming. The availability of procedures, inspections and records is required to minimise the possibilities causing unsafe or off-quality end products. Therefore, the development of control systems for the analysis of a wide range of compounds that could affect the food performance became a crucial demand. The market for diagnostics for agrifood application grows each year, taking in consideration that millions of assays are performed on milk, fish and meat to evaluate the freshness and presence of contaminants. Indeed, the diagnostic industry claims several commercial analytical systems for water and food control including both multi-analysers, bench-top portable instruments and one-shot disposable tests. These devices can provide specific, sensitive, rapid and inexpensive analyses of food components and contaminants, both under laboratory conditions but also at on-site locations. In Table 3, a selection of companies involved in the last 20 years in the commercialisation of (bio)sensors and their components for food quality control is reported together with the target analytes.

The determination of sugars is the most performed, as depicted in Table 3, due to its high occurrence in a wide number of food products. In particular, the most used biosensor configuration for glucose detection entails glucose oxidase as biocomponent and an amperometric electrode as transducer. In combination with other enzymes, a similar configuration allows the determination of other sugars, using oxidases, dehydrogenases and microorganisms in

TABLE 3 Companies Commercialising Biosensors for Food Quality Control

Company	Country	Target Analyte
Fuji Electric Co.	Japan	Glucose
IBA GmbH	Germany	Sucrose, glucose, alcohol
GeneScan Europe AG/Scil Diagnostic GmbH	Germany	NutriChip, DNA detection with array technology
Oriental Electric Co.	Japan	Freshness metre: Degradation products of ATP
Pegasus Biotechnology	Canada	Degradation products of ATP
Biosentec	France	Lactate
Analox Instruments	UK–USA	Ethanol, methanol, glucose, ethanol, lactate, glycerol
Sensolytics	Germany	Glucose, lactate, sucrose, ethanol, glutamate
BioFutura s.r.l.	Italy	Glucose, lactate, malate
Tectronik	Italy	Glucose, lactate, malate
Chemel AB	Sweden	Glucose, sucrose, lactate, ethanol, methanol
YSI Inc.	USA	Glucose, sucrose, lactate, lactose, ethanol, methanol, glutamate
Inventus BioTec	Germany	Ascorbic acid in juice, fruits and vegetables

combination with different types of transducers (eg, optical or potentiometric), and with sampling systems, such as flow injection analysis.

Among companies listed in Table 3, some of them rearranged their R&D strategies in the last years by commercialising different products, while other companies are nowadays widely involved in (bio)sensor development for food and beverage analysis. For example, *Biosentec* commercialises an electrochemical biosensor within the Microzym analyser for lactate determination in collaboration with the Laboratoire de Genie Chimique et d'Electrochimie (Toulose), GIP Exercice (C.H.U. Hopital de Bellevue, Sant Etienne), and the Laboratoire de Recherche 'Transfert industrielle en Biotechnologie' (I.N.S.E.R.M. Tolouse), demonstrating how the collaboration between academics, industries and laboratories involved in a technology transfer is a crucial assumption for marketable successes. The Microzym instrument is able to perform 100 analyses with each electrode showing high suitability in agrifood application, as demonstrated by its good analytical performances: (1) detection threshold of 0.05 mM, (2) linear range between 0.05 and 1.5 mM, and (3) short time response between 5 and 10 s. The analysis is also cost-effective (0.5 Euro for each test considering only the cost of the electrode without reagents). However, this biosensor is not characterised by a high robustness in term of stability, since a storage stability of 15 days at 4°C is indicated by the company.

Similarly, *Tectronik* Company claims collaboration with academic partners, including the Universities of Padua and Venice (Italy), and won the Innovation Award granted by the Veneto Region, in the development of a novel biosensor (Senzytec) to measure ethanol in the agrifood sector with specific application in wine, oranges and apples. The Senzytec instrument is a reliable and quick system which, thanks to the enzymatic-based technology, allows the user to run low cost tests in a very short time (3−6 min) and on-site, on a small sample volume (0.1 mL). Senzytec2 system is the evolution of Senzytec, allowing for the detection of a wide range of analytes (ethanol, malic acid, lactic acid, glucose and fructose) (Fig. 6).

Glucose detection is also provided by *Analox Instruments* that sells a series of analysers (GM8, GM10 and GL6) to be applied in beverages, based on a very fast electrochemical enzymatic assay, in the very short time of 20 s, and a working range of 0.05−20% W/V (for GM8, GL6) and 0.20−40% W/V (for GM10). In the technical bulletin of the instruments, the company highlights the advantages of the electrochemical detection rather than spectrophotometric, being able to work in complex matrices with high opacity or turbidity.

In Table 4, a selection of companies involved in the last 20 years in the commercialisation of biosensors and their components for food safety control is reported, together with the target contaminants. As for food quality testing, the market for food safety, concerning detection of pathogen, GMOs, toxins and pesticides, is increasing as well, being projected to reach a value of US$16.1 billions by 2020, at a CAGR of 7.4% from 2015 [13]. Traditional

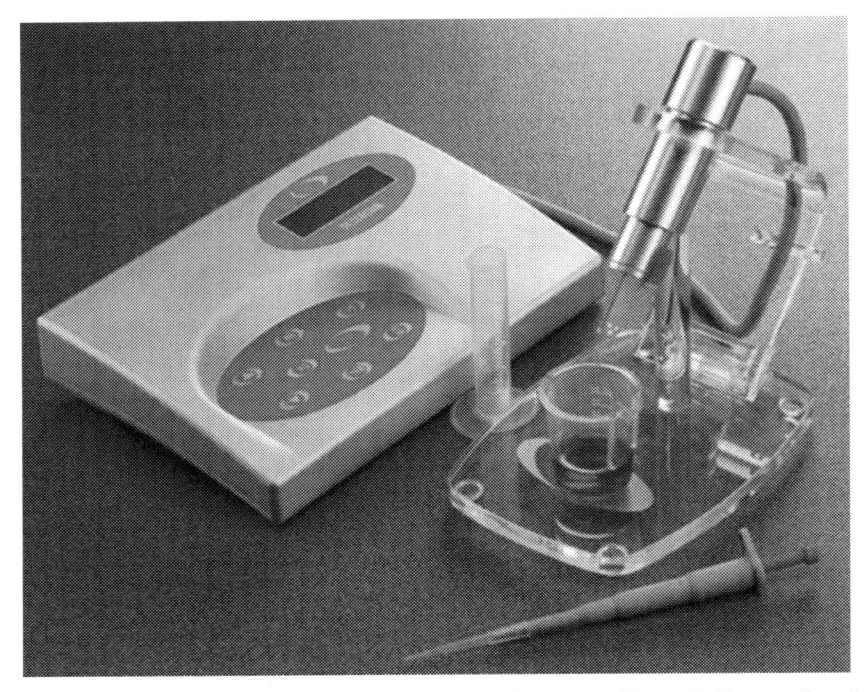

FIGURE 6 Senzytec instrument by *Tectronik*. *Photograph courtesy of Tectronik Company (http://www.tectronik.it)*.

methods of food safety testing embrace visual inspection, microbial assays and culturing methods, which require complex procedures and extended time (ie, days) to provide results. In particular, the response time is decisive, due to the short shelf life of some foods, such as milk and ready-to-eat vegetables and fruits. Thus, the technological focus from conventional food safety testing methods has been shifting to rapid testing methods able to ensure timely analysis of a larger sample size. Among the wide variety of contaminants, the total microbial detection is one of the most promising areas for biosensor application in food analysis. There are major players in pathogen identification and treatment market including *Abbott Laboratories, Achaogen Inc., Alnylam Pharmaceuticals Inc., Becton, Dickinson and Company, Biome´rieux Sa, Charles River Laboratories Inc., Siemens Healthcare* (subsidiary of Siemens AG), *Sigma—Aldrich Corporation, Qiagen N.V., Thermo Fisher Scientific Inc., Bio-Imaging Technologies Inc., Bio-Reference Laboratories Inc., Quest Diagnostics, Neogenomics, RadNet Inc.* [14].

Several other companies developed in the last decades a number of sensing systems for food contaminants. *Molecular Circuitry Inc.* launched, after more than 1-year development, a novel biosensor (Detex Pathogen Detection System) able to simultaneously test multiple pathogens, including *E. coli* O157

TABLE 4 Companies Commercialising Biosensors for Food Safety Control

Company	Country	Target Analyte
Molecular Circuitry, Inc.	USA	*Escherichia coli* O157, *Salmonella*, *Listeria* and *Campylobacter* in foodstuffs within 24 h
Ambri Ltd	USA	*Salmonella* and *Enterococcus*
Biosensor Systems Design	USA	Microorganisms and toxic substances
Biosensores S.L.	Spain	Online monitoring of the concentration of total microorganisms in water samples in 15 min
Motorola	Japan	Microorganisms and genetically modified organisms
Biacore GE	Worldwide	Folic acid, biotin, antibiotics in cereals, meat, milk, infant food, honey
Texas Instruments Inc.	USA	Sensors for gas and pH determination in food industry
Don Whitley Scientific Ltd.	UK	Sensors and components for microorganism detection
Integrated Genetics	USA	DNA probes for detection of microbial contamination (actually, the company has changed its R&D strategies towards other research products)
Molecular Devices Corporation	USA	Sensors and components for microbial determination

(including H7), *Salmonella*, *Campylobacter* and *Listeria* in food samples. This system has received AOAC Research Institute's performance tested method approval for the Detex 24 h *E. coli* O157 (including H7) assay. The system is a patented electro-immunoassay biosensor that shows high accuracy by utilizing unique mixtures of antibodies, as well as enhanced sensitivity and improved specificity thanks to the lower cross reactivity compared to most commercial antibodies. In addition, this instrument provides fast results on individual tests and also offers simultaneous testing for multiple pathogens. *Ambri Ltd.* obtained a worldwide exclusive licence for sensing applications for the Ion Channel Switch (ICS) technology from the Cooperative Research Centre for Molecular Engineering and Technology. This device was the first purpose-built nanomachine operating with moving parts that are nanometres in dimension, and has application in the on-site, fast detection of bacteriological contamination, with potential use in food testing to quickly identify dangerous organisms such as *Salmonella* and *Enterococcus*. This Australian company is also working with United States—based Biosensor Enterprises LLC (BEL), a joint venture between Dow Corning Corporation and Genencor International, to increase the range of applications for this platform technology.

More recently, *Biacore GE* proposes, among several products, Qflex kits (Fig. 7) to run specific assays with a Biacore Q system delivering rapid, reliable and automated quantitation of vitamin content (biotin, folic acid, pantothenic acid, vitamin B12) in food matrices, including coloured or opaque samples. Sample preparation and analysis times are significantly reduced compared to traditional techniques, also providing high flexibility required for routine assays. However, this instrumentation needs a laboratory set-up, and it is not useful for in field analysis.

2.2 Biosensors for Process Control

Continuous monitoring systems of industrial processes to detect in real-time possible errors in the production chain are a crucial requirement to address decision-making. Among these processes, food fermentation is used worldwide to produce a broad variety of beverages (wine, beer, cider and kefir) and

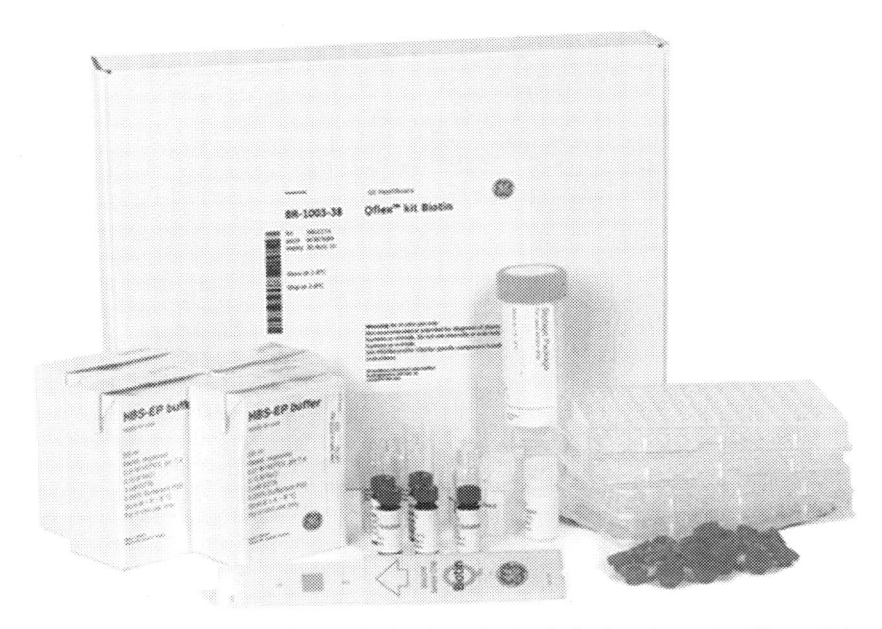

FIGURE 7 Qflex kit Biotin for quantitative determination in food products using Biacore Q by *Biacore GE. Photograph courtesy of GE Healthcare (www.biacore.com).*

food products (cheese, yogurt, bread, sausage, vinegar, pickled cucumbers and soy sauce olives).

Several analytical methodologies are involved to analyse the quality and the freshness of the produced food, controlling parameters such as pH, temperature, pressure, but also in online determining and quantifying various compounds of great importance to control the productivity and the quality of the process. Typically, central analyses entail the detection of contaminants (pesticides, heavy metals and nitrites), pathogenic microorganisms, toxins, antibiotics, allergens, and hormones, and the determination of food composition (sugars, amino acids, vitamins, fermentation products), including:

1. sugars, which are carbon sources and limiting factors of yeast growth involved in fermentative processes (glucose); or they can indicate excessive thermal treatment of milk during pasteurization processes (lactulose);
2. alcohols, which can be limiting factors in alcoholic fermentation of beer and wine, inhibiting enzymatic reactions when the amount exceeds 14% (ethanol);
3. amino acids, which can be used as supplements in animal feed (lysine), or to control the acidity and formation of cheese crust (lactic acid).

Table 5 reports the main companies commercialising biosensors for process control.

Among other analytical methods, (bio)sensors are of high relevance in process control being able to avoid productivity and economic losses. Furthermore, sensing devices can be integrated into the HACCP (Hazard Analysis Critical Control Points) to verify the successfulness of the processes. However, a limited exploitation of (bio)sensors occurs in process control, due to the short lifetime of enzyme sensors, the need for frequent calibration, the lack of reliable responses to different concentrations and changing physicochemical conditions.

As listed in Table 5, a number of companies realised biosensing systems for online detection of several analytes of interest in process control. *Yellow Springs Instruments* is the foremost company in the development of systems for monitoring and optimization of bioprocesses. The company constructed an integrated system consisting of the YSI 2900M Online Monitoring/Control System featuring the YSI 2960 Online Sampler, which automatically draws fluids from the bioreactor and directly delivers samples to the analyser for testing. This integrated system provides accurate results in less than 1 min, assuring sterility and detecting a broad range of analytes within the typical concentration range of bioprocesses (eg, glucose, lactate, glutamate, glutamine, glycerol, xylose, choline, hydrogen peroxide, sucrose, ethanol, methanol, lactose and galactose) (Fig. 8).

Gwent Sensor Ltd. provides an assortment of one-shot disposable biosensors (Fig. 9) for agrifood analysis, including biosensors for:

TABLE 5 Commercial Biosensors for Process Control

Company	Country	Target Analyte
Universal Sensors	USA	Ethanol, methanol, glucose, sucrose, lactose, L-amino acids, glutamine, ascorbic acid, oxalate
Yellow Springs Instruments	USA	Glucose, sucrose, lactose, L-lactate, galactose, L-glutamate, ethanol, H_2O_2, starch, glutamine, choline
BioFuture Srl	Italy	Glucose, fructose, malic acid, lactic acid
Chemel AB	Sweden	Glucose, saccharose, ethanol, methanol, lactose
Analox Instruments	UK, USA	Ethanol, methanol, glucose, lactate, glycerol
Nova Biomedical	USA	Glucose, L-lactate, L-glutamate, L-glutamine, alchol, sucrose, methanol, ammonia
Gwent Sensor Ltd.	UK	Sucrose, L-glutamate, alcohol
York Scientific technology	USA	Glucose, lactose, sucrose, galactose, lactate, ethanol/methanol, L-glutamate, L-amino acids
Biometra	Germany	Glucose, ethanol
Colora Messtechnik GmbH	Germany	Glucose, lactate, ethanol
NEC	Japan	Glucose, alcohol, L-lactate, glycerine
Provesta Corporation	USA	Glucose, lactate, lactose, alcohol

1. ammonia (based on glutamate dehydrogenase and NADH drop-coated onto sensors, giving response in water solutions in the range 1−10 ppm in water samples),
2. lactate (based on lactate oxidase and peroxidase, measuring concentrations between 0.04 and 0.5 mM in water samples),
3. glucose (based on glucose oxidase and an electrochemical mediator, measuring concentrations between 1 and 30 mM in food samples),
4. pyruvate (based on lactate dehydrogenase and NADH drop-coated onto sensors, giving response in the range 1−10 mM in onion).

2.3 Biosensors for Intelligent Food Packaging

A new application of (bio)sensors in the food sector is represented by monitoring systems integrated in food packaging to analyse and preserve the products

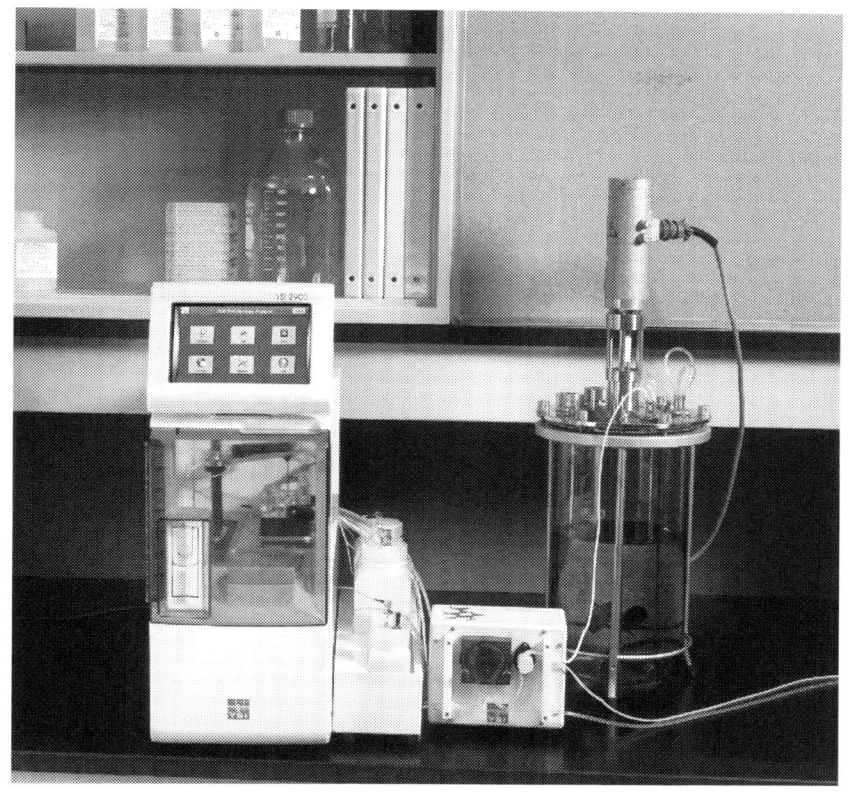

FIGURE 8 YSI 2900M Online Monitoring/Control System by *Yellow Springs Instruments. Photograph courtesy of YSI Company (https://www.ysi.com/lifesciences).*

along the food supply chain. Indeed, one of the most important issues to be considered regarding food sustainability is the management of food excess. Nowadays, about 30% of the total world food production is wasted [15].

Intelligent or smart packaging refers to packaging that monitors the conditions of food, senses freshness properties, and checks the integrity of the packages by exploiting biosensors during transport, storage and displays in markets [16]. Many intelligent packagings involve (bio)sensors as monitoring systems to measure physical parameters (humidity, pH, temperature, light exposure), to reveal gas mixtures (eg, oxygen and carbon dioxide), to detect pathogens and toxins, or to control freshness (eg, ethanol, lactic acid, acetic acid) and decomposition (eg, putrescine, cadaverine).

Sensors for food packaging are demanded to operate in specific physicochemical conditions; thus their application in the food chain is restricted by a number of limitations. As an example, these devices work in a wide temperature range (from -20 to $+30°C$) and/or high exposure to light (eg, UV/retail

FIGURE 9 One-shot disposable biosensors by *Gwent Sensor Ltd.* *Photograph courtesy of The Gwent Group (http://www.gwent.org).*

display lighting). For this reason, many research and development efforts have been spent to ensure their effectiveness and stability under all food storage and distribution conditions. Fast responses are also required (in some case as low as tenths of milliseconds) for real-time, online, and high-throughput control of food packages, to rapidly identify improperly sealed units and provide to their removal.

Several sensing systems have been realised for their integration in intelligent packaging, but few of them have been commercialised with the aim to improve the performance of the entire food supply chain and consequently to contribute to a worldwide food waste reduction.

Kerry et al. [17] have accurately reported some examples:

1. OxySense (Fig. 10) is an optical sensor system for a noninvasive oxygen measurement for sealed packages, tailored for the needs of the food, beverage and pharmaceutical industries. This system is able to measure oxygen with sensitive, rapid (5 s) and not-destructive measurements, both in headspace as well as in dissolved liquids. Its optical system can measure oxygen concentration within packages that are transparent, semitransparent and translucent [18].
2. ToxinGuard (Toxin Alert, Ontario, Canada) is a biosensing system able to detect pathogens (*Salmonella* sp., *Campylobacter* sp., *E. coli* 0517 and

FIGURE 10   OxySense Portable Oxygen Analyzer by *OxySense*. *Photograph courtesy of Oxy-Sense Company (http://www.oxysense.com).*

Listeria sp.) by the use of antibodies directly applied in a polyethylene-based plastic wrap in which food is packaged [19]. Any contaminant binds to a detector antibody that is suspended in a nutrient gel; this antibody is set on the packaging film in an easily recognized icon such as an X. As the contaminant collects in the icon area, a coloured or luminescent chemical, also attached to the detector antibody, makes the X visible on the plastic, warning of the presence of a toxin. A global market of $24 billion has been indicated by the company for their Toxin Guard technology, in targeting *Salmonella, Listeria, Campylobacter* and *E. coli.*

3. Food Sentinel System (SIRA Technologies, California, USA) is an immuno-biosensor system patented on 2000 (EP 1018013 A1), occurring in part of a barcode, able to continuously detect contaminating bacteria. As shown in Fig. 11, a barcode can be used on refrigerated food products which incorporate an ink that is invisible when safe conditions exist (*left*), while the ink will turn red in contaminated conditions (*right*) and the barcode will be rendered incapable of transmitting data when scanned. This indicator ink includes a first layer impermeable support, a second capillary layer and a third impermeable cover film layer. The second layer includes an absorbent material providing for the directional flow of juices, an immunobead solution pad and a detection area [20].

An alternative approach to preserve food packages and thus assure quality is the use of noninvasive indicator systems, which are able to provide qualitative or semiquantitative information about, for example, marker metabolites for freshness, or deliver data on time and temperature.

Through visual colorimetric changes or comparison with standard references, indicators are promising tools in ensuring package integrity throughout

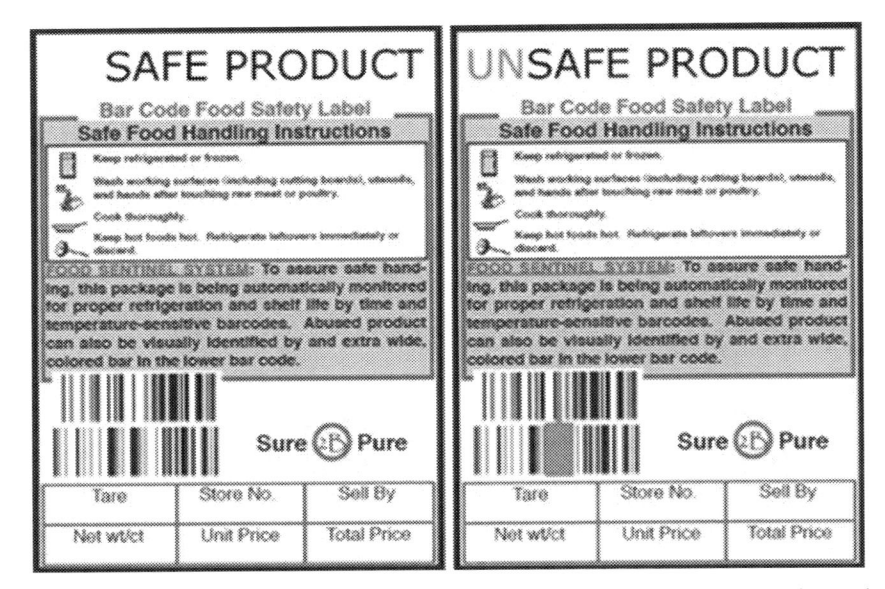

FIGURE 11 Food Sentinel System by *SIRA Technologies* (http://www.google.com/patents/EP1018013A1?cl=en).

the production and distribution chain. Several companies have commercialised indicators to sense oxygen in packs, including *Ageless Eye*, *Vitalon* and *Samso-Checker*. Their high sensitivity (indicators are susceptible to oxygen concentration, as low as 0.1%, thus their response could depend from residual oxygen) and reversibility (undesirable where an increased oxygen content, due to a leak, is consumed during subsequent microbial growth) are some of the disadvantage of these devices, which restrain their wide commercialisation.

Table 6 reports the main companies commercialising indicators for food packaging.

Indicators for gas measurement may employ specialized pigments inserted into plastic packaging, showing colour changes if a modified atmosphere occurs that can damage food. These kinds of systems allow advising manufacturers on possible breakages of package integrity to remove damaged foods from the supply chain before reaching supermarkets.

Colour indicator sensings are also employed as freshness indicators, detecting the production of volatile amines, which generate the typical 'fishy odour' common to all seafood, or indicating fruit ripeness by reacting to particular produced aromas. Time—temperature indicators estimate how long a product remained above a certain threshold temperature, within typical ranges between 2 and 8°C or 15—31°C. Moreover, certain sensitive labels indicate time—temperature data adapted to toxin formation of *Clostridium botulinum*.

TABLE 6 Commercial Indicators

Company	Country	Target Analyte
Novas Insignia Technologies	UK	Gas — modified atmosphere
O2Sense	Switzerland	Gas — amount of oxygen
FreshTag	The Netherlands	Freshness — volatile amines
RipeSense	New Zeland	Freshness — aromas released by the fruit as it ripens
3M MonitorMark	USA	Time—temperature — to estimate how long a product was above a threshold temperature (ranging between 15 and 31°C)
Timestrip Complete	UK	Time—temperature — to monitor high and low threshold temperature breaches outside selected ranges (2—8°C)
Fresh-Check	USA	Time—temperature — to monitor temperatures higher or lower than selected ones
CheckPoint	Palestine	Time—temperature — to monitor temperature ranges adapted to toxin formation of *Clostridium botulinum*
CoolVu Food	Switzerland	Time—temperature — to monitor the storage temperature
Innolabel Timestrip	Belgium	Time—temperature — to provide accurate shelf life guide for fresh fruit and vegetables
Chromatic Technologies, Inc.	USA	Thermochromic inks Reversible colour changes depending on temperature
Matsui International Company, Inc.	USA	Thermochromic inks Reversible colour changes depending on temperature

2.4 Biosensors for Smart Agriculture

Smart agriculture concept entails the integration of cross-disciplinary technologies, including crop grow control, soil analysis, remote sensing, yield mapping and positioning systems, to optimize the sustainability of agricultural processes and reduce their environmental impacts. Among these technologies, (bio)sensors for dynamic measurement of soil physical (ie, pH) and chemical (ie, nutrients) parameters, as well as for the determination of residual pesticides in crop and soil, are decisive to foster a sustainable farming.

For example, several optical sensors and laser scanning devices have been realised for real time in situ sensing of photosynthetic activity of plants. Nitrogen uptake prediction can be also provided using laser-induced chlorophyll fluorescence measurements. Sensor systems have been developed to monitor the presence of pests, pathogens, or pesticides to tune the amount of insecticides to be employed for crop productivity management. In addition, nanosensors integrated in platforms can monitor the controlled release of fertilizers and pesticides via nanoscale carriers avoiding their overdose, improving productivity and reducing waste. Networks of nanosensors located throughout cultivated fields will promise a real-time and comprehensive monitoring of crop growth, furnishing effective high-quality data for best management practices [21].

Moreover, data remotely sensed through satellites or drones can provide information on soil conditions, plant growth and weed infestation, with cost-effective and reliable measurements of crop emittance and reflectance [22]. The automation of irrigation systems is also a crucial requirement of smart agriculture. In this regard, sensor technology has the potential to maximize the efficiency of water use.

Several companies invested their resources in the development of sensing systems for sustainable farming. Among others, *PrecisionHawk* [23] is a leader company in the construction of open platform sensors designed on the requirements of end users, capable of plant health measurement, water quality assessment, vegetation index calculation, and plant counting, in addition to aerial mapping and imaging, and water temperature detection. Many companies have invested on the realisation of unmanned aerial systems (UAVs) or drones to support farmers to adopt an effective use of products for plant protection and fertilisation, as well as to provide data on soil parameters, to increase productivity, reduce inputs, and maximize yield potential.

Among others, *Ascending Technologies* [24] develops, in collaboration with *e-volo* (http://e-volo.com), systems based on precision positioning technology; *Lehmann Aviation* [25] utilises imagery from drones equipped with camera and sensors to provide crop information (Fig. 12); *MicroDrones* designs micro-UAVs equipped with several sensors and microcontrollers [26].

3. CONCLUSIONS AND FUTURE PERSPECTIVES

Nowadays, advances in (bio)sensor development are reasonably increasing with the aim to accomplish fast, sensitive and cost-effective food analysis. However, despite the straightforward investments to support R&D efforts, very low success rates are achieved in biosensor commercialisation. Indeed, there is

FIGURE 12 UAV/drone by *Lehmann Aviation. Photograph courtesy of Lehmann Aviation (www. lehmannaviation.com).*

not a huge market for (bio)sensors, except for glucose sensors, since the market entry of a new device is very difficult because of some drawbacks inherent to sensing technology. Limited lifetime of the bioreceptors, integration of the components, mass production as well as practicability in handling are some of the main obstacles to tackle with.

Last and future trends in biotechnology, biomimetic chemistry, nanotechnology, and material science guarantee innovation in biosensors and thus allowing their penetration into new application fields including agrifood, environmental monitoring, process industry, and security/biodefense, besides the consolidated biomedical field. The recent advances in synthetic biology and biomimetic chemistry resulted in tailor-made biosensors with desired features of robustness, sensitivity and specificity, by producing synthetic entities mimicking key properties of natural one, or engineering them to provide new desired features. Indeed, the design of artificial molecules, such as aptamers, biomimetics, molecular imprinting polymers, peptide nucleic acids and ribozymes, provided a trustworthy support to biosensing technology in solving some of the problems related to specificity, sensitivity and stability [27]. Associated to them, progress made in the development of new protocols for bioreceptor immobilisation helped to provide an enhanced robustness of the biosensors, in terms of both storage and working stability. Many physical and chemical methods for immobilizing bioreceptors, such as adsorption and entrapment within membranes, deliver complications to the conformational structure affecting the functional activity and result in less sensitivity and short lifetime. Thus, new strategies for the immobilization, orientation and molecular organization of biomolecules have been conceived by designing vanguard materials to construct well-defined microenvironments, mimicking biological membranes for a better orientation of the bioreceptors, and thus for enhancing their stability and activity.

To this regard, nanotechnology emerged in the last years helped to improve the biosensor performance thanks to the special physical, chemical, mechanical, magnetic and optical features of nanomaterials (eg, strong absorption band in the visible region, high electrical conductivity and good mechanical characteristics). Indeed, the use of nanomaterials to functionalise and immobilise the bioelements can improve their stability and specificity, yielding also reproducibility and reliability [28,29]. As an example, carbon nanotubes together with nanoparticles (gold, platinum, copper) have been widely exploited to boost detection sensitivity and facilitate biomolecule immobilization, as well as to promote electron-transfer reactions between bioreceptors and electrodes.

Transfer of (bio)sensor production from the research laboratory bench to large-scale manufacture is also crucial for their marketability. Production of high quantities and at low cost demands high levels of automation, while sensor development burdens a high requirement for manual processing. Among the alternatives to adapt sensors for commercial application, there is the possibility to develop multiplexed sensing systems through microfluidics and lab-on-chip technologies, which hold promise for automation of sampling and screening of a wide range of analytes with minimal consumption of reagents in portable devices. Moreover, progresses in microscale engineering plays an essential role in design (bio)sensors by scaling down systems and providing controllable microenvironments, offering also the possibility of economic mass production [30]. Indeed, novel miniaturised electrodes, micro-field effect transistors and micro-transducers obtained by the last advances in microengineering technology allow (bio)sensors to be integrated into disposable chips, capable of simultaneous determination of multiple food components/contaminants as well as of on-site monitoring of food processes.

The simultaneous detection of analytes in complex matrices is also a fundamental requisite in food analysis, since the main threat of food safety involves hazards caused by both biological and chemical contaminants. The recent results in material science highlight the potential of several materials (eg, silicon, glass, paper, plastic, other polymer) in multiplexed (bio)sensor design because of their miniaturization, integration and automation. Among these materials, paper has attracted significant attention in the last years for the development of simple, easily fabricated, disposable and low-cost devices. In addition, the combination of paper with microfluidics provides a mechanism for multiple detection, separating particulates from fluids and avoiding interference among different chemical species [31,32].

Finally, sample treatment in food analysis is also a critical concern to develop (bio)sensors for in field use. Generally, treatment procedures require several steps, including sampling, extraction of targets from the samples, clean-up and analysis. Considering the high complexity of food matrices and the presence of interfering species, it is hard to provide a simple extraction procedure to be accomplished on site without affecting recovery of the target analytes. For this reason, further research efforts are required to design (bio) sensor configurations for real application, entailing easy and effective procedures of sample treatment.

Actually, these above-mentioned technologies, including biomimetic chemistry, material science, nanotechnology, and microengineering, are still in their infancy, but they have the highest probability of making an impact in (bio)sensor design in the near future. Surely, their cross-cut disciplinary character will lead (bio)sensors to be commercialised and extensively used in several application field, including agrifood, in the next 20 years.

REFERENCES

[1] S.J. Updike, G.P. Hicks, Nature 214 (1967) 986—988.

[2] Yellow Springs Instruments — www.ysi.com.

[3] Medisense Inc. — www.medisenseinc.com.

[4] Intechno Consulting 2016 — www.intechnoconsulting.com.

[5] PR Newswire — www.prnewswire.com.

[6] A.C. Mongra, J. Acad. Indus. Res. 1 (6) (2012) 310—312.

[7] A.C. Mongra, A. Kaur, Int. J. Biomed. Adv. Res. 03 (07) (2012) 519—530.

[8] G. Spizzirri (Ed.), Food Analysis: Innovative Analytical Tools for Safety and Quality Assessment, 2016 (in press).

[9] Report "Food Safety Testing Market by Contaminant (Pathogen, GMO, Toxin, Pesticide), Technology (Traditional & Rapid), Food Type (Meat & Poultry, Dairy, Fruit and Vegetable, Convenience Food) & Region — Global Trends and Forecast to 2020" — http://www.marketsandmarkets.com/Market-Reports/food-safety-365.html.

[10] J.H. Luong, K.B. Male, J.D. Glennon, Biotechnol. Adv. 26 (5) (2008) 492—500.

[11] L.D. Mello, L.T. Kubota, Food Chem. 77 (2) (2002) 237—256.

[12] V. Scognamiglio, F. Arduini, G. Palleschi, G. Rea, TrAC 62 (2014) 1—10.

[13] Report "Food Safety Testing Market by Contaminant (Pathogen, GMO, Toxin, Pesticide), Technology (Traditional & Rapid), Food Type (Meat and Poultry, Dairy, Fruit and Vegetable, Convenience Food) and Region Global Trends and Forecast to 2020" — http://www.marketsandmarkets.com/PressReleases/food-safety-testing-market.asp.

[14] Report "Pathogen Identification and Treatment Market: Global Industry Analysis and Forecast 2016—2022" — http://www.persistencemarketresearch.com/market-research/pathogen-identification-treatment-market.asp.

[15] X. Jiang, D. Valdeperez, M. Nazarenus, Z. Wang, F. Stellacci, W.J. Parak, P. Del Pino, Part. Part. Syst. Charact. 32 (4) (2015) 408—416.

[16] M. Vanderroost, P. Ragaert, F. Devlieghere, B. De Meulenaer, Trends Food Sci. Technol. 39 (1) (2014) 47—62.

[17] J.P. Kerry, M.N. O'grady, S.A. Hogan, Meat Sci. 74 (1) (2006) 113—130.

[18] OxySense® — http://www.oxysense.com.

[19] ToxinGuard™ — http://www.prnewswire.com/news-releases/toxin-alert-wins-its-fifth-patent-for-toxin-guardtm-immunoassay-72658577.html.

[20] Food Sentinel System™ — https://www.adazonusa.com/blog/barcode-industry/sira-technologies-food-sentinel-system.

[21] M. El Beyrouthya, D. El Azzi, Adv. Crop. Sci. Tech. 2 (2014) e118.

[22] P. Monda, M. Basu, P.B.S. Bhadoria, Am. J. Exp. Agric. 1 (3) (2011) 49—68.

[23] PrecisionHawk® — www.precisionhawk.com.

[24] Ascending Technologies® — www.asctec.de.

[25] Lehmann Aviation (www.lehmannaviation.com).

[26] Microdrones (https://www.microdrones.com).

[27] V. Scognamiglio, A. Antonacci, M.D. Lambreva, S.C. Litescu, G. Rea, Biosens. Bioelectron. 74 (2015) 1076−1086.

[28] V. Scognamiglio, Biosens. Bioelectron. 47 (2013) 12−25.

[29] F. Arduini, S. Cinti, V. Scognamiglio, D. Moscone, Micro Chim. Acta (2016) (in press).

[30] B. Derkus, Biosens. Bioelectron. 79 (2016) 901−913.

[31] A.W. Martinez, S.T. Phillips, M.J. Butte, G.M. Whitesides, Angew. Chem., Int. Ed. 46 (8) (2007) 1318−1320.

[32] Y. Zhang, P. Zuo, B.C. Ye, Biosens. Bioelectron. 68 (2015) 14−19.

Section V

New Revolutionary Frontiers in Biosensor Technology

Chapter 11

Laser Scanning Approaches for Crop Monitoring

D. Hoffmeister

University of Cologne, Cologne, Germany
E-mail: dirk.hoffmeister@uni-koeln.de

Chapter Outline

1. INTRODUCTION

Precision agriculture (PA) is a holistic approach, comprising new sensors, information systems and smart machinery, aimed at securing food supply [1]. Improvements of nutrient use efficiency and development of better stress and disease management are necessary as agriculture generally faces the challenges of a growing world population, climate change, environmental protection and resource reduction. Thus, the remote detection of the contemporary status of a crop's state — height, density, vitality and biomass — is important for improving management and optimizing yield and inputs [2]. An additional research area with similar aims is lab- and field-based crop phenotyping [3,4]. This area investigates the monitoring of specific phenotypic traits, as the result

Comprehensive Analytical Chemistry, Vol. 74. http://dx.doi.org/10.1016/bs.coac.2016.02.018

of newly developed genotypes, and plants' adaptation responses to the environment, to select the phenotype with the best resource efficiency and stress tolerance. Likewise, crop height, crop growth and biomass distribution are also important parameters for environmental modelling approaches [5]. Destructive methods usually used to reach the previously described goals are often laborious and time-consuming. Thus, more efficient and nondestructive techniques are needed.

Laser ranging (LR) and laser scanning (LS) are highly accurate state-of-the-art remote sensing methods for data acquisition based on light detection and ranging (LiDAR) [6]. In particular LS, with an automated deflection of single laser measurements is used in numerous applications (eg, topographic surveys, forestry, and documentation of buildings and cultural heritage) [7]. Both methods have also been applied in agronomy in terms of PA goals and in phenotyping approaches. In these cases, LiDAR-based measurements allow a fast and objective determination of specific features of plants, such as crop height and leaf area. Another important research area that extensively uses LiDAR is forestry, which is not the focus of this chapter, but is briefly mentioned. General overviews of these applications are given in Kaartinen et al. [8], Hyyppä et al. [9], as well as van Leeuwen and Nieuwenhuis [10].

Generally, LS provides direct, highly accurate and dense 3D point measurements of objects [11], in contrast to photogrammetric approaches, which indirectly calculate 3D points. These systems are able to facilitate measurements on different scales and areas. The technology can be used on airborne platforms, known as airborne laser scanning (ALS), on mobile platforms (cars, boats, humans and so on), known as mobile laser scanning (MLS), as well as on terrestrial platforms, known as terrestrial laser scanning (TLS). Triangulation-based scanners, mainly mounted on scan arms, are exploited for plant phenotyping in the laboratory. The result of these different methods is a 3D point cloud with a specific coverage, density and accuracy, which can be used for further manual or automatic analysis.

In this chapter, an overview of these LS approaches for crop monitoring will be given and the overall advantages and disadvantages will be discussed. The chapter starts with LS basics. In the following section, the main applications of LS with regard to crop monitoring, including management enhancements, single plant measurements, canopy analyses at field level and the relationship to spatial variability, will be described. Multispectral LS as a new possibility, mainly driven by forestry, is presented afterwards. All applications are discussed, and a short conclusion finalizes the chapter.

2. LASER SCANNING BASICS

In general, all LS systems deliver polar 3D coordinates for each measured point [11]. Therefore, the horizontal and vertical angles for each laser shot are recorded in addition to the single-range measurements. In contrast to single measurements being produced, for example, by tachymetric surveying and LR, LS mechanisms record a huge amount of single measurements, resulting in so-called point clouds.

Every single-range measurement is conducted by the time-of-flight (TOF) principle, a phase-shift detection or triangulation [6,12]. For measurements with the TOF principle, the range to a target is calculated by measuring the time delay between each singly emitted laser beam and the respective backscattered echo. As laser light is emitted with near-light velocity, half of the time delay gives the range to the target. In contrast, for measurements with the phase-shift detection method, a modulated continuous laser beam is emitted, and the range is derived from the phase angle shifts between the emitted and the received signal.

Devices incorporating the basic triangulation method show a high resolution and accuracy, which is used particularly for close-range measurements in industrial applications, reverse engineering and medical research. A beam is projected onto an object and partially reflected to the known position of a sensitive receiver. Similarly, structured light scanners project a reference light pattern and derive 3D coordinates by detecting the distortion of this pattern caused by the recorded object. A further method to create 3D data or images with depth is range imaging, or flash LiDAR, which directly records an entire scene in 3D and detects the reflected signal with image sensors [11].

The previously mentioned LiDAR methods use a laser (light amplification by stimulated emission of radiation), which is highly intense, monochromatic, directional and coherent [6]. The active materials which generate the laser light can be crystals or gases, such as rubies, CO_2, helium, neon and argon. Two types of laser are commonly considered for range detection: solid-state and semiconductor lasers. Solid-state lasers (usually type: Nd:YAG, neodymium: yttrium aluminium garnet) are mainly used for the previously described TOF method, whereas semiconductor lasers are mostly used for the continuous wave method [11].

The wavelength of the laser light depends on the active material used, which is normally in the range of short infrared light (830−1350 nm). Frequency doubling methods, which halve the wavelength, allow laser light to be emitted in the visible light spectrum (∼530 nm). This is, for instance, important for bathymetric measurements, due to their ability to penetrate water. All lasers are classified from class 1 to 4 according to their emitted energy and their wavelength. Classes 3 and 4 are hazardous to humans.

To cover an area of interest by the previously described LiDAR methods, automatic deflection methods for each emitted laser pulse are applied (as depicted in Fig. 1). This is the difference between LR and automatic LS. The different methods of deflection have various effects on the resolved ground pattern, eg, in lines or in circles [13]. This is particularly important for MLS and ALS approaches, where single laser beams are deflected in one direction and aerial coverage is achieved by the movement of the carrier. In this case, the trajectory of

FIGURE 1 Terrestrial laser scanning (TLS) measurements are conducted by measuring single points with a given point spacing and an increasing footprint diameter. After a scan line is finished, the next scan line is measured by automatically tilting the head of the scanner. The actual footprint for this scanner (type: Riegl LMS-Z420i) is 25 mm, and point spacing is 7 mm over a 100 m distance. The device captures 11,000 points per second. The scanner is additionally equipped with a digital camera (with a known and fixed orientation) and an RTK-GPS receiver.

the sensor and its orientation need to be calculated, which is typically recorded by inertial measurement systems, which combine inertial measurement units with recordings of global navigation satellite systems (GNSS).

In contrast, TLS is conducted from fixed positions (Fig. 1), which are accurately determined by measuring highly reflective targets. The position is then calculated by a resection. Additionally, the position of the sensor itself can be measured. These measurements can be conducted by tachymetric surveying or GNSS measurements. Typically, RTK-DGPS (real time kinematic differential global positioning system) measurements with the highest possible accuracy of about $1-2$ cm are used [14]. This additionally allows the simultaneous georeference of the recorded point clouds.

For all LS surveys, occlusion effects need to be taken into account. All LiDAR measurements are only capable of deriving information from their current viewshed. Thus, points behind occlusions are not recorded, and another scan position or an according trajectory for ALS and MLS operation must be used to fully cover an object or area of interest.

In the post-processing step (depicted in the overall workflow in Fig. 2), the position information and the raw point clouds are normally combined, which is called registration. Registration can be conducted or enhanced by the iterative closest point algorithm [15]. The initial registration of approximately registered point clouds is enhanced by iteratively searching the closest points of similar regions until the Euclidean distance is within a specified threshold, where a transformation is conducted at every step. A further similar algorithm was developed, where a surface is used as the master fitting area [16].

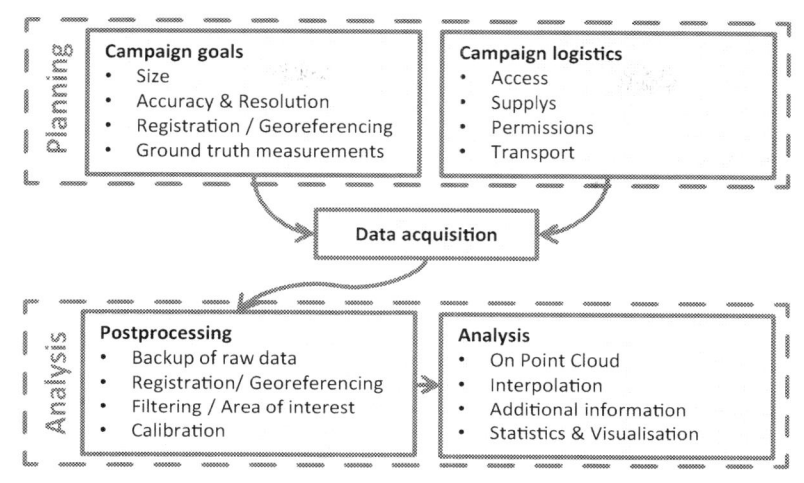

FIGURE 2 The workflow depicts the most important parts for planning, conducting and analysing laser scanning surveys.

As a result, all of the previously mentioned LS methods achieve raw 3D point clouds with an accuracy of millimetre to centimetre, with a similar resolution (point spacing) containing billions of points. These raw point clouds can already be displayed in different occurrences, for instance, coloured by height, with assigned RGB-picture values or reflected intensity values (see Fig. 3). These unstructured huge point clouds need, as opposed to single measurements, further post-processing steps, such as the selection of the area of interest, filtering, calibration and further analysis (see Fig. 2). They are comparable to methods of photogrammetry or radar procedures, in particular concerning the derivation of digital elevation models (DEM) [12].

Beside the detailed geometric representation in a point cloud, a radiometric backscatter of the scanned object is also received for each point [17] (see

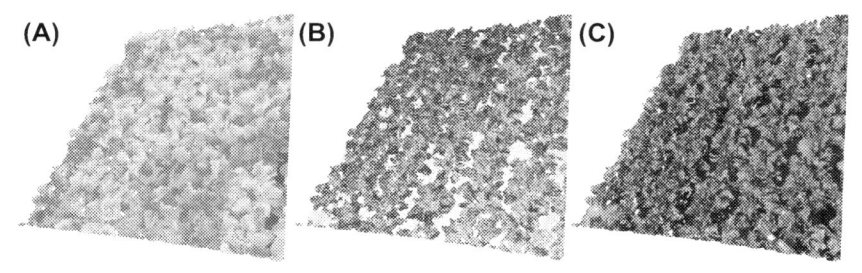

FIGURE 3 The basic 3D point cloud can be displayed in several ways: (A) coloured by height from blue (grey in print versions) to red (light grey in print versions); (B) coloured by intensity from black (low intensity) to light grey (high intensity) and (C) in 'true' RGB-colours collected by an additional camera. All figures display the same area and perspective of a sugar beet plot.

Fig. 3B). This information is valuable, as different objects show different surface characteristics and, accordingly, different backscatter signals. However, the backscatter signal is differently converted and amplified in available systems [18]. Thus, the intensity of the measurements needs to be radiometrically calibrated. In this case, different laser scanners show a different behaviour, in particular regarding the influence of range to a target or surface. A data driven range correction of the intensity signal allows to achieve 3D data, which can be used for single plant detection [19] and can be related to nitrogen content [20], as shown in Section 3.3.4.

Recently developed sensors record the full waveform (FW) of the reflected signal, as depicted in Fig. 4. In contrast to discrete echo measurements, the digitized FW information can be decomposed to derive further specific characteristics, such as the amplitude or echo width [21]. The latter FW record was first used in ALS [22] and is also applied on the ICESat satellite [23−25]. The most recent TLS devices are also capable of deriving this information (such as the Riegl VZ-series, http://riegl.com/nc/products/terrestrial-scanning/produktdetail/product/scanner/27/).

As a further information source, images from various cameras (eg, RGB, thermal, or hyperspectral) can be used to combine information and to extend feature information on the data level [26]. This is depicted by the mounted camera in Fig. 1 and the resulting coloured point cloud in Fig. 3C. Therefore, the orientation between both sensors needs to be known, which also can be derived from highly reflective targets. Likewise, several systems are in development which allow multi- or hyperspectral active LR or LS on the physical level [4,27]. This will be discussed later in this chapter.

3. APPLICATIONS OF LASER SCANNING FOR CROP MONITORING

Applications of the previously presented methods for crop monitoring on different scales are described in the following paragraphs. For clarity, these paragraphs are subdivided into (1) management enhancements (eg, seeding, harvest and postharvest monitoring); (2) measurements of single plants for lab-based phenotyping; (3) monitoring of current plant canopy structures and vitality on field level and (4) spatial variability research for yield mapping and prediction.

3.1 Management Enhancements

Enhancing and simplifying management tasks such as seeding, harvesting and all steering-related issues are key elements of PA. For example, LR sensors can be used to determine crop edges and forage swaths [28]. Similar sensors are already available, such as the Laser Pilot from Claas (http://www.claasofamerica.com/product/precision-farming/guidance/laser-pilot), and

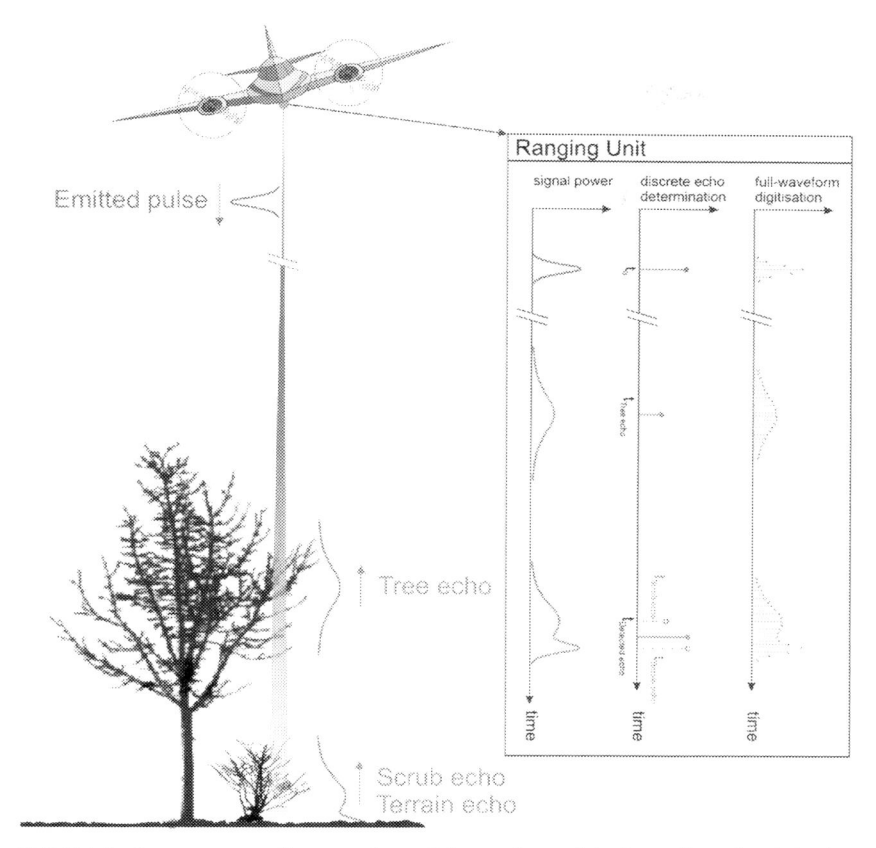

FIGURE 4 In contrast to discrete echoes, full waveform digitations allow the derivation of further parameters, such as echo width for statistical analysis. *Reprinted from M. Doneus, C. Briese, M. Fera, M. Janner, J. Archaeol. Sci. 35 (4) (2008) 882–893, Fig. 1, with the permission of Elsevier.*

SmartSteer from Case-New Holland (http://www.agriculture.newholland.com/germany/de/PLM/Guidance_systems/Crop_Guidance/SmartSteerSystem/Pages/PLM_overview.aspx). However, enhanced sensors can also be used to adjust the driving speed of a harvester according to crop density, since a constant rate of input is of importance [29]. In this study, two kinds of LS devices were tested for different densities of wheat and varying driving speeds. The computation of crop density by the detection of the ground and top canopy was obtained based on a linear model ($R^2 = 0.81 - 0.96$). The low-frequency LS showed less accurate results. With a similar set-up, the effects of straw output settings were investigated, as straw quality is an additional variable for farming [30]. The swath is automatically detectable by an LS system and can be used for auto-steering a harvester [31]. In addition, problematic postharvest growth can be detected by the analysis of the geometric and range-corrected radiometric properties of an LS [32].

Dimensional information of orchards from LS measurements is also taken into account to adjust spraying amounts [33]. A similar approach is implemented for a stop and go sensor on a small cart for maize plant localization [34]. Likewise, recent research has used laser scanners information for the automatic guidance of autonomous vehicles (robots) for mapping single tree trunks in orchards in real time [35,36]. A robotic system was also presented for autonomous navigation in maize [37]. Real-time adjustments of the position of spraying arms are possible [38], as well as accurate on-the-fly determination of foliage surface to optimize the pesticide and fertilizer application [39].

3.2 Single Plant Measurements for Phenotyping

The accurate and fast determination of phenotyping traits, such as the number of leaves, plant height, leaf size and internode distances, is important to select promising genotypes for crop breeding [4]. Several systems exist which allow comprehensive phenotyping. Some systems are based on LiDAR, while others rely on imagery, flash LiDAR [40] or structured light [41] (see Section 2).

For these type of measurements, high-resolution devices with high accuracy are usually used, such as triangulation-based devices. However, every system which is capable of deriving 3D information is applicable [42]. One major issue is the fully automatic and fast feature extraction of traits from these measurements. Therefore, plant organs can be separated by applying histogram-based classification algorithms, and subsequent fitting of planes and cylinders into the separated areas (leaves and stems) of a virtual plant represented by a point cloud [43]. The segmentation process is fully automated, as unsupervised clustering methods are applicable and were successfully tested on grapevines, wheat and barley [44]. However, the measurements of plants by specific LiDAR-based devices using a specific wavelength (660 nm, the absorption range of chlorophyll) might cause inaccurate results, due to refraction or penetration effects of the signal on the leaf tissue, depending on exposure time [45]. Sensor fusion enables to combine measurements from different sensors and accordingly allows the enhancement of the geometric information gathered by LiDAR measurements. 3D monitoring by LS and fluorescence imaging of herbicide effects on a single plant was also conducted [46]. The study revealed that the uptake processes of herbicide are depending on structural parameters, such as leaf inclination.

3.3 Field-Level Canopy Measurements

3.3.1 Direct Canopy Measurements

Early crop measurements on barley by LR were conducted in 1989 [47]. A thorough investigation was conducted under field conditions by exploiting a triangulation-based LR device with two different inclination angles [48]. This study showed a very high coefficient of determination of LR-based crop heights (oilseed rape, winter rye, winter wheat and grassland) for fresh and dry biomass. Likewise, the RMSE was smaller than 0.76 kg/m^2.

As an advancement, the previously described ranging device and a TOF-based laser scanner were used in combination with a specific height guiding device to keep the sensors at a constant distance above the ground [49]. Additionally, these LR devices were attached to a swivelling device to achieve a sinusoidal line of measurements. Both devices provided similar results. The TOF measurements showed a higher measurement frequency and lower noise values compared to the triangulation-based LR device. Likewise, a close correlation was also found over time at different growth stages and for leaf area index (LAI) estimation [50]. In contrast to these short-range sensors (<2.5 m), a TOF laser scanner for driving assistance was also tested with different driving speeds and inclination angles, which showed no major influence on the results [51]. However, an overestimation of plant height with increasing distance was detected. This overestimation is the result of laser beam geometry and a gap fraction, which emphasizes higher plants. A model that is able to partly resolve these effects needs to be established individually for every crop type and device [52].

The previously stated promising results were confirmed by other studies on crop height and crop growth. For instance, a similar low-cost laser scanner for automotive purposes was used in a static mode and by driving for yield prediction of *Miscanthus giganteus* [53]. An average error of plant height detection of <5% was achieved for both methods after applying a correction algorithm for the sloping of the instrument. In addition, the authors found a high correlation between plant height and stem mass ($R^2 = 0.86$). Similar results in crop height detection accuracy ($R^2 = 0.88$) with a low-cost LS mounted on a quad were also reported [54]. In this case, weed infestations in maize plants were detected at a rate of 78%.

Similar results were achieved by a study with a more expensive phase shift—based scanner mounted on a 3-m rack [55]. The manually measured height of differently fertilized barley, oat and wheat plots showed a strong correlation ($R^2 > 0.88$) with the detected plant height and harvesting results. In addition, a strong correlation of the detected single ears with grain yield was reported. Several studies have additionally been undertaken in orchards to accurately characterize crop canopy [56–61].

3.3.2 Indirect Canopy Characterization

In contrast to the previously mentioned direct plant height measurements, the height can also be indirectly calculated by the comparison of high-resolution, multi-temporal crop surface models (CSMs) from different time steps, as shown in Fig. 5 [62]. In addition, the proposed method allows the compilation

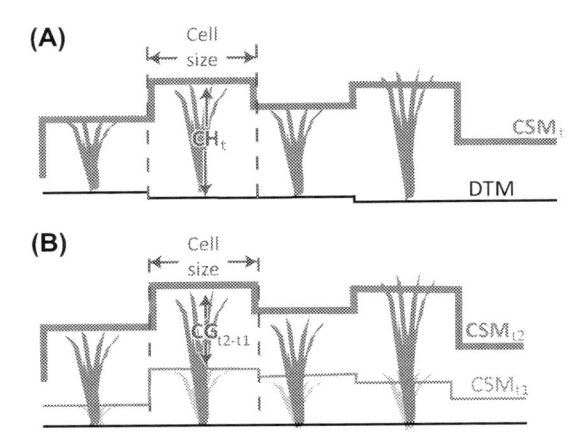

FIGURE 5 The use of crop surface models (CSM) for the derivation of crop height (CH) for a certain time step (t) by the use of an initial digital terrain model (DTM) (A) is depicted. Likewise, crop growth (CG) can be derived by including two consecutive CSMs (B). *Reproduced from D. Hoffmeister, G. Waldhoff, W. Korres, C. Curdt, G. Bareth, Precis. Agric. 17 (3) (2016) 296–312, Fig. 2, with the permission of Springer.*

of corresponding crop height and crop growth distribution maps for larger areas (see Fig. 5), and the combination with other geodata for statistical analysis.

Therefore, a comparison of the initial digital terrain model (DTM) and the multi-temporal CSM has to be conducted. The initial DTM can be created based on the first survey taken before crop emergence. Likewise an artificial, estimated DTM can be selected. The successive CSMs were taken from those surveys conducted at a later stage of crop development. To achieve the spatial distribution of crop height (CH), a CSM of a certain time step (t) is compared to the initial DTM from the first measurement before crop emergence, as depicted in Eq. (1) and shown in Fig. 5A.

$$CH_t = CSM_t - DTM \tag{1}$$

The resulting CH depicts the current, absolute crop height of the assigned survey date, in accordance with a normalized (n-) DSM or canopy height model in forestry applications. An example is shown in Fig. 5. In contrast, the spatial distribution of crop growth (CG) can be established by comparing two CSMs from successive dates to retrieve the relative crop growth between those dates, as shown in Eq. (2) and Fig. 5B.

$$CG_{t2-t1} = CSM_{t2} - CSM_{t1} \tag{2}$$

This approach was first used on sugar beet and wheat fields [62] and successfully transferred to paddy rice [63,64], barley [65] and maize [66]. The results show a high correlation to manual plant height measurements

($R^2 > 0.72$) and biomass ($R^2 > 0.65$). Similarly, UAV-based dense imaging is able to derive such accuracies [67]. However, multi- and hyperspectral indices hardly enhance the relationship between these measurements and biomass [65]. Hämmerle and Höfle [68] tested artificially reduced data sets and the 90th-percentile elevation for CSM generation to reduce the scanning time of low-cost LS systems. Twenty-five percent of the initial high-resolution data set still allows for a high coverage with a small difference to manually measured crop heights ($<4\%$).

3.3.3 Leaf Area Index and Crop Density Estimation

One further key parameter to describe vegetation, besides crop height, is the LAI. This is particularly important for modelling plant-atmosphere processes. The LAI is the ratio of leaf area to ground area. Several forestry studies exist, where the LAI is derived by ALS data. LAI can also be measured with high accuracy ($R^2 = 0.724$, RMSE $= 0.449$) by FW-ALS data for maize crops [69]. A similar approach also using intensity-corrected ALS data and derived canopy cover established a correlation with the fraction of absorbed photosynthetically active radiation, which is an important parameter that describes the physiological state of a plant [70].

Besides the previously reported direct use of the 3D point cloud and 2.5D raster information of crop height, point clouds can also be transferred to a voxel (a 3D pixel) representation. For instance, a triangulation-based scanner was used in a study for the voxel-based estimation of plant area density and plant area index (similar to leaf area density and LAI) of wheat, with few errors ($<10\%$) [71]. This analysis was conducted by counting single laser beam rays reaching or passing through a voxel, detecting the corresponding angle, and relating these values to horizontal layers of vegetation. Thus, horizontal areas with more measurements had a more intense density. This density was then compared to manually performed measurements and was related to the plant area index by the vertical integration of each horizontal layer. It was noted that, depending on row spacing, canopy height and inclination angle, the canopy measurements can be misleading (similar to the findings of Ref. [52]). Thus, the approach was successfully transferred to paddy rice by implementing a mirror above the dense canopy to accurately estimate the previously stated parameters, which are important for the collection of data concerning, for example, transpiration, photosynthesis, habitat characteristics and carbon stock [72]. In contrast to the approaches described in Section 3.3.2, this approach exploits the full 3D structure, even if it needs a high point density.

3.3.4 Nitrogen Content Detection

In addition to the highly accurate geometric information derived by LS, the intensity of the backscattered signal with a specific wavelength is of interest, as shown in Section 2. Passive remote sensing methods such as multi- and hyperspectral imagery are affected by solar illumination and atmospheric conditions. In addition, spectral reflectance sensors are limited to canopy closure. Active remote sensing, such as LS, is independent of these problems.

As a first approach, single leaves of different tree types were investigated for a relationship between recorded intensity values and nitrogen content [73]. Therefore, a standard reflectance target was measured at different distances and angles by a scanner exploiting a wavelength in the green-visible light spectrum (532 nm). These recorded values were used for an intensity range correction. Additionally, a specific intensity threshold was used to remove edge returns. A good correlation between measured chlorophyll a+b and laser intensity ($R^2 = 0.778$) was found, as well as a high correlation between manually measured leaf angle and leaf area. This lab-based approach was then used in a field application, where the N status of wheat in an early growth stage and corrected and filtered intensity values showed an accurate determination of N ($R^2 = 0.68$) in contrast to conventional chlorophyll measurements ($R^2 = 0.36$) and spectral indices ($R^2 < 0.41$) [20]. The geometric and radiometric information can be further used to derive the nitrogen nutrition index (NNI) [74]. This index shows the relationship between current and minimum nitrogen concentration for optimal growth, where the nitrogen concentration is a relation between nitrogen and crop volume. A strong correlation between estimated and measured crop volume was found ($R^2 > 0.72$), and a good relationship to N-concentration and intensity was used to calculate NNI ($R^2 = 0.45-0.54$). Overall, LS might help to actively derive NNI and thus improve N-application rates. Similarly, temporal changes in thermal energy dissipation at the leaf scale caused by stress can be related to backscattered intensity [75].

3.4 Spatial Variability

The measurements of a crop's status as described in Section 3.3 are the result of management, topography, diseases, soil and weather [76]. The quantification of the effects from these numerous and very variable factors affecting crop development can be conducted by statistical analysis of yield and other selected factors [76–78]. For instance, depending on precipitation, yield is correlated to topographic factors. In years with lower precipitation, the well-known effect of higher yields in low-lying areas, which benefit from soil moisture and better soil properties, can be proven. However, in years with high precipitation or strong precipitation events, the pattern is reversed. This knowledge facilitates the establishment of yield stability maps, which can be used for adjusting management practices. To derive topographic factors, LiDAR data can be used. A high-density LiDAR DEM in combination with yield data allows the statistical derivation of yield stability maps [77]. However, in another study, high resolution maps of crop height (depicted in Fig. 6)

FIGURE 6   Plant height distribution on a single field is derived indirectly from multi-temporal terrestrial laser scanning surveys (see Section 3.3.2). Each pixel has a size of 0.04 m². Sugar beet plant height on 24 July 2008, about 115 days after sowing, shows varying development across the field. *Reproduced from D. Hoffmeister, G. Waldhoff, W. Korres, C. Curdt, G. Bareth, Precis. Agric. 17 (3) (2016) 296–312, Fig. 3, with the permission of Springer.*

and crop growth derived by LS surveys were compared to several attributes related to topography [79]. These selected topographic attributes, as well as soil and soil moisture attributes, showed only minor correlations ($r < 0.44$, $R^2 < 0.31$) with the investigated variation of crop height and crop growth.

4. MULTISPECTRAL LASER SCANNING

It was shown in Section 3.3.4 how the backscattered intensity of an LS device using a specific single wavelength could be used on plants. As a further development, LiDAR with dual and multiple wavelengths will be presented. Some of these devices are based on a super-continuum laser [80], others use several different laser diodes emitting at specific wavelengths.

The application of the Salford Advanced Laser Canopy Analyser instrument as a dual wavelength system (1063 and 1545 nm) has been exploited for forestry applications in the field [81]. The selected wavelength enables the determination of the equivalent water thickness ($R^2 > 0.8$), which shows plant moisture distribution as an early drought stress indicator [82]. A similar design is shown for the Dual Wavelength Echidna LiDAR (DWEL), emitting at 1064 and 1548 nm, with the aim of distinguishing leaves and trunks [83]. In a field test, ranges over 70 m were shown, and the results revealed a very impressive distinction between leaves and trunks [84].

A multiwavelength canopy LiDAR was developed [85], capable of measuring four different wavelengths (555, 670, 700 and 780 nm) emitted by four semiconductor laser diodes. Specific indices were developed, and a support vector machine classification was shown [86]. In contrast, Wallace et al. [87] presented the design and evaluation of another prototype multi-spectral LiDAR system based on a super-continuum laser source. This device is capable of measuring at four different wavelengths (670, 780, 531 and 571 nm). The first pair of wavelengths was used to derive the normalized difference vegetation index (NDVI) and the second pair was implemented to derive the photochemical reflectance index. The measurements in a high geometric resolution demonstrated a significant agreement with in situ NDVI measurements.

A super-continuum source is also used by the Finnish Geodetic Institute. The proposed device records eight different channels between 542 and 981 nm [88]. These measurements can be used to determine chlorophyll concentration, which is related to plant nitrogen content. Among several computed vegetation indices, the chlorophyll absorption ration index provided the best results ($R^2 = 0.85$) [89]. The system can be enhanced to record more channels by the implementation of a better detector array. This possibility is shown by a 25-band device working between 1080 and 1620 nm [27].

Overall, the active emission of laser light allows the quick and accurate detection of objects and delivery of additional information at different wavelengths. These different wavelengths are used to distinguish different objects and, in addition, are able to discriminate between different plants and plant health levels by the incorporation of vegetation indices. With these solutions, the registration of different sensor images is unnecessary. Each point in 3D space will be unambiguously associated with a piece of spectral information.

5. DISCUSSION

There is general consensus in the literature that nondestructive detection by an active sensor is the key advantage of LiDAR. In addition, LiDAR is not generally dependent on weather conditions or dust and can be performed in most farming circumstances. In opposite, wind may cause altered measurements in particular for TLS-based approaches. Also, shadows of surrounding objects, like trees or other machinery, are not an issue. This advantage of active measurements is particularly shown by examples of already implemented or successfully tested applications in research for the enhancement of management tasks, like adjusted harvesting, automatic steering, spraying control and guidance and detection by robots.

Furthermore, LiDAR is able to partly penetrate the canopy by entering through small gaps in the vegetation, allowing direct measurement of the ground soil and the top canopy [29]. All these key features are promising for real-time applications in PA. In contrast, passive sensors, such as multi- and hyperspectral remote sensing [90], measurements of chlorophyll fluorescence

[91] and photogrammetric approaches conducted from UAVs [67,92] or balloons [93] need more post-processing tasks and computation time.

However, the acquisition of LiDAR data for most of the presented studies (see mainly Section 3.3.3) is complex and time-consuming. This complexity applies in particular to high-resolution data sets by TLS, as several scan positions are needed, which also require additional measurements for registration purposes. Thus, there is a need to practice the whole workflow [94]. In addition, the equipment itself is expensive. Hämmerle and Höfle [68] showed that artificially reduced data sets are also capable of crop detection. Likewise, a lot of research effort was undertaken to test low-cost scanners on field machinery, eg, by Ehlert et al. [51], and small scanners with new measuring principles have been developed [95]. Automatic TLS systems with low-cost parts may be a solution for constant monitoring of field-based phenotyping experiments [96]. However, depending on the crop type and parameters needed, a high density of the data might be required, as shown in Section 3.2. An automated workflow could be able to directly deliver plant development results from every experiment site to the office.

As a further advantage, the work regarding range-corrected intensity values for single plant detection [19], research regarding nitrogen content and further radiometric capabilities by a scanner in the visible green light wavelength [20,75] and the numerous studies in developing dual and multiple wavelength LiDAR (Section 4) [4], show that additional radiometric information for each point in the point cloud is available. This additional information can be used for a more accurate and easier distinction of features [82]. Furthermore, this added information allows the reuse or enhancement of all research of vegetation indices, overcoming the disadvantages of previously used passive remote sensing [89].

The data processing is likewise time and cost intensive, and new workflows and algorithms need to be developed for this newer remote sensing method. For the data derived by this method, a lot of computing power is needed to display and work with the data. Data sets can consist of billions of points, with subsequent versions of the original point cloud. However, studies indicate that once the resources are provided, suitable data are available. This is shown on all different scales, starting from the automatically derived features of single plants [43,44] to the field scale used by the CSM approach [63], which can be used for further statistical analysis.

The chapter demonstrated that in terms of PA, accurate plant height detection and several other parameters at different stages of crop development can be accurately derived from the 3D data sets. Plant height is mostly closely connected to plant volume or biomass [63,64], which is valuable information related to the health and nutrient requirements of a crop. In addition, plant biomass variation can be related to variation in yield. Thus, nondestructive plant biomass measurements by LS enable the estimation of yield, as well as the use of specific counter measures to prevent yield loss. Maps of crop height and crop growth can be applied to detect areas where additional management tasks or adjusted processes are necessary [79]. Furthermore, it was shown at the single plant scale that high-throughput phenotyping is possible (see Section 3.2).

6. CONCLUSION AND OUTLOOK

In summary, LiDAR, as an active remote sensing method is able to facilitate PA in several directions, such as management, plant height and plant vitality detection. Destructive measurements, in particular those used in crop phenotyping, become superfluous. In addition, the high-resolution monitoring data of detected crop variability is important for statistical analysis in terms of adjusted management and yield prediction. Newer sensors are smaller, cheaper and adjustable in many variables, such as the wavelength. The latter contributes to a combined measurement of crop height and vitality by an active sensor. Faster and smaller hardware, as well as automatic procedures for feature analysis, might allow LiDAR to be widely distributed. Thus, LiDAR will play an important role in future PA and hence contribute to sustainable food supply.

REFERENCES

[1] R. Gebbers, V.I. Adamchuk, Science 327 (5967) (2010) 828−831.
[2] D.J. Mulla, Biosyst. Eng. 114 (4) (2013) 358−371.
[3] F. Fiorani, U. Schurr, Annu. Rev. Plant Biol. 64 (2013) 267−291.
[4] Y. Lin, Comput. Electron. Agric. 119 (2015) 61−73.
[5] K. Schneider, Int. J. Remote Sens. 24 (14) (2003) 2959−2980.
[6] G. Petrie, C.K. Toth, in: J. Shan, C.K. Toth (Eds.), Topographic Laser Ranging and Scanning, Taylor & Francis Group, Boca Raton, Florida, USA, 2009, pp. 1−29.
[7] G.L. Heritage, A.R.G. Large, in: G.L. Heritage, A.R.G. Large (Eds.), Laser Scanning for the Environmental Sciences, Wiley-Blackwell, Chichester, UK, 2009, pp. 1−20.
[8] H. Kaartinen, J. Hyyppä, X. Yu, M. Vastaranta, H. Hyyppä, A. Kukko, M. Holopainen, C. Heipke, M. Hirschmugl, F. Morsdorf, E. Næsset, J. Pitkänen, S. Popescu, S. Solberg, B.M. Wolf, J.-C. Wu, Remote Sens. 4 (12) (2012) 950−974.
[9] J. Hyyppä, H. Hyyppä, D. Leckie, F. Gougeon, X. Yu, M. Maltamo, Int. J. Remote Sens. 29 (5) (2008) 1339−1366.
[10] M. van Leeuwen, M. Nieuwenhuis, Eur. J. For. Res. 129 (4) (2010) 749−770.
[11] J.-A. Beraldin, F. Blais, U. Lohr, in: G. Vosselman, H.G. Maas (Eds.), Airborne and Terrestrial Laser Scanning, Whittles Publishing, Dunbeath, UK, 2010, pp. 1−42.
[12] E.P. Baltsavias, ISPRS J. Photogramm. Remote Sens. 54 (2−3) (1999) 199−214.
[13] G. Vosselman, H.-G. Maas (Eds.), Airborne and Terrestrial Laser Scanning, first ed., Whittles Publishing, Dunbeath, UK, 2010, p. 311.
[14] S.-G. Mårtensson, Y. Reshetyuk, L. Jivall, J. Appl. Geod. 6 (1) (2012) 25−32.
[15] P.J. Besl, N.D. McKay, IEEE Trans. Pattern Anal. Mach. Intell. 14 (2) (1992) 239−256.
[16] Y. Chen, G. Medioni, Image Vision Comput. 10 (3) (1992) 145−155.
[17] B. Höfle, N. Pfeifer, ISPRS J. Photogramm. Remote Sens. 62 (6) (2007) 415−433.

[18] N. Pfeifer, B. Höfle, C. Briese, M. Rutzinger, A. Haring, ISPRS Archives — Volume XXXVII Part B5, Proc. XXI. ISPRS Conference (July 03—11, 2008), 2008.

[19] B. Höfle, IEEE Geosci. Remote Sens. Lett. 11 (1) (2014) 94—98.

[20] J.U.H. Eitel, L.A. Vierling, D.S. Long, E.R. Hunt, Agric. For. Meteorol 151 (10) (2011) 1338—1345.

[21] W. Wagner, A. Ullrich, V. Ducic, T. Melzer, N. Studnicka, ISPRS J. Photogramm. Remote Sens. 60 (2) (2006) 100—112.

[22] M. Doneus, C. Briese, M. Fera, M. Janner, J. Archaeol. Sci. 35 (4) (2008) 882—893.

[23] V.H. Duong, R. Lindenbergh, N. Pfeifer, G. Vosselman, Int. J. Remote Sens. 29 (5) (2008) 1453—1473.

[24] M. Hayashi, N. Saigusa, H. Oguma, Y. Yamagata, ISPRS J. Photogramm. Remote Sens. 81 (2013) 12—18.

[25] M.A. Lefsky, D.J. Harding, M. Keller, W.B. Cohen, C.C. Carabajal, F. Del Bom Espirito-Santo, M.O. Hunter, R. de Oliveira, Geophys. Res. Lett. 32 (22) (2005) 1—4.

[26] S.J. Buckley, T.H. Kurz, J.A. Howell, D. Schneider, Comput. Geosci. 54 (2013) 249—258.

[27] M.A. Powers, C.C. Davis, Appl. Opt. 51 (10) (2012) 1468—1478.

[28] T. Chateau, C. Debain, F. Collange, L. Trassoudaine, J. Alizon, Comput. Electron. Agric. 28 (3) (2000) 243—257.

[29] W. Saeys, B. Lenaerts, G. Craessaerts, J. De Baerdemaeker, Biosyst. Eng. 102 (1) (2009) 22—30.

[30] B. Lenaerts, B. Missotten, J. De Baerdemaeker, W. Saeys, Comput. Electron. Agric. 85 (2012) 40—44.

[31] T. Coen, A. Vanrenterghem, W. Saeys, J. De Baerdemaeker, Comput. Electron. Agric. 63 (1) (2008) 57—64.

[32] K. König, B. Höfle, M. Hämmerle, T. Jarmer, B. Siegmann, H. Lilienthal, ISPRS J. Photogramm. Remote Sens. 104 (2015) 112—125.

[33] P.J. Walklate, J.V. Cross, G.M. Richardson, R.A. Murray, D.E. Baker, Biosyst. Eng. 82 (3) (2002) 253—267.

[34] Y. Shi, N. Wang, R.K. Taylor, W.R. Raun, Comput. Electron. Agric. 112 (2015) 92—101.

[35] N. Shalal, T. Low, C. McCarthy, N. Hancock, Comput. Electron. Agric. 119 (2015) 254—266.

[36] J. Gimenez, D. Herrera, S. Tosetti, R. Carelli, Comput. Electron. Agric. 116 (2015) 88—100.

[37] S.A. Hiremath, G.W.A.M. van der Heijden, F.K. van Evert, A. Stein, C.J.F. ter Braak, Comput. Electron. Agric. 100 (2014) 41—50.

[38] J. Palacin, T. Palleja, M. Tresanch, R. Sanz, J. Llorens, M. Ribes-Dasi, J. Masip, J. Arno, A. Escola, J.R. Rosell, IEEE Trans. Instrum. Meas. 56 (4) (2007) 1377—1383.

[39] A. Osterman, T. Godeša, M. Hočevar, B. Širok, M. Stopar, Comput. Electron. Agric. 98 (2013) 175—182.

[40] Y. Chéné, D. Rousseau, P. Lucidarme, J. Bertheloot, V. Caffier, P. Morel, É. Belin, F. Chapeau-Blondeau, Comput. Electron. Agric. 82 (2012) 122—127.

[41] T.T. Nguyen, D.C. Slaughter, N. Max, J.N. Maloof, N. Sinha, Sensors 15 (8) (2015) 18587—18612.

[42] S. Paulus, J. Behmann, A.K. Mahlein, L. Plumer, H. Kuhlmann, Sensors 14 (2) (2014) 3001—3018.

[43] S. Paulus, J. Dupuis, S. Riedel, H. Kuhlmann, Sensors 14 (7) (2014) 12670—12686.

[44] M. Wahabzada, S. Paulus, K. Kersting, A.K. Mahlein, BMC Bioinform. 16 (2015) 248.

[45] S. Paulus, T. Eichert, H.E. Goldbach, H. Kuhlmann, Sensors 14 (2) (2014) 2489—2509.

[46] A. Konishi, A. Eguchi, F. Hosoi, K. Omasa, Funct. Plant Biol. 36 (10—11) (2009) 874—879.

[47]   P.J. Walklate, Agric. For. Meteorol. 46 (4) (1989) 275−284.

[48]   D. Ehlert, H.J. Horn, R. Adamek, Comput. Electron. Agric. 61 (2) (2008) 117−125.

[49]   D. Ehlert, R. Adamek, H.-J. Horn, Precis. Agric. 10 (5) (2009) 395−408.

[50]   R. Gebbers, D. Ehlert, R. Adamek, Agron. J. 103 (5) (2011) 1532−1541.

[51]   D. Ehlert, M. Heisig, R. Adamek, Precis. Agric. 11 (6) (2010) 650−663.

[52]   D. Ehlert, M. Heisig, Comput. Electron. Agric. 93 (2013) 10−16.

[53]   L. Zhang, T.E. Grift, Comput. Electron. Agric. 85 (2012) 70−76.

[54]   D. Andújar, A. Escolà, J.R. Rosell-Polo, C. Fernández-Quintanilla, J. Dorado, Comput. Electron. Agric. 92 (2013) 11−15.

[55]   J. Lumme, M. Karjalainen, H. Kaartinen, A. Kukko, J. Hyyppä, H. Hyyppä, A. Jaakola, J. Kleemola, ISPRS Archives − Volume XXXVII Part B5, Proc. XXI. ISPRS Conference (July 03−11, 2008), 2008. Beijing, China.

[56]   J. Llorens, E. Gil, J. Llop, A. Escola, Sensors 11 (2) (2011) 2177−2194.

[57]   R. Sanz-Cortiella, J. Llorens-Calveras, A. Escola, J. Arno-Satorra, M. Ribes-Dasi, J. Masip-Vilalta, F. Camp, F. Gracia-Aguila, F. Solanelles-Batlle, S. Planas-DeMarti, T. Palleja-Cabre, J. Palacin-Roca, E. Gregorio-Lopez, I. Del-Moral-Martinez, J.R. Rosell-Polo, Sensors 11 (6) (2011) 5769−5791.

[58]   S.D. Tumbo, M. Salyani, J.D. Whitney, T.A. Wheaton, W.M. Miller, Appl. Eng. Agric. 18 (3) (2002) 367−372.

[59]   K.E. Keightley, G.W. Bawden, Comput. Electron. Agric. 74 (2) (2010) 305−312.

[60]   J. Arnó, A. Escolà, J.M. Vallès, J. Llorens, R. Sanz, J. Masip, J. Palacín, J.R. Rosell-Polo, Precis. Agric. 14 (3) (2012) 290−306.

[61]   F.A. Auat Cheein, J. Guivant, R. Sanz, A. Escolà, F. Yandún, M. Torres-Torriti, J.R. Rosell-Polo, Comput. Electron. Agric. 118 (2015) 361−371.

[62]   D. Hoffmeister, A. Bolten, C. Curdt, G. Waldhoff, G. Bareth, in: Proceedings of SPIE-7840, 6th International Symposium on Digital Earth: Models, Algorithms, and Virtual Reality (September 09−12, 2009), Beijing, China, 2010.

[63]   N. Tilly, D. Hoffmeister, Q. Cao, S. Huang, V. Lenz-Wiedemann, Y. Miao, G. Bareth, J. Appl. Remote Sens. 8 (1) (2014) 083671.

[64]   N. Tilly, D. Hoffmeister, Q. Cao, V. Lenz-Wiedemann, Y. Miao, G. Bareth, Agriculture 5 (3) (2015) 538.

[65]   N. Tilly, H. Aasen, G. Bareth, Remote Sens. 7 (9) (2015) 11449−11480.

[66]   N. Tilly, D. Hoffmeister, H. Schiedung, C. Hütt, J. Brands, G. Bareth, Int. Arch. Photogramm. Remote Sens. Spatial Inf. Sci XL-7 (2014) 181−187.

[67]   J. Bendig, A. Bolten, S. Bennertz, J. Broscheit, S. Eichfuss, G. Bareth, Remote Sens. 6 (11) (2014) 10395−10412.

[68]   M. Hämmerle, B. Höfle, Sensors 14 (12) (2014) 24212−24230.

[69]   S. Nie, C. Wang, P. Dong, X. Xi, Remote Sens. Lett. 7 (2) (2015) 111−120.

[70]   S. Luo, C. Wang, X. Xi, F. Pan, Opt. Express 22 (5) (2014) 5106−5117.

[71]   F. Hosoi, K. Omasa, ISPRS J. Photogramm. Remote Sens. 64 (2) (2009) 151−158.

[72]   F. Hosoi, K. Omasa, ISPRS J. Photogramm. Remote Sens. 74 (2012) 11−19.

[73]   J.U.H. Eitel, L.A. Vierling, D.S. Long, Remote Sens. Environ. 114 (10) (2010) 2229−2237.

[74]   J.U.H. Eitel, T.S. Magney, L.A. Vierling, T.T. Brown, D.R. Huggins, Field Crops Res. 159 (2014) 21−32.

[75]   T.S. Magney, S.A. Eusden, J.U. Eitel, B.A. Logan, J. Jiang, L.A. Vierling, New Phytol. 201 (1) (2014) 344−356.

[76]   A.N. Kravchenko, G.P. Robertson, K.D. Thelen, R.R. Harwood, Agron. J. 97 (2) (2005) 514−523.

[77] J.M. McKinion, J.L. Willers, J.N. Jenkins, Comput. Electron. Agric. 70 (1) (2010) 187−198.

[78] A. Heuer, M.C. Casper, M. Herbst, Z. Geomorphol. Suppl. Issues 55 (3) (2011) 169−178.

[79] D. Hoffmeister, G. Waldhoff, W. Korres, C. Curdt, G. Bareth, Precis. Agric. 17 (3) (2016) 296−312.

[80] C. Gmachl, D.L. Sivco, R. Colombelli, F. Capasso, A.Y. Cho, Nature 415 (6874) (2002) 883−887.

[81] R. Gaulton, F.M. Danson, F.A. Ramirez, O. Gunawan, Remote Sens. Environ. 132 (2013) 32−39.

[82] F.M. Danson, R. Gaulton, R.P. Armitage, M. Disney, O. Gunawan, P. Lewis, G. Pearson, A.F. Ramirez, Agric. For. Meteorol. 198−199 (2014) 7−14.

[83] E.S. Douglas, A. Strahler, J. Martel, T. Cook, C. Mendillo, R. Marshall, S. Chakrabarti, C. Schaaf, C. Woodcock, L. Zhan, Y. Xiaoyuan, D. Culvenor, D. Jupp, G. Newnham, J. Lovell, Geoscience and Remote Sensing Symposium (IGARSS), 2012 IEEE International, 2012.

[84] E.S. Douglas, J. Martel, L. Zhan, G. Howe, K. Hewawasam, R.A. Marshall, C.L. Schaaf, T.A. Cook, G.J. Newnham, A. Strahler, S. Chakrabarti, Geosci. Remote Sens. Lett. IEEE 12 (4) (2015) 776−780.

[85] W. Gong, S.L. Song, B. Zhu, S. Shi, F.Q. Li, X.W. Cheng, ISPRS J. Photogramm. Remote Sens. 69 (2012) 1−9.

[86] W. Gong, J. Sun, S. Shi, J. Yang, L. Du, B. Zhu, S. Song, Sensors 15 (9) (2015) 21989−22002.

[87] A.M. Wallace, A. McCarthy, C.J. Nichol, R. Ximing, S. Morak, D. Martinez-Ramirez, I.H. Woodhouse, G.S. Buller, IEEE Trans. Geosci. Remote Sens. 52 (8) (2014) 4942−4954.

[88] T. Hakala, J. Suomalainen, S. Kaasalainen, Y. Chen, Opt. Express 20 (7) (2012) 7119−7127.

[89] O. Nevalainen, T. Hakala, J. Suomalainen, S. Kaasalainen, ISPRS Annals of Photogrammetry, Remote Sensing and Spatial Information Sciences, II-5/W2, 2013, pp. 205−210.

[90] F. Li, Y. Miao, S.D. Hennig, M.L. Gnyp, X. Chen, L. Jia, G. Bareth, Precis. Agric. 11 (4) (2010) 335−357.

[91] M. Meroni, M. Rossini, L. Guanter, L. Alonso, U. Rascher, R. Colombo, J. Moreno, Remote Sens. Environ. 113 (10) (2009) 2037−2051.

[92] T. Oberthür, J. Cock, M.S. Andersson, R.N. Naranjo, D. Castañeda, M. Blair, Comput. Electron. Agric. 58 (1) (2007) 60−77.

[93] T. Murakami, M. Yui, K. Amaha, Comput. Electron. Agric. 89 (2012) 70−75.

[94] G.L. Heritage, A.R.G. Large, in: G.L. Heritage, A.R.G. Large (Eds.), Laser Scanning for the Environmental Sciences, Wiley-Blackwell, Chichester, UK, 2009, pp. 21−34.

[95] M. Bosse, R. Zlot, P. Flick, IEEE Trans. Robotics 28 (5) (2012) 1104−1119.

[96] J.U.H. Eitel, L.A. Vierling, T.S. Magney, Agric. For. Meteorol. 180 (2013) 86−96.

Chapter 12

Robotic-Based Agriculture for Rural Renaissance: Drones and Biosensors

K.C. Siva balan
IIAT, Trichy, Tamil Nadu, India
E-mail: shiv_balan@yahoo.com

Chapter Outline

Comprehensive Analytical Chemistry, Vol. 74. http://dx.doi.org/10.1016/bs.coac.2016.04.017

1. INTRODUCTION

Drones (in technical terms unmanned aerial vehicles (UAVs)) are aircraft either monitored or controlled by 'pilots' from the ground or autonomously following a preprogrammed air travel and landing. Drones are equipped with low-cost aerial camera platform. Based on the budget and requirement, drones may be designed like miniature aeroplanes or quadcopters and multibladed small helicopters. These aircrafts are equipped with an autopilot using the Geographic Positioning System and also use software for predesigned aerial mosaic flying. The high-resolution aerial camera records aerial shots which could be interpreted using suitable software like Image J. Drones ensure optimal farm input management by giving ad hoc information for applying exact dose of fertilizers and chemicals. By drone technology, the frequency of farm visits and travel time for farmers to visit their respective farms could be considerably reduced. Although drones were invented for military purpose (UAV called 'Queen Bee' produced by the British military in 1934), people subsequently started using them for recreational purposes. The main advantage of drone technology is the approach specificity to the target and collection of information at a micro level [1,2] with lower cost compared to satellites while a detailed field overview is also possible.

According to the assessment report by the International Institute for Strategic Studies in 2014, the United States of America ranks first in the officially declared count of drones and other variants of UAVs. Although the Chinese and Japanese UAV manufacturers rule, market individuals as well as institutions worldwide are in progress of making drones, network balloons, biosensors and geo-tagging for serving the clientele needs.

1.1 Application of Drone Technology in Agriculture

In developing countries like India, traditional farm and crop scouting are done by farmers themselves which often consume much more time and manpower. Moreover, farmers could identify major crop issues only if they visited the farm regularly. The delay in diagnosis of crop pest and diseases and crop protection measures may lead to crop damage with a reduction in yield and profitability. By aerial vehicles, real-time diagnosis of crop systems is possible. Aerial vehicles could perform geo-referenced farm scouting, with a considerable reduction in time and manpower cost. The detailed three-dimensional (3D) mapping and colour spectrum charts could be developed over surveys by drone devices. These databases would be useful in the line of measurement of stratospheric ozone levels, measurement of air quality, groundwater discharge monitoring, early warning systems on earthquakes and volcanoes, meteorology warnings on cloud tsunami (cloud bursting) avalanche patrols, status of glacier and ice sheet thickness and surface deformation, open pit mine surveys and others. These mobile survey drones can be easily displaced into areas whenever the need arises irrespective of the topographical zones. The advantage of flying aerial space is that drones could reach even difficult topographical conditions compared to wheeled farm vehicles. Drones being machines are not influenced by weather conditions and can operate in cloudy weather and even on rainy days. The infrared camera attached to drones can take multispectral images from the sky which can suitably be decoded using software [4]. Based on the data recorded, the soil and crop stress status could be identified. Sensors attached to drones are able to monitor the vegetative index of the crops using a laser scanning system [5]. By comparing with the standard parameters, quicker farm decisions regarding the time of application of fertilizers, spraying of chemicals and weeding operations could be provided.

Apart from crop scouting, UAVs could be deployed for a variety of agricultural operations from spraying of pesticides and fertilizers to monitoring soil and water quality, surveillance of erosion and maturity stage of the crops [6]. Almost 25 years ago in the late 1980s, Japanese were the first to apply drone technology in the farm segment. Today, unmanned aircrafts like quadcopters and helicopters carry out farm operations in more than 10% of Japan's paddy area [7]. At present, Japanese manufacturing giants Fuji Heavy Industries and Yamaha Corporation manufacture and sell fully autonomous sprayers extensively [3]. The Yamaha Corporation commercialised a remote-controlled helicopter (UAV) that can be used to spray chemicals or spread granules. By aerial spraying of inputs, including fertilizers, fungicides and pesticides, considerable reduction in labour costs could be achieved as well as timely farm operations. By reduction of movement for wheels and tracks on the field, there is no need for *pakka* roads (a levelled terrain or roads) and crop space could also be increased. Drones for agricultural purpose could be of light weight, which require less fuel and maintenance costs in comparison with military or logistic utility models. Plant protection operations such as spraying involving manual labour could result in allergy, eye irritation, skin disorders and diseases like cancer in the long run, which could be avoided with aerial spraying with machines. Concerning safety issues, the risk of manned survey and operation vehicles can be also evaded using remote controlled drones by following preprogrammed destinations by a remote pilot [8].

At present, precision farming practices are gaining momentum across the globe. Agricultural survey drones can monitor and take crop quality visuals, which help researchers to advocate technological solutions and farmers to carry out various precision agriculture applications. Since the cost of the drone is cheaper than a tractor or any other farm vehicle, individual farmers in developing countries could purchase a drone for their own use (easy-to-use agricultural drones are available online for less than $1200). That the unmanned aircraft consumes less fuel and has lower maintenance cost compared with manned farm vehicles is an added advantage. With preloaded job commands, the agriculture drones or UAVs could be operated at ease even by a non-technocrat.

1.2 Economic Impact of Drone Technology

Partial budgeting is the direct measure of financial impact of any new resources implemented (Eg. farm machinery) in the farm business. It is also a systematic approach by which farm owners and managers could decide the alternate use of resources. In Asia, the cost of cultivation of paddy over the years has consistently been rising owing to the increased costs of inputs and labour. Based on the financial gain by adopting drones for spraying plant protection chemicals, with lesser manpower, we could strongly promote drone technology for farming. The partial budgeting analysis was done for paddy growers since labour shortage issues (for farm operation and plant protection) were reported often for paddy, more than for any other crop.

In reality, many farmers cannot afford to own a drone device. Considering the initial investment cost, they would prefer pay and user mode. Farmers using drones for spraying agrochemicals in rental model operations may incur service charges for every spraying. In the same line, nonusers of drones for chemical spraying have to spend extra cost for labour for spraying during the demand time or peak seasons. That is why the expenditure incurred by the nonusers of drone technology with respect to the labour cost was higher than the users of drones for crop protection (Table 1).

TABLE 1 Paddy Crop — Cost and Return Analysis Among Users and Nonusers of Drone Technology

No	Costs/Returns (INR/Acre)	Users	Nonusers
1.	Nursery preparation	2400	2400
2.	Farm Yard Manure (FYM)	5000	5000
3.	Main field preparation	4800	4800
4.	Seed	2500	2500
5.	Transplanting	1500	1500
6.	Fertilizer	1457	3381
7.	Irrigation	1160	1160
8.	Plant Protection (PP) Chemicals	1750	1750
9.	Human labour	6511	9961
10.	Machine labour	4980	3890
11.	Harvesting	3100	3100
12.	Yield (kg)	2400	2250
13.	Selling price (Rs/kg)	20	20

Author's estimation based on 2015 survey data.

TABLE 2 Results of Partial Budgeting Analysis for Using Drone Technology in Paddy Cultivation (INR/Acre)

No.	Debit (A)	Value	Credit (B)	Value (INR)
1.	Added costs	2000[a]	Reduced costs	5374
2.	Reduced returns	—	Added returns	3000
	Total	2000[a]	Total	8374
Net gain = (B − A)				6374

[a]Approximate price for rental model/conditions applied.

The comparative cost—benefit analysis is presented in Table 2. The approximate rental cost has been worked out from inputs from rental vendors. On an average, $30 (INR 2000) would be the added cost in the rental model operation of drones for spraying plant protection chemicals. For a smallholder farmer even if he uses rental drones for chemical spraying, he could expect a net gain to the tune of $100 (INR 6374). The considerable reduction of farm operation timings (the total spray period could be within 20—30 min) and flexibility of usage of drones even during night hours are the other intangible benefits. The complete details of spread mapping spread tracking analytics and reports are also possible since the device can be controlled by a tablet or smartphone. Along with the plant protection measures, farm surveillance is also possible by using drone technology. Thus, drone interventions have the potential to realize a better profit margin as well as to reduce the direct cost in every stage of the crop value chain.

1.3 Drone Technology for Nonagricultural Key Areas

1.3.1 Real-Time Forestry and Vegetation Coverage

Drones could be used for observing the changes in the gross forestry and vegetation pattern thereby helping tropical and tribal communities to better manage and conserve their natural resources [9,12]. The population count of wild animals could be carried out with accuracy with the help of thermal infrared imaging technology, data telemetry by the drone technology. Entry of wild animals inside farms and human habitats in tribal areas is also possible; thereby man—animal conflicts could be avoided. In India, an effort for tracking a killer leopard in Uttarakhand state of Northern India was successfully done by multifunctional drone 'Netra UAV' [3,10]. A similar attempt for detection of coconut wilt disease outbreak was carried out with drone technology. Illegal wood logging could be monitored and forest surveillance by drones can better prevent illegal logging and poachers in the dense forests, bio parks and wildlife sanctuaries [11].

1.3.2 Coastal Management and Fishing Zone Mapping

With the proliferation of drone technology, tracking changes in coastal vegetation, coral reefs and nutrients of coastal zones, surveillance of coastal areas for intrusion and management of resources are possible [13,14]. The drones could measure sea surface temperature accurately, related to the concentration of fish population. Daily or periodical information regarding the fish population could alert the fishermen for potential areas of fish catch.

1.3.3 Global Natural Resources Monitoring Systems

The detailed 3D mapping and colour spectrum charts developed over survey by drone devices would be useful in the line of measurement of stratospheric ozone levels, measurement of air quality, groundwater discharge monitoring, early warning systems on earthquakes and volcanoes and meteorology warnings on cloud tsunami (cloud bursting).

1.3.4 Hawk Eye in the Sky for Law and Order Maintenance

In many countries, police and military departments have started deploying drones and UAVs equipped with high-definition cameras and sensors to monitor national borders as well as public gatherings and sensitive communal situations. Cross-border intruders could be caught by the latest night vision drones [15,16]. The drones could support the military personnel in disaster management, to provide relief measures during natural disasters when reaching the affected places and people in need is a cumbersome task.

1.4 Ethical Issues in Using Drone Technology

Although the unmanned surveillance has multifaceted utility, the prime outcry is about breach of privacy. For instance, the Indian Information Technology (Amendment) Act of 2008 raises questions on equipment ownership and security of data and violation of privacy [17]. The misuse of drones for private observational tasks may lead to pandemonium and calls for user regulatory mechanisms. The civil aviation and border security departments already caution on the regulations for the privately owned drones which could be a threat to national security. In the United States of America, this low-altitude view of around 120 m is the regulatory ceiling for unmanned aircraft operating without special clearance from the US federal aviation administration. In India, there is no clear doctrine on rules and regulations for flying drones either for academic or for personal purposes [3,18]. Apart from the lawful stand, other technical problems like chance of mechanical failures and crashes, unclear or low-quality visuals, operator skills and affordability of devices and software also need to be considered and resolved.

2. APPLICATION OF BIOSENSORS IN AGRICULTURE, HEALTH AND FOOD PRESERVATION

When resolving food security challenges, food safety and protection is also a matter of concern in the midst of environmental issues. The food safety concept entails the production and commercialisation of foods free from any allergenic or toxic substances, pathogens, pesticides, fertilizers, heavy metals, organic compounds, toxins and pathogen microorganisms, which can have a high toxic effect on human health and environment. However, during the postharvest transits, the quality of food grains may deteriorate due to storage temperature conditions and exposure to sunlight. Often, components such as vitamins, minerals and antioxidants might alter during the processing and value addition of food products. As a result of the catalytic and enzymatic processes, organoleptic indicators, texture, colour, taste, smell and aroma may be distorted. According to the World Health Organization (WHO), each year millions of people around the world are affected by food-borne diseases. Food poisoning and infections have a strong economic and social impact, causing irreparable loss of productivity [19]. The traditional means of food quality testing involves detection and isolation of potential pathogenic microorganisms which necessitate long response times (ranging from 48 h to 5 days), and also involving high identification costs [20]. In this context, biosensors could play an important role in the agricultural and food sectors to ensure food quality and safety by detecting food-borne pathogens and chemical contaminants. A biosensor can be defined as a quantitative or semiquantitative analytical instrument containing a sensing element of biological origin, which is either integrated within or is in intimate contact with a physicochemical transducer [21]. Biosensors have been useful for the evaluation of food quality and composition, particularly in food products from plant and animal origin. Biosensors could also be used as indicators of food freshness by monitoring the pH, temperature, volatile substances rancidity and maturity of different food components. Biosensors could harness specificity and sensitivity of biological systems by in situ analysis of pollutants in crops and soils, detection and identification of concentrations of herbicides, pesticides and heavy metals, diseases in crops and livestock. Along with food safety, food security measures like evaluation of important food processing and packaging parameters could be possible. Biosensor technologies are most beneficial due to the potential for rapid, specific and sensitive detection of biological hazards [22,23].

2.1 Role of Biosensors in Detecting Food Pathogens

Biotoxins produced by inborn bacteria, viruses, pesticide residues and chemical substances added in the food products either as adulteration or contamination could generate polycyclic aromatic hydrocarbons or mutagenic agents like tryptophan. This may cause quality deterioration of food products as well as ill effects on the consumers' health. Colourimetric, refractometric, chromatographic and spectrophotometric tests are commonly used for the detection of food freshness and quality of alcoholic beverages. These methods involve high-end laboratory procedures, pricey tools and considerable time for laboratory testing. The advanced and emerging analytical technology like electronic or bioelectronic tongues and nose recognized as taste sensors, could be well utilized for testing beverage quality in the fermentation stage itself and fungal contamination of food grain with substantial reduction of cost and time for analysis.

Gram-negative bacteria *Salmonella* are commonly found in the gastro-intestinal tract of warm-blooded animals and humans [24,25]. The least quality concern about the purified water for food processing industries leads to biological contamination hazards by *Salmonella* bacteria [26]. The piezoelectric antigen—antibody biosensors are used for the early detection of *Salmonella*. They are also used in milk industries including milking utensils, farms or livestock firms and food processing industries. Another bacterium flagellate microaerophilic coccobacillus called *Listeria mono-cytogenes* could cause disease resembling influenza or meningitis leading to miscarriage. Causing severe muscle pain and diarrhea in the later stages it may affect the central nervous system of humans. The bacterium infects fresh and processed foods such as meats, shellfish, unpasteurized milk and vegetables [27]. According to the Food and Agriculture Organization (FAO)/WHO report 2004, *L. monocytogenes* has had negative consequences at the productive, economic and social levels of the community. Fibre optic and surface plasmon resonance (SPR) biosensors help preventing the spread of food-borne illnesses and ensure food safety in processed and minimally processed foods [28]. Although SPR sensors are more expensive than fibre optic sensors, they are suitable for minimal time frame (5—15 min) detection [29].

Among children, the gram-negative bacterium *Campylobacter* causes gastroenteritis with symptoms of fever, abdominal pain, vomiting and diarrhoea. Campylobacteriosis illness is mainly spread by eating contaminated food. The disease also has side effects including reactive arthritis and muscle pain due to alterations in the immune system [30]. Conventional microbiological methods for the identification of *Campylobacter* require 3—4 days. The optical SPR biosensors could be implemented for early detection. The gram-negative bacillus *Escherichia coli* normally populates the intestine of humans and warm-blooded animals [31]. The infection caused by *E. coli* results from consuming fresh fruits and vegetables. Infections may also be caused by eating poorly cooked animal foods cleaned with contaminated water [32]. *E. coli* detection could be done in a shorter time using amperometric biosensors, through the detection of hydroxyl radicals produced by *E. coli* during aerobic metabolism [33].

2.2 Role of Biosensors in Detecting Pesticide Residues and Contaminants

Over the years, the indiscriminate use of inorganic inputs such as pesticides and fertilizers has led to biological, chemical and physical threats to food grains and the quality of value-added food products. The imbalanced fertilization is primarily attributed to indiscriminate use of nitrogenous fertilizers and exclusion of other macro- and micronutrients. It is considered a serious concern which is rapidly deteriorating the soil health. The excessive nitrogen fertilizer use affects groundwater and also causes eutrophication in aquatic ecosystems. Biosensors could precisely detect and deliver the correct quantity of nutrients required by crops in a suitable proportion that promotes productivity while ensuring environmental safety. OPH, an organophosphotriester hydrolysing enzyme could be used in a biosensor for the detection of paraoxon and parathion [34]. Acetyl cholinesterase could detect monocrotophos, malathion, metasystox, lannate, chlorpyrifos, malathion, carbofuran, methomyl and carbendazim [35]. Immunochemical techniques, including piezoelectric immunosensors, are gaining acceptance as alternative or complementary methods for the analysis of pesticides. Immunosensors have great potential for monitoring herbicides in drinking water [36,37].

2.3 Biosensors for Detection of Heavy Metal Deposits in Soil and Water Bodies

Soil fertility and water bodies such as rivers, streams and drinking water ponds are commonly polluted by leaching from industrial and consumer wastages. The ingestion of plant- and animal-based foods from the polluted landscapes leads to cardiovascular and respiratory problems, skin allergy, hormonal imbalance, fertility issues and malfunction of vital organs. The protection of soil health and the environment requires rapid, sensitive detection of pollutants and pathogens with molecular precision. Accurate sensors are needed for in situ detection, as miniaturized portable devices, and as remote sensors, for the real-time monitoring of large areas in the field. The sensors are now used for the identification of toxic chemical compounds at ultralow levels (parts per million and parts per billion) in industrial products, chemical substances, water, air and soil samples, or in biological systems [38]. The detection of cadmium, copper, chrome, nickel and zinc by the urease biosensor and detection of silver and cadmium by invertase and glucose oxidase biosensor were reported by Verma et al. [39] and Bagal-Kestwal et al. [40], respectively. The heavy-metal nanosensor technology (nano-contact sensor) could monitor heavy metals in drinking water and alert the stakeholders in governance of water bodies [41].

2.4 Product Identity Preservation and Tracking

With appreciable global demand for organic agriculture, the proper monitoring of production system through nanotechnology will be very appropriate to promote quality and make clear distinction with organic grown produces. The identity preservation (IP) system could deliver information about practices and activities used to produce a particular crop. The certifying inspectors can take advantage of IP for documenting, verifying and certifying agricultural practices at ease. The nanoscale IP holds a possibility of the continuous tracking and recording of the history which a particular agricultural product experiences. The timely detection and in situ control over online process control is feasible by relatively small biosensor apparatus. The nanoscale monitors linked to recording and tracking devices would evaluate food safety and identity quality parameters of agricultural products.

3. CONCLUSION AND FUTURE PERSPECTIVES

In the complex agricultural environment, before advanced robotics application becomes a reality, deployment of drones and biosensors could be possible at least in micro crop cluster levels [42]. With direct effects like farm labour reduction and better input management, other intangible benefits of drones are monitoring air quality, land-use change such as soil erosion, surveillance of wildlife, illegal poaching and wood logging. Soil quality testing, preparation of location-specific soil maps and nutrient management measures could be possible through drone technology so that tailor-made farm advisories could be delivered to farmers. In future, the development of artificial intelligence systems and services should be carried out in a larger scale keeping in mind the heterogeneity of clienteles [43]. The proper legal framework for manufacturing and using small drones could increase the user population. Suitable legislative amendment should be carried out, taking into account individual privacy and developmental objectives as well as punishments for trespassers of regulation.

In the era of globalization, food safety protocols ensuring international quality standards are warranted. The proficient detection of microorganism and contamination could produce considerable effects on food quality improvement and consumer health. Apart from food quality testing, biosensor technologies including optical, microcantilevers, immunoDNA and whole cell—based biosensors are acquiescent for analysis of new emerging pollutants such as surfactants, hormones and antibiotics. For successful deployment of the drones and biosensor technologies, networked analytical stations should be established at multiple locations for agriculture, veterinary and environment domains. Integrated research complex should be initiated for developing multiarray biosensing systems with lesser physical space. The complex data processing methods should be revamped with latest user-friendly techniques, thereby users can use these robotics techniques without much training and skill. Moving the drone and biosensor technology to address the niche markets is the need of the hour, which warrants multi-stakeholder approach. Interdisciplinary research on biosensor technologies in the field of agriculture, horticulture and food science at the global level involving both public and private players could be initiated. Apparently, policy formulation and execution of public—private partnership should be initiated. According to FAO Save Food key findings, 2015, the global quantitative food losses and waste per year are roughly 30% for cereals, 40—50% for root crops, fruits and vegetables, 20% for oil seeds, meat and dairy and 35% for fish products. Therefore strengthening of the supply chain management through skill training for farmers and entrepreneurs of food package industries could reduce the food wastage as suggested by the United Nations Environment Programme (quick facts 2013). Hence seeking opportunities for more automations and innovations in food safety and agriculture is crucial. Thus robotics and biosensor design in agriculture could be a game changer in fighting against food insecurity, poverty and hunger.

ACKNOWLEDGEMENTS

The author acknowledges and expresses his sincere thanks to Mr Syed Nazir Razik of Avere consulting, Chennai, India, for the on-going drone for agriculture research support and Dr (Mrs) Nithila, Asst. Professor (Crop Physiology), Tamil Nadu Agricultural University, Coimbatore, India, for rendering help in field-level investigation and encouragement.

REFERENCES

[1] A. Rango, A. Laliberte, J.E. Herrick, C. Winters, K. Havstad, C. Steele, D. Browning, J. Appl. Remote Sens. 3 (2009) 1—15.

[2] H. Xiang, L. Tian, Biosyst. Eng. 108 (2011) 174—190.

[3] R. Swaminathan, Observer Research Foundation Report, 2015.

[4] M.A. Druy, R.A. Crocombe (Eds.), Next-Generation Spectroscopic Technologies V, Proc. SPIE: Baltimore, Maryland, USA, 8374, 2012, pp. 1—9.

[5] L. Johnson, S. Dungan, B. Lobitz, D. Sullivan, R. Slye, S. Herwitz, International Symposium on Remote Sensing of the Environment, 2003, pp. 10—14.

[6] N. Kondo, K.C. Ting, Artif. Intell. Rev. 12 (1998) 227—243.

[7] L.R. Newcome, American Institute of Aeronautics and Astronautics, 2004, pp. 127—130.

[8] H. Geer, C. Bolkcom, Unmanned Aerial Vehicles: Background and Issues for Congress, Washington, 62, 2005.

[9] A. Rango, A. Laliberte, C. Steele, J.E. Herrick, B. Bestelmeyer, T. Schmugge, A. Roanhorse, V. Jenkins, Environ. Pract. 8 (2006) 159—168.

[10] http://www.thehindu.com/news/national/drones-to-guard-indias-forest-and-wildlife/article 6286830.ece. Retrieved on January 26 2015.

[11] http://www.timesofindia.indiatimes.com/india/Dromes-to-keep-eye-on-panna tigers/article show/28653645.cms. Retrieved on 18 June 2015.

[12] T. Knowles, M. McCall, M. Skutsch, L. Theron, Perspectives Series, UNEP Risø Centre, Roskilde, Denmark, 2010.

[13] Louisiana Agriculture, Assuring our future through scientific research and education, La. Agric. Exp. Stn. 57 (2014) 8−10.

[14] J. Ratti, G. Vachtsevanos, J. Intell.Robotic Syst. 65 (2011) 437−455.

[15] E. Miasnikov, Threat of Terrorism Using Unmanned Aerial Vehicles: Technical Aspects, MIPT, Dolgoprudny, (2005).

[16] http://www.firstpost.com/india/on-modis-us-wishlist-drones-for-india-defence-internal-security-1583025.html. Retrieved on 23 November 2015.

[17] http://cis-india.org/internet-governance/resources/section 66 An information technology - act. Retrieved on 1 December, 2015.

[18] http://www.dot.gov.in/sites/default/files/itbill2000_0.pdf. Retrieved on 10 November, 2015.

[19] G.V. de Plata, Salud UIS 35 (2003) 48−57.

[20] J. Meng, M. Doyle, Microbes Infect. 4 (4) (2002) 395−397.

[21] N. Turner, S. Subrahmanyam, S. Piletsky, Anal. Chim. Acta 632 (2) (2009) 168−180.

[22] S. Pathirana, J. Barbaree, B. Chin, M. Hartell, W. Neely, V. Vodyanoy, Biosens. Bioelectron. 15 (3−4) (2000) 135−141.

[23] K. Ropkins, A. Beck, Trends Food Sci. Technol. 11 (1) (2000) 10−21.

[24] B. Nowak, T. Müffling, S. Chaunchom, J. Hartung, Int. J. Food Microbiol. 115 (3) (2007) 259−267.

[25] Y. Lu, W. Yang, L. Shi, L. Li, M. Alam, S. Guo, S. Miyoshi, J. Health Sci. 55 (5) (2009) 820−824.

[26] D. White, S. Zhao, S. Simjee, D. Wagner, P. McDermott, Microbes Infect. 4 (4) (2002) 405−412.

[27] J. Sánchez, S. Jiménez, R. Navarro, M. Villarejo, Ediciones Díaz de Santos, España, 2009, p. 209. ISBN 978 84 7978 922 0.

[28] T. Geng, M. Morgan, A. Bhunia, J. Appl. Environ. Microbiol. 70 (10) (2004) 6138−6146.

[29] P. Van der Merwe, Surface Plasmon Resonance, May 2011. Available from: http://users.path.ox.ac.uk/∼vdmerwe/Internal/spr.PDF.

[30] WHO fact sheet 255, Campylobacter, World Health Organization, 2000. Available from: http://www.who.int/mediacentre/factsheets/fs255/en.

[31] N. Darnton, L. Turner, S. Rojevsky, H. Berg, J. Bacteriol. 189 (5) (2007) 1756−1764.

[32] WHO Factsheet No 125, Enterohaemorrhahic *Escherichia coli* (EHEC), World Health Organization, 2005. Available from, http://www.who.int/mediacentre/factsheets/fs125/en/.

[33] H. Tang, W. Zhang, P. Geng, Q. Wang, L. Jin, Z. Wu, M. Lou, Anal. Chim. Acta 562 (2) (2006) 190−196.

[34] M. Schöning, M. Arzdorf, P. Mulchandani, W. Chen, A. Mulchandani, Sensors 3 (6) (2003) 119−127.

[35] A. Hildebrandt, R. Bragos, S. Lacorte, J. Marty, Sens. Actuators, B 133 (1) (2008) 195−201.

[36] K. Yokoyama, H. Ikebukuro, E. Tamiya, I. Karube, N. Ichiki, Y. Arikawa, Anal. Chim. Acta 304 (2) (1995) 139−145.

[37] A. Székács, N. Trummer, N. Adányi, M. Váradi, I. Szendrö, Anal. Chim. Acta 487 (1) (2003) 31−42.

[38] A. Biswas, B.K. Sharma, J.L. Willett, A. Advaryu, S.Z. Erhan, H.N. Cheng, J. Agric. Food Chem. 56 (14) (2008) 5611−5616.

[39] N. Verma, S. Kumar, H. Kaur, Biosens. Bioelectron. 1 (2010) 102.

[40] D. Bagal Kestwal, M. Karve, B. Kakadeb, V. Pillai, Biosens. Bioelectron. 24 (4) (2008) 657−664.

[41] T. Lin, M. Chung, Using monoclonal antibody to determine lead ions with a localized surface plasmon resonance fiber-optic biosensor, Sensors 8 (1) (2008) 582−593.

[42] S.Y. Nof (Ed.), Springer Handbook of Automation, Springer Berlin Heidelberg, 2009, pp. 1095−1128.

[43] C. Anderson, Agricultural Drones. http://www.technologyreview.com/featuredstory/526491/agricultural-drones. Retrieved on 23 October 23 2015.

Chapter 13

Intelligent Food Packaging

S. Otles[1,*] and B.Y. Sahyar[1,2]
[1]*Ege University, Izmir, Turkey;* [2]*Indesit Company, Manisa, Turkey*
Corresponding author: E-mail: semih.otles@ege.edu.tr

Chapter Outline

1. INTRODUCTION

In developing world, needs and expectations change day by day, due to technological developments and life style diversities, and, consequently, food knowledge of consumer is increased [1,2]. As a result, food packaging is one of the technologies subjected to huge progresses in the last years. Packaging is a kind of preservation system suitable for food industry to protect food, inhibiting contaminations and avoiding breaking up or leaking. Packaging can be classified in active and intelligent packaging. Active packaging refers to the incorporation of additives into packaging systems with the aim of maintaining or extending food product quality and shelf life. These systems include oxygen scavengers, carbon dioxide scavengers and emitters, moisture control agents and antimicrobial packaging technologies. Intelligent packaging comprises carbon dioxide absorbers/emitters, odour absorbers, ethylene removers and aroma emitters [3]. These systems monitor the condition of packaged foods to give information on the quality of the packaged food during transport and storage.

Several packaging systems been developed in the last years to monitor food products and give important information about their chemical and/or microbiological status, as well as about nutritional content and guidelines for food preparation. In addition, these systems are also able to ensure traceability of

Comprehensive Analytical Chemistry, Vol. 74. http://dx.doi.org/10.1016/bs.coac.2016.04.010

the products in every step of the food supply chain by providing communications between food, food packaging and atmosphere.

General food packaging materials can show antimicrobial function, gas barrier (O_2, CO_2), mechanical properties, vapour barriers, optical properties, aroma barriers, thermal properties, and they are environmental friendly. Traditional and basic packaging materials (paper/paperboard, plastic, glass, metal and a combination of materials with various physico-chemical features) are used to fulfil the requirements and functions of packaged foods depending on their type. Several efforts have been gained to improve packaging system by developing new smart materials devoted to gain additional functions, enhancing packaging effectiveness and maintaining food quality (eg, nanoparticles, which are able to block sunlight and/or UV light) [4]. Nanopolymers have been also exploited providing improved mechanical, barrier and antimicrobial properties to packaging, with many advantages in terms of stiffness and strength properties, oxygen and moisture barrier and flexibility [5–7]. These polymers show a number of economic, safety and environmental advantages including reduced migration of components between food matrices and food packaging, energy inputs for production, transport and storage, CO_2 emissions, increase of biodegradability and barrier protection to gases and light [5–17].

Finally, sensors, nanosensors and/or indicators can be combined with or integrated into packaging materials to furnish information to the customers about freshness and/or spoilage of food, as well as to monitor and trace food condition during transport and storage [5]. The potential of sensor technologies, indicators (including integrity, freshness and time–temperature (TTI) indicators), and radio frequency identification (RFID) have been extensively exploited in many food product packaging. The use of those sensors and indicators inside or outside the covers can be considered as smart or intelligent packaging. Indeed, smart packaging refers to packaging that utilizes chemical sensor or biosensor to monitor the food quality and safety from the producers to the costumers [18].

Recognition of the benefits of active and intelligent packaging by the food industry, development of economically viable packaging systems and increased consumer acceptance are necessary for commercial realization of these technologies [19].

2. INTELLIGENT PACKAGING EXAMPLES

There are lots of studies, innovations and designs on intelligent food packaging. Firstly, many researches are interested in developing bio-based polymers aimed to reduce the dependence on fossil fuel and move to a sustainable material basis. These bio-based polymers open an opportunity for the use of new, high performance, lightweight green nanocomposite materials making them suitable to replace conventional nonbiodegradable petroleum-based plastic packaging materials. The most studied bio-nanocomposites for packaging applications are starch and cellulose derivatives, polylactic acid, polycaprolactone, poly-(butylene succinate) and polyhydroxybutyrate.

There are also many studies about nanomaterials and their use in packaging. The most promising nanoscale fillers are layered silicate nanoclays such as montmorillonite and kaolinite. Moreover, several nanostructures can be useful to ensure active and/or smart properties to food packaging systems, as exemplified by antimicrobial properties, oxygen scavenging ability, enzyme immobilization or indication of the degree of exposure to some detrimental factors such as inadequate temperatures or oxygen levels.

Challenges remain in increasing the compatibility between clays and polymers and reaching complete dispersion of nanoparticles. Successful technical development of polymer nanomaterials for food packaging (PNFP) has to overcome barriers in safety, technology, regulation, standardization, trained workforce and technology transfer in order that commercial products can benefit from the global market potential and requires therefore a high degree of multidisciplinary. Moreover, because of its enormous growth application potential, the emerging technology of PNFP will be a major provider of new employment opportunities, based upon growing international commercial success combined with ecological advantages. This technology can inform with a visible indicator the supplier or consumer that foodstuffs are still fresh, whether the packaging has been breached, kept at the appropriate temperatures throughout the supply chain or has spoiled. Key factors in their extensive application are cost, robustness and compatibility with different packaging materials. Foremost developments were based on devices, which were incorporated with the product in a conventional package with the aim to monitor the package integrity, the time—temperature history of the product and the effective expiration date [20,21].

Metallic-based micro- and nano-structured materials have been also used into food contact polymers to enhance mechanical and barrier properties, and to prevent the photodegradation of plastics. Additionally, heavy metals are efficient antimicrobials in the form of salts, oxides and colloids, complexes like silver zeolites or as elemental nanoparticles. They can be used to inhibit food contamination and also beware of contamination of environmental surfaces in food and related industry. Similar active and intelligent packaging with these properties involves ethylene oxidation or oxygen scavenging, which could be also used to extend food shelf life. As antimicrobial activity of nanomaterials, silver is the most commonly used in commodities, while titanium, copper, and zinc nanomaterials are indicating promises in food safety and quality. Nanozinc oxide is a safe, cost-effective and innovative material thanks to its antimicrobial properties for sensorial and intelligent packaging systems. While nanocopper has been denoted to be an effective sensor for humidity, nanotitanium oxide has UV blocking properties and resistive to abrasion. The migration of cations from the polymer matrices could be the key point to determine their antimicrobial effectiveness. However, this cation migration might affect legal status of the polymer as a food contact material [22].

A new standard (IEEE 1451.5) is currently being developed to bridge wireless sensors with the 'smart transducer' concept to produce intelligent systems that combine sensing, computing and communication. Using this standard, intelligent sensors and actuators can be connected to a common network through both wired and wireless transmissions to perform sophisticated functions. RFID has been considered the most important identification tool to establish an effective 'traceability system' [23]. Compared with the traditional barcode method, RFID allows an 'intelligent tag' assigned to each individual product to be read at any position without physical contact with the readers. Furthermore the intelligent tag can be updated along the entire supply chain to provide complete archives of information on the growth, processing, packaging, transportation, distribution, storage, shelving and recycling. When combined with wireless sensors, the RFID system can also record environmental parameters and specific quality/safety attributes of the product along the chain. It can be predicted that deployment of RFID and wireless sensors in traceability systems will experience a great boom in the near future [24].

3. INTELLIGENT PACKAGING INCLUDING SENSOR MECHANISMS

As previously described, intelligent packaging can incorporate sensors and indicators, which can communicate information about the food to the consumer or react to the information and change conditions within the packaging to delay spoilage/contamination [25,26]. When a system is called as sensorial or sensor, this means that there are some properties which can imitate sensorial properties of living organisms, thus giving a chance to the system to be intelligent. Sensing mechanism can be physical, biological and/or biochemical to detect target compounds such as gasses, minor and major food components and/or contaminants. There are numerous kind of sensors and indicators attached to the packaging materials or packaging materials directly including sensorial or indicating properties to extend the shelf life of food and inform consumer about food items.

For example, there has been an increasing interest to develop nontoxic and irreversible oxygen sensors to assure oxygen absence in oxygen-free food packaging systems, such as packaging under vacuum or nitrogen. Among them, UV-activated colourimetric oxygen indicators involve TiO_2 nanoparticles to photosensitize the reduction of methylene blue by triethanolamine in a polymer encapsulation medium, using UVA light. Upon UV irradiation, the sensor bleaches and remains colourless, until it is exposed by oxygen, when its original blue colour is restored. The rate of colour recovery is proportional to the level of oxygen exposure. Nanocrystalline SnO_2 was used as a photosensitizer in a colourimetric O_2 indicator with the colour of the film varying depending on the O_2 exposure.

An oxygen indicator, formulated from a combination of electrochrome, titanium dioxide and EDTA, was evaluated in modified atmosphere packaging. These polyviologen electrochromes showed a faster reduction than methylene blue after exposure to UV light. Thionine and 2,2'-dicyano-1,1'-dimethylviologen dimesylate, which have more anodic reduction potentials compared to methylene blue, can be used to produce oxygen indicators with decreased sensitivity to oxygen. Different electrochromes were formulated into an oxygen indicator that was triggered by UV light by mediated reduction via EDTA and titanium dioxide. The oxygen indicator based on poly(pxylylviologen dibromide) was more efficient in the trigger step compared to methylene blue, determined by colour contrast after activation. Poly(p-xylylviologen dibromide) can therefore be used in the oxygen indicators when package contents are sensitive to UV exposure associated with the photoreduction step. These indicators can be used to detect oxygen even when O_2 levels increase up to 4.0% [27].

Intelligent packaging can also help to trace a product history through critical points of the food supply chain. Generally, occurrence of elevated CO_2 gas level is an important indicator of food spoilage in packed foods and also its maintenance at optimal levels is essential to avoid spoilage in foods packed under modified-atmosphere packaging conditions. Consequently, a CO_2 sensor incorporated into food package is crucial for efficiently monitoring product quality. Although much progress has been made so far in the development of sensors monitoring CO_2, most of them are not versatile for food packaging applications and suffers from limitations such as high equipment cost, bulkiness and energy input requirement, including safety concerns. Thus, the development of efficient CO_2 sensors that can intelligently monitor gas concentration changes inside a food package and specific to food packaging applications is essential [28]. An example of gas indicator is the chitosan-based carbon dioxide (CO_2) indicator developed to indicate freshness or quality of packaged foods during their storage, since elevated CO_2 levels inside a package is one of the prime indicators of microbial spoilage of food [40].

Additionally, an alginate polymer can been applied to prevent dyes from leaching out of colourimetric gas indicator films, which enable people to notice the presence of gas in the package in an economic and simple manner. However, these dye-based indicators film suffer from dye leaching upon contact with water, thus water-resistant indicators are successfully exploited [29].

Also, pH indicators based on organically modified silicate nanoparticles have been recently introduced [30−32]. pH-sensitive dyes are also valid alternatives which can be used to develop sensors and also as indicators.

Another study indicates that, freshness indicators monitor the quality of the packed food by reacting to changes that take place in the fresh food product as a result of the microbiological growth. The freshness sensor has to be able to react to the presence of these metabolites with the required sensitivity. The indication of freshness is based on a colour change of the indicator tag due to the presence of the microbial metabolites produced during spoilage. It is to be noted that the formation of the different metabolites depends on the nature of the packed products spoilage flora and type of packaging. The embedded sensors in a packaging film must be able to detect food-spoilage organisms and trigger a colour change to alert the consumer that the shelf life is ending/ended [33].

Some examples could be given including colour-indicating tags named 'FreshTag', consisting of a small label attached to the outside of the packaging film. They are used to monitor the freshness of seafood products, and consists of a reagent-containing wick embedded within a plastic chip. As the seafood ages, spoils, and generates volatile amines in the headspace, these are allowed to contact the reagent, causing the wick in the tag to turn bright pink [18,34]. Other indicator samples could be given as TTI called Fresh-Check (LifeLines Technology, USA), consisting of a small polymer circle which is surrounded by a pH indicator (The Vitsab TTI indicator, Vitsab Sweden AB, Sweden) through enzymatic reactions. The American company 3M Packaging Systems Division developed several TTI, which can thoroughly be applied to the package, including 3M (MonitorMarkTM) samples used for different types of ready-made meat and dairy foods [35].

A recent study about antimicrobial activity of the active packaging containing cinnamon or oregano as essential oils was evaluated against *Escherichia coli* and *Staphylococcus aureus*. The vapour phase activity and the direct contact between the antimicrobial agents themselves (or once incorporated in the packaging material) and the microbial cells have been studied. The results indicated that these techniques ensure information about the antibacterial activity of cinnamon and oregano in direct contact as well as in the vapour phase. The antimicrobial packaging indicated a fast efficiency supporting its suitable application as a food packaging material [36]. On the other hand, there are also biosensors samples that detect *Listeria monocytogenes* [37,38].

Food packaging materials are also capable of releasing nanoscale antimicrobial compounds, antioxidants and/or flavours, which would improve the shelf life or sensory characteristics of a food, termed as 'active', 'intelligent' or 'smart'. These materials will need to have all the required safety data to ensure their place on the positive list of food additives before their use. For example, food-contact-materials-bound active particulate nanomaterials are being used in commercial products such as commercially available nano-Ag embedded baby bottles. These materials impart an active effect but are not ingested with the food [39].

In addition, nanosensors can be incorporated into food packaging matrices having the ability to identify specific microbial and chemical contaminants or environmental conditions. These nanosensors can respond in a way that alerts the consumer about any contamination, being useful to detect pathogen, chemical spoilage and/or contaminants [25]. These technologies are also able of changing an environment in response to an actuator. Actuators might include a specific pH, pressure, presence of gas, liquids or products of microbial metabolism, or spoilage accelerators such as temperature or light intensity [26,40].

Thanks to the many research concerns about sensors and indicators for packaging, a number of commercial systems are nowadays available for the food sector. Among them, *e-nose* is a kind of sensor that mimic the mammalian olfactory system that allows identification and classification of aroma mixtures present in the odour. The e-nose system has been successfully used for quality evaluation of fresh yellowfin tuna and vacuum-packed beef [41−43]. As an example, an e-nose was realized to determine volatile compounds produced during tomatoes ripening stages [41,44]. While another e-nose was designed to evaluate the quality of modified atmosphere-packed broiler chicken cuts [41,45].

'Wondersensor' is another type of sensor which shows oxygen concentrations $>0.5\%$ by the development of a blue colour, while oxygen concentration less than 0.1%, by the development of a pink colour. 'Wonderkeep' is also an oxygen absorber. 'Wondersensor and Wonderkeep' combination can be used in products (Aw between 0.2 and 0.9) such as dry food, pasta, peanuts and biscuits. Other commercially available visual system sensors are 'Food Sentinel System' and 'Toxin Guard', which alert consumer by a colour change in the food package that food must not be consumed. The 'Food Sentinel System' is declared as detecting the presence of harmful microorganisms through immune reactions that make the barcode unreadable [46,47].

4. THE BENEFITS AND LIMITATIONS OF INTELLIGENT PACKAGING TECHNOLOGY

As discussed previously, there are a number of benefits in using and applying intelligent packaging technology. However, there are also some limitations, one of these regarding the unknown effects of nanomaterials such as migration, safety, quality and movement principle. There are some studies referred to this open point trying to lead ways to overcome these unbeknown. Until 2004, there was a legislative lack in Europe for intelligent and/or active packaging decreasing their penetration in the EU market. After that, some regulations have been legislated including Regulation 1935/2004/EC and Regulation 450/2009/EC, aimed to set new legal basis for their correct use, safety and marketing. Additionally, due to its deliberate interaction with the food and/or its environment, the migration of substances could represent a food safety concern. Definitions stated in these regulations regard active materials, specifically 'materials and articles that are intended to extend the shelf-life or to maintain or improve the condition of packaged food'. They are designed to deliberately incorporate components that would release or absorb substances into or from the packaged food or the environment surrounding the food [48–50].

There are also some benefits and limitations about sensor usage on intelligent packaging. The sensors usage provide an alternative to the time consuming, expensive and laborious analytical techniques which are currently applied to monitor a packaged food product and its environment throughout the entire supply chain. On the other hand, some obstacles need to be overcome, including the reduced size and rigidity of the systems, the high production costs, and the low robustness and sensitivity, to meet strict legislations and taking into account food safety considerations [50].

5. CONCLUSION

Traditional packaging is a kind of preservation system which protect food from contamination and undesirable components such as humidity, oxygen, carbon dioxide, microorganisms, undesirable flavour and taste. Last trends on intelligent and active packaging systems showed additional advantages to enhance food storage time, increase safety and quality of food and maintain long period freshness to food. In addition, devices and/or nano-devices can be integrated in packaging, to control packaging medium, give information to the customers about food and packaging medium and to guarantee food quality and safety control along with the food chain. Moreover, the use of nano-materials in packaging can be a promising strategy to provide several benefits to the food industry by reducing food waste, foodborne illness, spoilage and deterioration of food products. Although several research efforts have been obtained on packaging technology, there are still open points and gaps about its applications to agrifood sector due to regulations concerns, thus further research is required.

REFERENCES

[1] K.L. Yam, P.T. Takhistov, J. Miltz, Intelligent packaging: concepts and applications, J. Food Sci. 70 (2005) R1–R10.

[2] M. Vanderroost, P. Ragaert, F. Devlieghere, B. De Meulenaer, Intelligent food packaging: the next generation, Trends Food Sci. Technol. 39 (2014) 47–62.

[3] D. Restuccia, U.G. Spizzirri, O.I. Parisi, G. Cirillo, M. Curcio, F. Iemma, F. Puoci, G. Vinci, N. Picci, New EU regulation aspects and global market of active and intelligent packaging for food industry applications, Food Control 21 (2010) 1425–1435.

[4] J.-W. Rhim, H.-M. Park, C.-S. Ha, Bio-nanocomposites for food packaging applications, Prog. Polym. Sci. 38 (2013) 1629–1652.

[5] C. Silvestre, D. Duraccio, S. Cimmino, Food packaging based on polymer nanomaterials, Prog. Polym. Sci. 36 (2011) 1766–1782.

[6] S.E.M. Selke, J.D. Culter, R.J. Hernandez, Plastics Packaging: Properties, Processing, Applications, and Regulations, second ed., Carl Hanser Verlag, Munich, 2004.

[7] J. Jordan, K.I. Jacob, R. Tanenbaum, M. Sharaf, I. Jasiuk, Experimental trends in polymer nanocomposites: a review, Mater. Sci. Eng. A 393 (2005) 1–11.

[8] R. Coles, D. McDowell, M.J. Kirwan (Eds.), Food Packaging Technology, Blackwell Publishing Ltd, Oxford, UK, 2003, pp. 65–94.

[9] C.I. Moraru, C.P. Panchapakesan, Q. Huang, P. Takhistov, S. Liu, J.L. Kokini, Nanotechnology: a new frontier in food science, Food Technol. 57 (2003) 24–29.

[10] A. Broady, Packaging by the numbers, Food Technol. 62 (2008) 89–91.

[11] J.M. Lagaron, D. Cava, L. Cabedo, R. Gavara, E. Gimenez, Improving packaged food quality and safety. Part 2: nanocomposites, Food Addit. Contam. 22 (2005) 994–998.

[12] G.L. Robertson, Food Packaging: Principles and Practice, second ed., CRC Press Taylor & Francis Group, Boca Raton, FL, 2006.

[13] P. Sanguansri, M.A. Augustin, Nanoscale materials developments– a food industry perspective, Trends Food Sci. Technol. 17 (2006) 547–556.

[14] J.M. Lagaron, E. Gimenez, M.D. Sánchez-García, M.J. Ocio, A. Fendler, Novel nanocomposites to enhance quality and safety of packaged foods, in: Food Contact Polymers, 19, Rapra Technologies, Shawbury, UK, 2007, pp. 1–5.

[15] Q. Chaudhry, M. Scotter, J. Blackburn, B. Ross, A. Boxall, L. Castle, R. Aitken, R. Watkins, Applications and implications of nanotechnologies for the food sector, Food Addit. Contam. 25 (2008) 241–258.

[16] M.C. De Azedero Henriette, Nanocomposites for food packaging applications, Food Res. Int. 42 (2009) 1240–1253.

[17] A. Arora, G.W. Padua, Review: nanocomposites in food packaging, J. Food Sci. 75 (2010) R43–R49.

[18] A. Pavelková, Intelligent packaging as device for monitoring of risk factors in food, Pavelková 2 (1) (2012) 282–292.

[19] J.P. Kerry, M.N. O'Grady, S.A. Hogan, Past, current and potential utilisation of active and intelligent packaging systems for meat and muscle-based products: a review, Meat Sci. 74 (2006) 113–130.

[20] J. Kerry, P. Butler (Eds.), Smart Packaging Technology for Fast Moving Consumers Goods, John Wiley and Sons Ltd, Chichister, UK, 2008.

[21] F.L. Yam (Ed.), The Wiley Encyclopedia of Packaging Technology, third ed., John Wiley and Sons Inc, New York, 2009.

[22] A. Llorens, E. Lloret, P.A. Picouet, R. Trbojevich, A. Fernandez, Metallic-based micro and nanocomposites in food contact materials and active food packaging, Trends Food Sci. Technol. 24 (2012) 19–29.

[23] E. Sahin, Y. Dallery, S. Gershwin, Performance evaluation of a traceability system: an application to the radio frequency identification technology, in: Proceedings of the 2002 IEEE International Conference on Systems, Man and Cybernetics, Vol. 3, Yasmine Hammamet, Tunisia, October 6–9, 2002, pp. 647–650.

[24] N. Wang, N. Zhang, M. Wang, Wireless sensors in agriculture and food industry—Recent development and future perspective, Comput. Electron. Agric. 50 (2006) 1–14.

[25] S. Neethirajan, D. Jayas, Nanotechnology for the food and bioprocessing industries, Food Bioprocess Technol. 4 (1) (2011) 39–47.

[26] M. Cushen, J. Kerry, M. Morris, M. Cruz-Romero, E. Cummins, Nanotechnologies in the food industry – recent developments, risks and regulation, Trends Food Sci. Technol. 24 (2012) 30–46.

[27] L. Roberts, R. Lines, S. Reddy, J. Hay, Investigation of polyviologens as oxygen indicators in food packaging, Sens. Actuat. B 152 (2011) 63−67.

[28] P. Puligundla, J. Jung, S. Ko, Carbon dioxide sensors for intelligent food packaging applications, Food Control 25 (2012) 328−333.

[29] C.-H. Thai Vu, K. Won, Novel water-resistant UV-activated oxygen indicator for intelligent food packaging, Food Chem. 140 (2013) 52−56.

[30] S.W. Lee, C. Mao, C.E. Flynn, A.M. Belcher, Ordering of quantum dots using genetically engineered viruses, Science 296 (2002) 892−895.

[31] A. Mills, D. Hazafy, Nanocrystalline SnO_2-based, UVB-activated, colourimetric oxygen indicator, Sens. Actuat. B 136 (2009) 344−349.

[32] S. Jurmanovíc, S. Kordíc, M.D. Steinberg, I. Murkovíc Steinberg, Organically modified silicate thin films doped with colourimetric pH indicators methyl red and bromocresol green as pH responsive sol−gel hybrid materials, Thin Solid Films 518 (2010) 2234−2240.

[33] M. Smolander, Freshness indicators for packaging, Food Sci. Technol. 18 (2004) 26−27.

[34] J.H. Han, Antimicrobial food packaging, Food Technol. 54 (3) (2000) 56−65.

[35] N. de Kruijfy, M. van Beesty, R. Rijky, T. Sipiläinen-Malm, P. Paseiro Losada, B. De Meulenaer, Active and intelligent packaging: applications and regulatory aspects, Food Addit. Contam. 19 (Suppl.) (2002) 144−162.

[36] R. Becerril, R. Gómez-Lus, P. Goñi, P. López, C. Nerín, Combination of analytical and microbiological techniques to study the antimicrobial activity of a new active food packaging containing cinnamon or oregano against *E. coli* and *S. aureus*, Anal. Bioanal. Chem. 388 (2007) 1003−1011.

[37] Y. Liu, S. Chakrabartty, E. Alocilja, Fundamental building blocks for molecular biowire based forward error-correcting biosensors, Nanotechnology 18 (2007) 1−6.

[38] B. Kuswandi, Y. Wicaksono, J.A. Abdullah, L.Y. Heng, M. Ahmad, Smart packaging: sensors for monitoring of food quality and safety, Sens. Instrumen. Food Qual. 5 (2011) 137−146.

[39] S.M. Alfadul, A.A. Elneshwy, Use of nanotechnology in food processing, packaging and safety e review, Afr. J. Food Agric. Nutr. Dev. 10 (6) (2010) 2719−2739.

[40] S. Otles, B. Yalcin, Intelligent food packaging, LogForum 4 (4) (2008) 1−9.

[41] B. Kuswandi, R.A. Jayus, R. Oktaviana, A. Abdullah, L.Y. Heng, A novel on-package sticker sensor based on methyl red for real-time monitoring of broiler chicken cut freshness, Packag. Technol. Sci. 27 (1) (2014) 69−81.

[42] Y. Blixt, E. Borch, Using an electronic nose for determining the spoilage of vacuum-packaged beef, Int. J. Food Microbiol. 46 (1999) 123−134.

[43] R. Dobrucka, R. Cierpiszewski, Active and intelligent packaging food research and development- a review, Pol. J. Food Nutr. Sci. 64 (1) (2014) 7−15.

[44] A.H. Gòmez, G. Hu, J. Wang, A.G. Pereira, Evaluation of tomato maturity by electronic nose, Comput. Electron. Agric 54 (2006) 44−52.

[45] T. Rajamäki, H. Alatomi, T. Titvanen, E. Skyttä, M. Smolander, R. Ahvenainen, Application of an electronic nose for quality assessment of modified atmosphere packaged poultry meat, Food Control 17 (2004) 5−13.

[46] D.A. Pereira de Abreu, J.M. Cruz, P. Paseiro Losada, Active and intelligent packaging for the food industry, Food Rev. Int. 28 (2012) 146−187.

[47] R.M. Goldsmith, C. Goldsmith, J.G. Woodaman, D.L. Park, C.E. Ayala, Food Sentinel System TM, World Intellectual Property Organization, 1999. WO/1999/014598 Patent.

[48] A.L. Brody, What's active in active packaging, Food Technol. 55 (2001) 104−106.

[49] A. Castle, Chemical migration into food: An overview, in: K.A. Barnes, C.R. Sinclair, D.H. Watson (Eds.), Chemical Migration and Food Contact Materials, Woodhead Publishing Ltd, Cambridge, 2007, pp. 1—14.

[50] J.D. Floros, L.L. Dock, J.H. Han, Active packaging technologies and applications, Food Cosmet. Drug Packag. 20 (1997) 10—17.

Chapter 14

(Bio)Sensor Integration With ICT Tools for Supplying Chain Management and Traceability in Agriculture

M. Durresi
European University of Tirana, Tirana, Albania
E-mail: durresim@gmail.com

Chapter Outline

Comprehensive Analytical Chemistry, Vol. 74. http://dx.doi.org/10.1016/bs.coac.2016.06.001

1. NEW CHALLENGES IN AGRICULTURE

In his book, *Full Planet Empty Plates: the New Geopolitics of Food Scarcity*, Lester Brown states "The world is in transition from an era of food abundance to one of scarcity. Over the last decade, world grain reserves have fallen by one third. World food prices have more than doubled, triggering a worldwide land rush and ushering in a new geopolitics of food. Food is the new oil. Land is the new gold" [1]. Indeed, one of the greatest human challenges today is surviving in the new modern environment created from industry and agriculture development, where air pollution, genetic engineering, toxic wastes and chemicals entering the water cycle are threatening food quality, safety and security. The existence of new regulatory measurements regarding the ecological problems and the environmental protection restrictions also bring new challenges [2]. In this context, the recent advances in information and communication technology (ICT) can tackle with these new challenges in helping to control the entire agriculture supply chain and thus produce safe food [3]. Among others, traceability is a key process to ensure quality of food production processes and thus safe products, by furnishing the history of a product or a process from the origin to the final destination [4]. To meet traceability requirements, several technologies have contributed to the flow of information about food production processes between farmers and markets [5], as shown in Fig. 1A.

Among these technologies, ICT tools and their integration in the agrifood chain can provide accurate food traceability, meeting the standards of each production country by exploiting the emerging technologies in mobile wireless systems, including mobile phones, radio frequency identification (RFID) systems, wireless sensor networks (WSNs) and global positioning systems (GPS) [3]. These technologies are able to monitor environmental and location-based variables and communicate them to databases for analysis [6]. Traceability systems are also useful to check the quality of livestock and meat exports. For example, a sensor and a tag are attached to the beef, and the product is monitored during the transport to check if it remains in a place for long time. Later the beef is placed in refrigerated trucks, and containers are sealed with a sensor bolt and a tag for identification. Shipments are also tracked to ensure that they do not have long transition times. In major stops during transportation, the tag will be read again to check the status of the food.

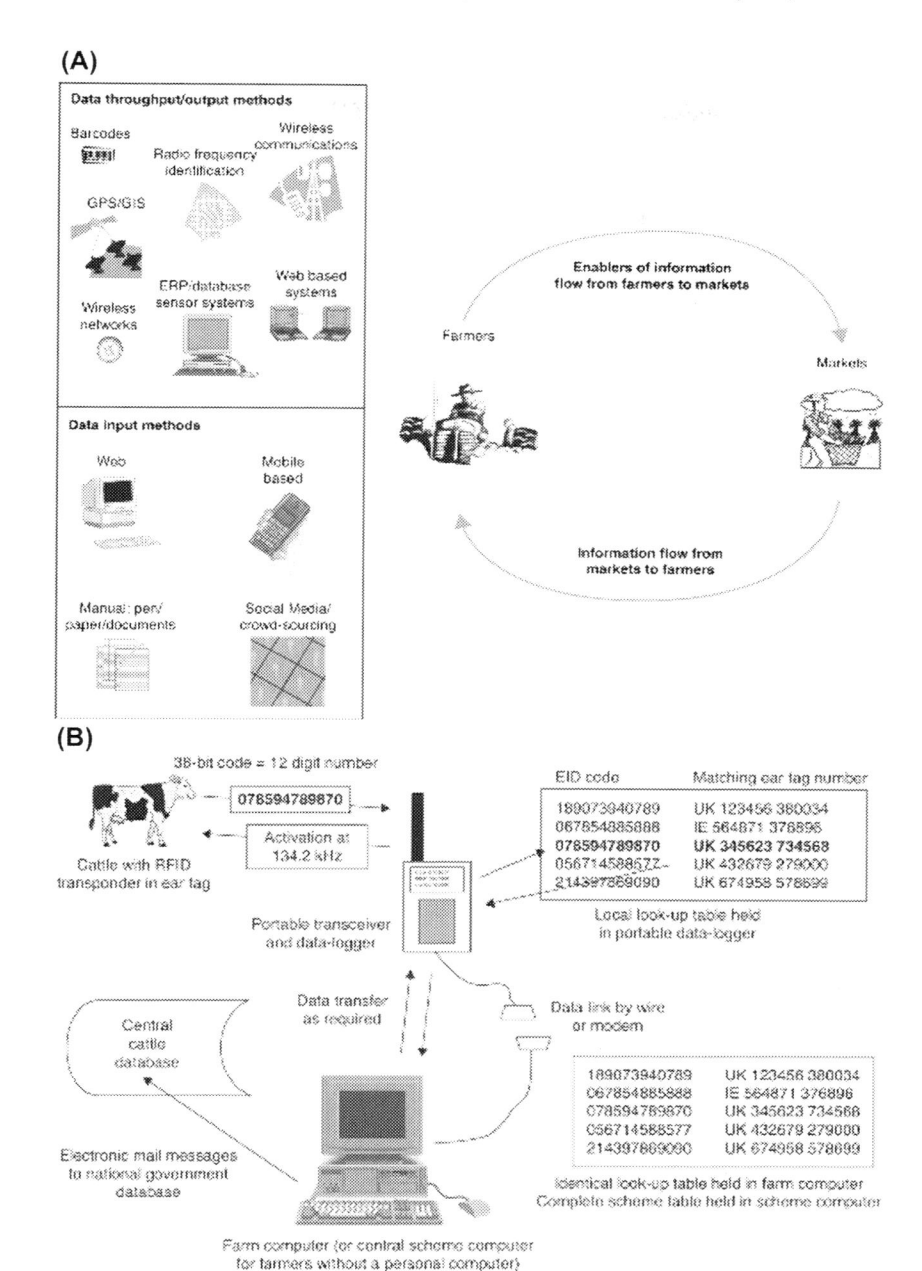

FIGURE 1 Cattle Tracing System.

A new data entry, generated from reading the data tag is added in each key point during transportation at the Central Cattle Database until the food reaches the final destination (Fig. 1B).

2. INFORMATION AND COMMUNICATION TECHNOLOGY TOOLS MEETING THE NEEDS OF AGRICULTURE

ICT has been introduced to the agriculture and food industry for more than 30 years now, starting with personal computers, Internet applications such as e-mail, mobile phone and wireless communication and multimedia networks. In the last decade, this technology is moving much faster due to the growing market demand for ICT implementations for agrifood application. Key advantages in using ICT are mobility, wireless communications and the use of small mobile devices, which are giving the most useful worldwide applications [6]. According to the application areas, the role of ICT can be divided into the following categories:

1. Tools that provide information (eg, geo-positioning systems, decision-support systems).
2. Data storage, data processing and data retrieval tools (eg, databases, data warehouse and data mining).
3. Analysis (eg, cost-benefit analysis, product prevision or risk analysis).
4. Sensors and communications networks (eg, in field or remote sensors, sensors systems connected and exchanging information, control and automation).

2.1 Information and Communication Technology Application in Today's Food and Agriculture

ICT developments continuously provide new applications in agriculture and food industry, including precision farming. Among the wide variety of new systems, which represent a convenient alternative of farming than traditional methods, digital technologies are being applied to improve the industrial structure of the food supply chain by increasing the number of industry contributors worldwide. Digital technologies may hold the key to successful coordination of a wider number of food systems [7]. ICT maps the entire production process digitally, from the crop breeding to harvesting, or from animal husbandry to product selling. In this way, these systems can increase efficiency and quality of the food production chain.

ICT is able to provide high quality of storage, processing and controlling of food creating new ways to fulfil the market requirements. For example, ICT can help in creating new products based on customer preferences, or in develop global economics in general and, especially, in agriculture. An indicator for globalization is the integration of several classification systems that were first used in North America and now are being used in many other countries. ICT brought standardization [8] in all the norms, requirements and weight unit that now cover all the transactions done worldwide. Moreover, the data about sizes and import/export economic volume are becoming internationally coordinated, providing more accurate, updated data for business intelligence and research purposes. The main concern is the food safety. Since there are so many fragile points in the food supply chain, it is very important to have a comprehensive trace of the food product. ICT implementations give us the possibility to receive data from the whole chain. This is useful because, as food safety standards increase and trade becomes more prevalent, the need to rapidly respond to a potential food illness or product safety recall will become more important [8]. Today agriculture is more complex than ever before in terms of globalization and the whole food industry processing systems. ICT implementations in agriculture are bringing its development to a completely new level by increasing the efficiency and the effectiveness of the agricultural supply chain.

2.2 Information and Communication Technology and Food Supply Chain Management

There are different definitions about the supply chain management, but the term that refers in the greatest manner to the management of multiple relationships and that underlie the supply chain as a concept is defined as follows: "Supply Chain Management (SCM) is the integration of key business processes from end users through original suppliers that provides products, services and information that add value for customers and other stakeholders" [9].

The integration of ICT tools together with mechanics, chemistry, biotechnology and logistics provides better communication between the various segments of the food supply chain in the food and agriculture sector. Such tools are able to adapt solutions to the preference of customers and open new markets for agrifood products. At the organizational level, collaboration among the supply chain actors such as producers, manufacturers, distributors and retailers is considered a key driver for innovation and may facilitate the sharing of tacit and explicit knowledge [10].

The food supply chain connects three important sectors: agriculture, food processing and distribution that together make more than 7% of European employment. Its performance affects customers, since food represents on average 16% of European household expenditures.

The modern food supply chain includes well organized and coordinated operations such as farms growing the food (including plant, meat and dairy products); complex food processing, manufacturing and packaging; bulk handling and food transportation; food distribution centres and coordination systems followed by complex, well-organized variations of retail and food service outlets (Fig. 2A) [7].

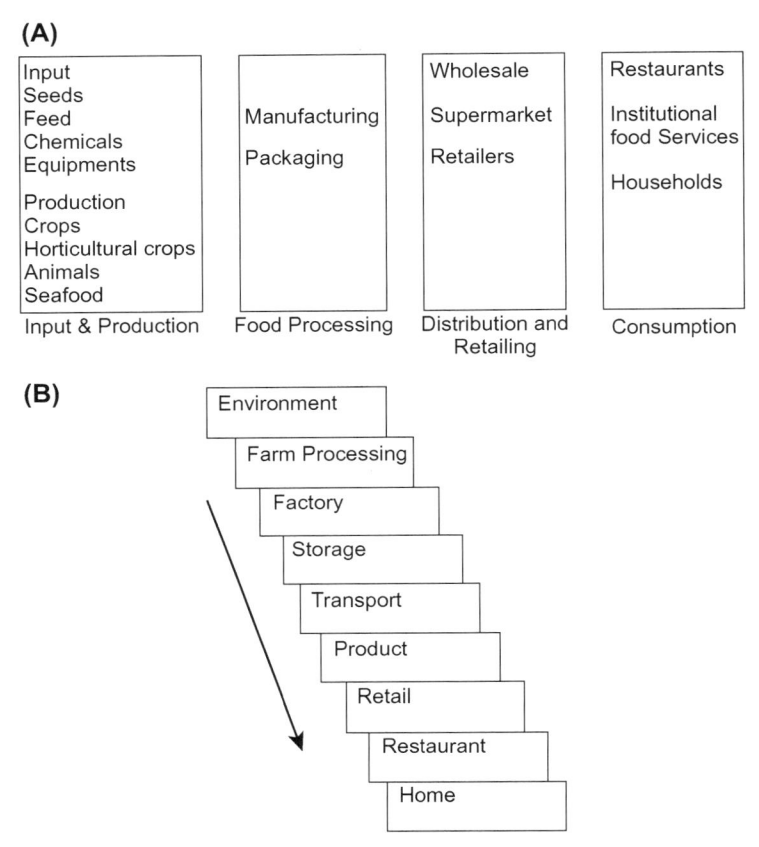

FIGURE 2 (A) Food supply chain, linear model, (B) Food supply chain.

Several segments are included in the food supply chain in agriculture such as input supply, post harvesting, storage, food processing, marketing, distribution, food services and food consumption. The final segment is related to the way the food is consumed as fresh or processed and/or furnished by a food service provider. Besides that, the food supply chain in agriculture interferes and depends on the external environment that includes other supply chains, such as geographical, political and organizational settings [11]. As a matter of fact, this relates to the new challenges set for the supply chain management in the food and agriculture sectors. The integration of new ICT technologies for food safety can help to redefine the whole agrifood scenario [9]. This will be done in three directions that are prevention, control and technological innovations. In fact, the benefit will be for the consumers and the marketing by reducing the extra processing of the products which will affect the operators and consumers by avoiding processing of products high in chemical and biological contaminants [7].

2.3 Information and Communication Technology and Food Security

2.3.1 Supply Chain Traceability

The new developments of food industries are causing new challenges to food safety and quality. The increasing complexity of the food supply chain, including the existence of outsourcing, contributes to enhance the vulnerability of food, which can be very detrimental to environment and human health. Nature equipped us with the senses that would be useful only for the very short food supply chain (time interval between the food production and consumption). Since today's food supply chain is much longer than before, the presence of so many fragile points and factors in the food supply chain would increase the risk to have contaminated and harmful food. While being processed through all the segments of food supply chain (Fig. 2B), food is exposed to different risks and can be contaminated with bacteria, parasites, allergens and heavy metals. Since food safety is the biggest concern of today's agriculture and food industry, several tests are developed based on the level of contamination and actions to be taken. The most common categorization [3] is as follows:

1. *Qualitative test* to verify the presence or absence of contaminant in a sample.
2. *Quantitative test* to quantify the amount of contaminant in a sample.
3. *Screening test* to give a strong indication on the presence of some form of chemical contaminant.
4. *Confirmatory tests* to give unequivocal proof on the presence of some form of chemical contaminant.

3. SENSORS AND WIRELESS SENSOR NETWORKS

Several questions need to be answered considering the manner by which these tests are done, at which point in the food supply chain, the way that contamination is detected and how the information is converted and transmitted. To answer all these questions ICT integration is a major factor. In fact, the exploitation of converging technologies in combination with ICT, including sensor technology and nanotechnology, can help to monitor food quality and safety in every food batch. Indeed, besides the genetics and information technology, nanotechnology is one of the triad of advancements driving the future transformation of supply chains worldwide [12]. One of the main tools created by the convergence of sensor technology and nanotechnology and used for food control quality is nanosensors. Nanosensors are smaller and faster than the regular sensors, require less power to run, have relatively simple design, have greater sensitivity, better specificity, are cost-effective and can be used remotely. But, in most of the applications there is the need to connect the separate sensors in a network to have large coverage areas, such as fields of crop, under control. When connected in a network, the separate sensors become much more powerful in their distribution in space and time, they can communicate with each other, collect, process and transmit their data to the data centres, giving a whole new meaning to their use.

However, nanotechnology is also applied to the management of supply chain processes associated with food quality, such as handling, packaging and safety. In the field of agricultural supply chains, nanotechnology deployment is already bringing potential benefits to farmers, the food industry and consumers alike, through innovations in agrifood production, processing, preservation and packaging [13].

3.1 Sensors Technology

The general term for defining sensors might be considered as an object whose purpose is to detect events or changes in its environment, and then provide a corresponding output. Sensors may provide various types of output, but typically use electrical or optical signals. For example, a mercury-in-glass thermometer converts measured temperature into expansion and contraction of a liquid, which can be read on a calibrated glass tube [14]. A *biosensor* is an analytical device used for the detection of an analyte, which combines a biological component with a physicochemical detector [15,16]. The sensitive biological component can be a tissue, a microorganism, an enzyme, an antibody, as well as a biomimetic component that recognizes and interacts with the target analyte. The transducer or detector element transforms the signal resulting from the interaction of the analyte with the biological element into another signal that can be more easily measured [17]. A reader device (associated with electronic components or other kinds of signal processors) will be responsible for the display of the results. Sensors and biosensors can be used to measure or detect a vast variety of physical, chemical and biological target analytes, including proteins, bacteria, chemicals, gases, light intensity, motion, position and sound.

3.1.1 Smart Sensors and Wireless Sensors Network

Smart sensors are sensors equipped with a digital interface. These interfaces have usually a serial configuration, or also some sensors might have parallel interfaces in case they need high data throughput, as might be the case of digital imaging sensors. The serial ones can be asynchronous or synchronous depending on the timing of how they produce the output. Serial interfaces have certain advantages over parallel ones, such as simpler wiring, the cables may be longer, and there is much less interaction (cross talk) among the conductors in the cable [18].

The most important feature of a smart sensor is the ability to communicate, which can be done by displaying the data directly to the user and transfer it over a wired interface wirelessly. These sensors can transmit a binary code by

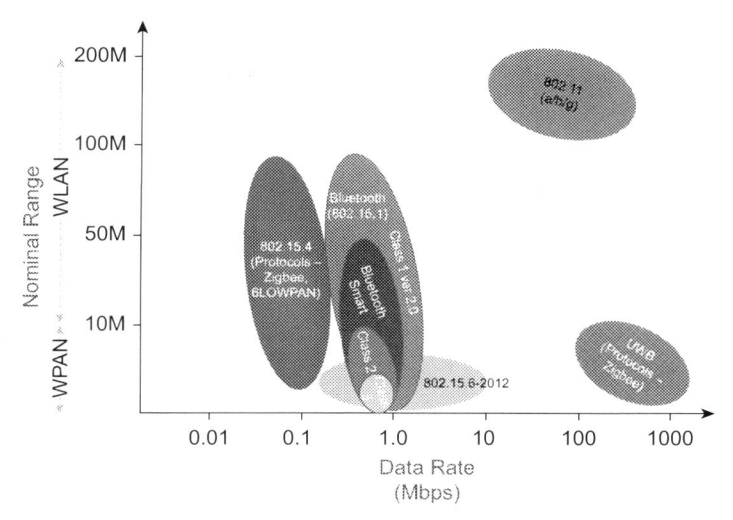

FIGURE 3   Common protocols used for wireless sensor applications.

using, for example, a diode LED (power on/off), colour changing indicator, or a liquid crystal display (LCD). The LCD screens can display characters, symbols or graphics, and they can be larger and more complex as the complexity of data to get transferred increases. Smart sensors are controlled by a simple microcontroller and if their data are the input to a graphical display, for example, to a smartphone, a graphical processing unit is used too. The wireless communication is provided according to several international standards that are Bluetooth [19], UWB, ZigBee and Wi-Fi. Each of these standards has different data throughputs and ranges, as shown in Fig. 3 [20]. In addition, the use of these standards depends on their purpose; for example, Bluetooth is used for shorter distance communications, while Zigbee or 802.15.4 are used for indoor and outdoor multinode network applications. These systems can be applied in agriculture and food industry, as for Wi-Fi it is usually applied in ambient applications where longer ranges, higher data rates and less signal attenuation are tolerated.

3.1.2 Sensor Network Topologies and Design Considerations

The Internet is the largest network in the world and together with the unlimited applications is named the Internet of Things (IoT). Because of the geographical setting of a farm or the food stored in a storage area, the data collected from smart sensors would need to be routed to a local network before getting through an Internet connection to a final destination. However, not always and all the sensor networks are connected to the Internet. Their data might be aggregated and displayed at a local aggregation point, such as a smartphone or a PC. Different applications require different topologies. Thus, a personal area network (PAN) may simply stream data from all sensors to a single central aggregator (star topology), or a home monitoring network may use a mesh topology. While the number of devices connected to Internet increases, the big data and their volume, velocity and variety rise likewise. Data generated from the smart sensors need to be transferred, stored and managed, and this process is even more expensive than data generation. The aggregator in sensor networks has a key role in aggregating data, that is choosing which data should be presented to the user or a remote network and which data should be discarded. In agriculture applications, sensors are distributed in distant locations and cover large surface areas, making their management more challenging. Recently the number of sensors connected to the cloud computing is increasing quickly because of their large capacity and fast speed.

These tools provide real-time data updates and the ability to remotely change a sensor configuration, data storage and visualization. Besides that, many popular cloud-based services can be integrated increasing their uses and effectiveness.

3.1.2.1 Sensor Network Elements

A sensor network is formed by a group of smart sensors that might be wired or wirelessly connected to another smart sensor or to a common aggregator. In such a network, each component that has a communication module is called a *node*. Each node that generates data is called a *source node* and each node that receives data is called a *sink* or *destination node*. A destination can be another sensor node on that network, a gateway to another larger network or a local aggregator. A source node can report monitoring data, an alert or maintenance data. Data dissemination and data collection are the main processes performed in the sensors network. *Data collecting* is used to capture and transfer the data from each sensor node to a destination [20]. The source sends data to the destination periodically or on demand, and the destination processes the data. For example, let us consider a storage area where the sensors are distributed randomly to check for a certain bacteria. In the case that they are programmed to send the status information every hour, the data will be gathered every hour towards the aggregation centre or the sink. The collected data needs to be disseminated to reach to the monitoring and control data centre that can be considered as the sink. Data dissemination is a two-step process. In the first step, the sink node describes the data it is interested in and broadcasts its detailed request, which is called 'interest', across the network. This might be the case when the sensors distributed in the food or storage area are packed in modules, including several kinds of sensors but not all of them proceed with their specific testing at a time depending from the descriptor broadcasted at that certain moment. Each node maintains an interest cache of all data to be reported. In the second step, the nodes that have data of interest (regarding the specific test required) send this data to the sink.

Sensor networks can include the same type of sensors distributed over a region, providing the same sensor data (a homogenous sensor network), or different sensors (a heterogeneous sensor network), which provide different sensor data to the system. Homogenous sensor networks can be used to extend the sensing region of a sensor. For example, a network of heavy metal sensors in a certain field can provide richer information than a single sensor placed in certain location. The positive feature of homogenous sensor networks is that they are fault tolerant and can be used in modules of several types of sensors.

WSNs are a special kind of sensor network consisting of a sink node, which is usually called a 'base station' and a number of wireless, battery-powered sensor nodes. The base station has usually much higher processing and data storage capabilities than the other nodes of the network. For example, a smartphone can act as a base station in a wireless personal area network (WPAN) even if it is battery powered. Also, the smartphone will have significantly more battery power than the rest of the nodes in the network and will be regularly charged. The lifetime of the sensor network depends on the balancing of their energy consumption during sensing and processing and the battery life. This is the reason why the selection of low-power radios and efficient network protocols are decisive factors in WSN's life span.

Sensor platforms depend on the microcontroller used (for example, Arduino) and allow users to connect sensor and communications modules to a base platform.

Smart sensors have the ability to interchange hardware radio modules, so users can change not only their communication protocol but also the network topology. For example, replacing a Wi-Fi module with an XBee 868 module allows the user to replace a star-based network with a self-healing mesh network that has a wider sensing range.

3.1.2.2 Aggregators, Base Stations and Gateways

Sensor data get collected, processed, stored or forwarded to other networks by going through their specific communication mechanisms, making them a very powerful technology [21]. Several computing devices, such as machine to machine (M2M) devices or PCs, can be configured to act as aggregators, gateways, bridges, base stations or coordinators. There are proprietary solutions such as Libelium Meshlium M2M device (www.libelium.com/products/meshlium) that can complete the connection between the sensor and the cloud. Those proprietary solutions are to interface with the manufacturer's own sensors using a predefined messaging protocol over a predefined radio as those are easier for the end user's use. On the other hand, smartphones can be used as sensor platforms because of their integrated sensors, processing power and communication abilities. They can act as aggregators collecting data from external sensors and other data sources.

After storing and analyzing the data, the smartphones can be interfaced with cloud computing services. One key advantage that is used in agriculture and the food industry is the use of the smartphones as 'app-enabled accessories', such as smart packages of warehouses smart panels. Those last ones can communicate with the smartphone wirelessly and by using a specific messaging protocol defined by the operation system. The Apple App Store and Google Play provide access to many software apps that can interface with the sensors and actuators. These apps are written from specialists that work in the field and are designated to interpret and process data from a specific kind of sensor. For long-term storing and tracking the apps can upload the sensor data to the cloud.

3.1.2.3 Sensor Network Topologies

As above described, sensors, smart sensors and sensor systems combine sensing, processing, communication and power subsystems in a single integrated system. In real-world applications, sensors are often integrated into higher-level topologies. These topologies can be different, such as a single node connected to an aggregator or even fully meshed networks distributed over a large geographical area. Sensor network topologies can have flat or hierarchical architectures. In a flat (peer-to-peer) architecture, every node in the network (sink node and sensor node) has the same computational and communication capabilities. In a hierarchical architecture, the nodes are situated relatively close to their respective cluster heads. As a result, nodes with low energy consumption simply capture the required raw data and forward it to their respective cluster heads. However, the cluster heads consume more as they have more processing and storage capacity than a simple sensor node. Some of the usual forms of sensor network topologies are as shown in Fig. 4. Sensor networks that are physically connected together commonly use star, line or bus topologies. But, often even the wireless sensors networks are built using star, tree or mesh topology configurations. These topologies are used in different applications even in food industry and agriculture sensor networks. The hierarchical schemes make their operation faster and more efficient. Each configuration has different characteristics.

Point-to-point topology links two endpoints, as shown in Fig. 4. This topology in a certain case can be permanent or switched. A permanent

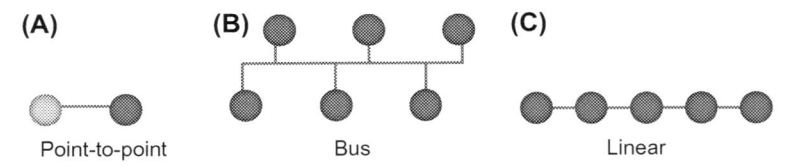

FIGURE 4   Point-to-point, bus, linear topology.

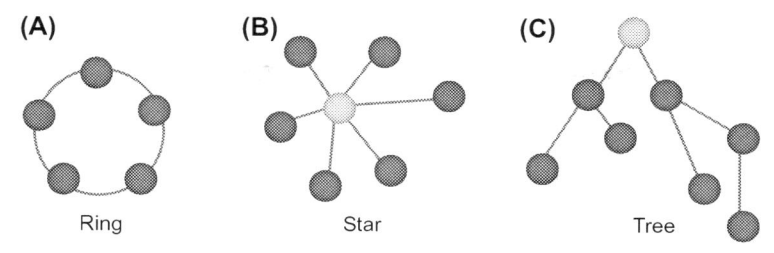

FIGURE 5 Ring, star, tree topology.

point-to-point topology can represent a hardwired connection between two points or a switched connection if the nodes are connected wirelessly.

Bus topology represents the topology where each node is connected to a shared communication bus, as in Fig. 4B. A signal can be transmitted in both directions along the bus until it reaches the destination. Specifically, bus systems need to include a collision-avoidance system to resolve issues when two nodes simultaneously send out data on the bus. These networks are simple and easy to install. The disadvantage of these networks is that they are not fault tolerant in a single point of failure: if the bus fails, the whole network fails.

As for the *linear topology* (Fig. 4C) there exists a two-way link between one node and the next one. In this topology there are two terminating nodes at the extremities of the network that have a single connection to their respective neighbours, and all other nodes are connected to two other nodes. Thus, this network is a fault tolerant one, because the nodes depend on each other to propagate a message to the next one.

Ring topology represents a network set up in a circular way, as shown in Fig. 5A. It is similar to a linear topology, with the exclusion that the end nodes are connected to each other. According to this configuration, each node connects to exactly two other nodes and data flows in one direction from the source to each node in between until it finds the destination node. Every node can be a point of failure and this represents the disadvantage of this type of network.

Star topology consists of a single 'central node', such as a hub or a switch and all the nodes in the network connect to that central node, as shown in Fig. 5B. This topology has a lot of advantages, is easy to design, implement and extend. All the data in the network goes toward the central node in both directions, making this node more intelligent than the rest of the nodes of the network. The central node represents a node of failure for the whole network. The star topology is one of the most used topologies used in the sensor networks.

Here we can mention the example of a WPAN, consisting of a smartphone connected to several wireless sensors.

Tree topology is a hierarchical connection model where the highest level of the hierarchy is a single 'root node'. This node is connected to one or many nodes in the level below, as shown in Fig. 5C. The tree topology can contain several levels of nodes. The processing power in nodes increases as the data moves from the branches toward the root so that the data can be processed close to where it is generated. This topology is scalable (expandable) and simple to be diagnosed in the case of a node failure. This topology can be used in food supply chain in the sensors used for tagging back batches of food production, etc.

Besides the hardware network topology in sensors networks, there is also the logical topology that represents the data flow in the sensors network. There are two types of logical topology: shared media and token-based topology. In the shared media topology, all nodes can transmit at any moment but this leads to collisions that need to be managed by a collision-avoidance protocol. The logical topology is used in bus, star or hybrid physical topology networks due to their shared data bus or shared node. In a token-based logical topology, a token is passed around the network. In the case a node needs to send data, it must get the token from the network. When the data gets to the destination, the token is released and continues to travel around the network until another node will get it as this last one needs to transmit. This logical topology is used in most ring-based topologies.

3.2 Sensor Network Applications

Sensor network applications are unlimited and they are getting used in areas and processes where one would not have imagined before. In terms of the application domain, sensor networks are usually described by their application type rather than the physical or logical topology of the network created. For example, a PAN that transfers personal data can have star or point-to-point physical topology, and it can use any of a number of low-power, short-range frequencies to communicate.

PANs connect many hosts (devices) such as laptops, tablets or smartphones to each other in short distance. The connection can be wired or wireless. Concern is security as the data usually is personal. According to their data throughput and power consumption, PAN can be categorized in three kinds [22].

High-data rate WPANs are used for real-time multimedia applications. Wide area network (WAN) is a network that covers a broad area (such as metropolitan, regional or national boundaries) using private or public network transports. Usually WAN can be used for commercial and government purposes as it might support communications between employees, clients, buyers and suppliers from distant locations. Even the Internet can be considered a WAN. WAN is used in food and agriculture to transmit the data from a field in Africa cultivating product ordered from a firm in Europe (Fig. 6) [20].

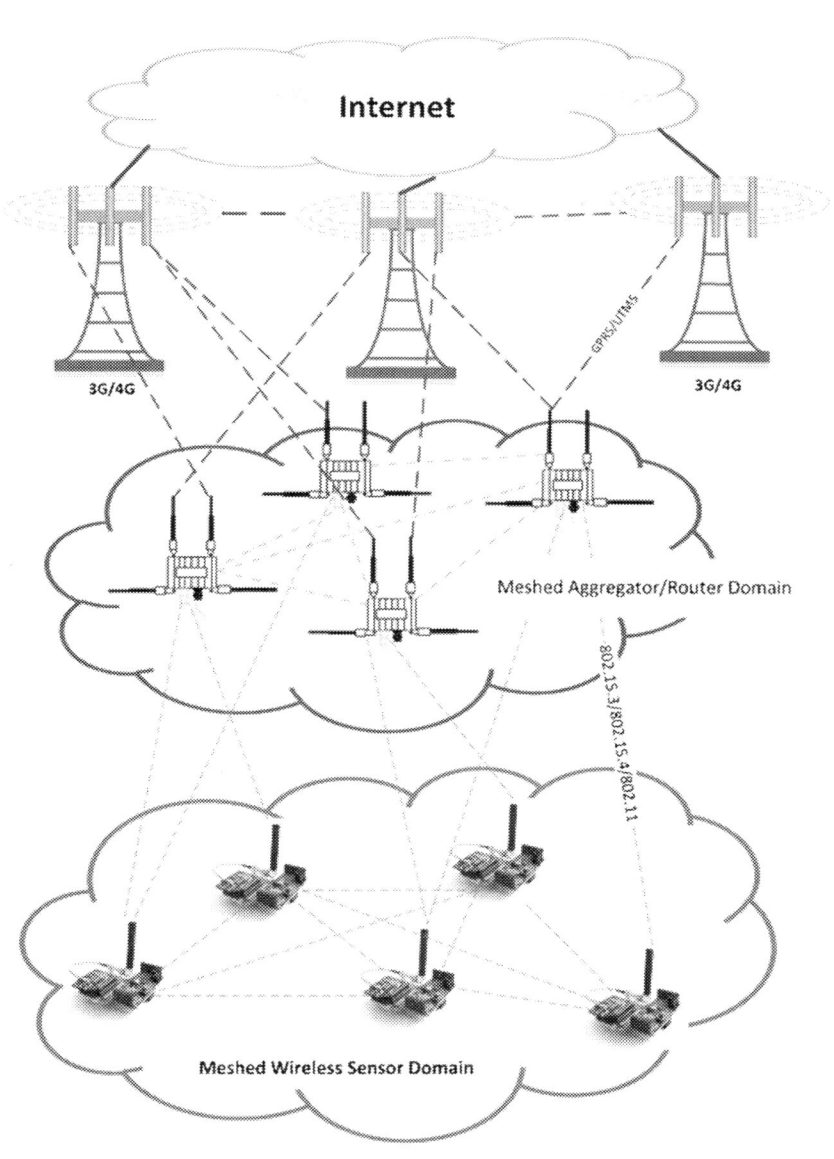

FIGURE 6 A wide area wireless sensor network.

Communication is a network where each device has the ability to transfer data to other devices in the sensor network hierarchy. The lowest order sink nodes (the end users) in the network are usually wireless and battery powered and must therefore implement a very low-power radio protocol. The *messaging* is the way in which their communication will be held and which protocol will be used. When a new device is added to the network, a new protocol will be defined and a gateway will be used to interpret both protocols.

Processing is the step for data transmission and storage. Since these steps are very expensive at every stage of a sensor network, it is important that the data be processed and reduced as close to where it is generated. Recently the cloud computing applications are making everything easier. These applications offer big storage and processing power using tools such as Hadoop (hadoop.apache.org).

Storage is the step in which data are stored and the memory required on a smart sensor device is dependent on its application. A sensor node that continually transmits raw data to a smartphone device requires minimal data storage. The databases are used for data storing and mining that is very helpful in the case of looking for a certain product, because they allow not only to store, but also to search for a certain doubtful hazard, such as a bacteria in a certain food product. The type of distributed database to be used depends on the size of data sets to be processed. So, a distributed database such as Cassandra (cassandra.apache.org) can be used for managing small to medium data sets, and MongoDB (mongodb.org) can be used for managing 'big data'.

The data in the databases can be queried or updated from the application device through Application Programming Interface (APIs).

A sensor network that cannot be remotely managed is a nonscalable sensor network. A number of cloud-based services, such as Xively and Device Cloud, have emerged in recent years to provide cloud-based device manageability. Both services offer libraries that can be installed on an IP-addressable sensor or gateway device, a method to register new devices on the cloud-based management console, and a manageability console that allows the sensor network manager to view status and remotely configure the devices on the network. These services also provide basic data storage, basic data analytics capabilities and APIs for application devices and services to interface to the data and the network.

3.2.1 Attacks on Sensor Networks

A challenge that the information technology is facing is the presence of malicious attacks, and those are becoming more sophisticated everyday. As every software-dependent (depends on) devices, even the smart sensors are endangered from those attacks. The main dangers are message spoofing or message replays. The threats of data altering can happen for many reasons even in the food and agriculture field. So, it might happen that small owners can have products not according to the standards and are interested on sending false data. Attacks can be internal or external. An external attack can be active or passive [21]. Attacks can also be classified based on the network layer which is used by the attackers. To avoid those attacks cryptography and intrusion detection are being used. There are many robust cryptographic implementations designed for WSNs, such as 128-bit Advanced Encryption Standard (AES)-based encryption with multiple keys. But, these solutions might be expensive as they bring significant computational overhead and power consumption.

3.2.2 Processing Sensor's Data

The integration of sensors into many aspects of daily life is generating enormous volumes of data. These data come from different directions, applications and are sent to cloud infrastructures for data storage, processing and visualization.

New methods and environments are being developed to process, interpret and display sensor data. One such example is the IoT. The 'Internet of Things' is the general idea of things, especially everyday objects that are readable, recognizable, locatable, addressable and controllable via the Internet — whether via RFID, wireless LAN, WAN or other means [23].

IoT is growing and includes a myriad of devices such as sensors, smart clothing, consumer electronics, utilities metres, cars, streetlights, advertising signs, buildings and more, everything that can be connected to the Internet. A very interesting fact is that the number of connected devices already exceeds the number of people on the planet, and Cisco estimated that the total number of connected devices will exceed 50 billion by 2020 [24]. A key element of the IOT is the usage of sensors and smart sensors: discrete sensors, such as those for environmental monitoring; body-worn sensors (such as those worn to animals being tagged), sensors embedded into devices, such as accelerometers in smartphones. In fact, many devices feature multiple sensors. For example, a module installed in food storage warehouse includes a set of sensors used for different purposes such as temperature, humidity, air quality control, etc. However, a GPS is added too to find the location. All these data streams can then be connected to the Internet via a smartphone. There are platforms such as Xively (https://xively.com) that are being used lately from hobbyists and companies to intuitively collect data from Internet-enabled devices, including sensors.

3.2.3 Sensors and the Cloud

The cloud computing is transforming the way in which businesses manage their data. This is because of advantages such as flexibility and scalability of computing, storage and application resources, optimal utilization of

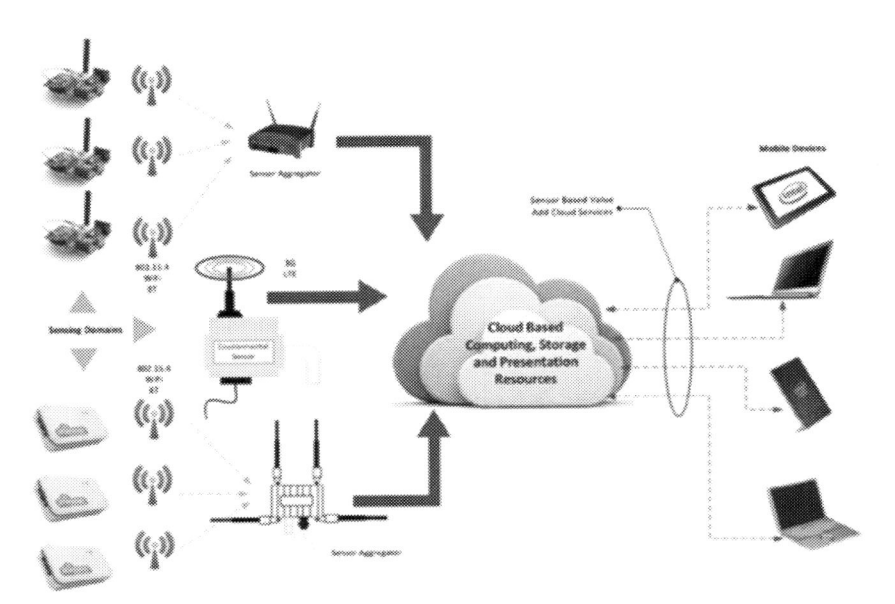

FIGURE 7 A cloud wireless sensor architecture.

infrastructure and reduced costs. As sensor can be a part of a geographically distributed network that generates highly dynamic data throughput, cloud computing represents a good solution for their storage, processing and visualization (Fig. 7) [20]. While using cloud-based functions, the end users and their applications do not need to be concerned about the physical location of sensing resources because they are virtualized. They can manage virtually sensor organization, activation and their termination when it is necessary. But, adding new sensors would require to register them through a registration process and also to describe sensor's data in a specific format (XML like), for example, SensorML. These might happen during the food supply chain monitoring even if the management might be in another continent.

3.2.4 Data Processing and Visualization

The measurements from all the sensors are not always accurate as they may contain noise, can be incomplete or not from the required locations. To improve the informational value of sensor's data, sensor fusion and virtual sensors are being used.

In sensor fusion, the sensor data or the derived sensor data is combined with data from other sensors or sources. In some cases, multiple sensors may be required to fully quantify the measurement of interest. This can be the case when a module of different sensors is used, but their data is related and cannot be interpreted without having every result.

3.2.4.1 Data Mining

Extracting useful and valuable data from raw sensor data is part of the data mining process that can be applied to the knowledge-discovery process on sensor data. This is a very effective process used to model relationships among sensor measurements to reveal nonobvious patterns among the data and to determine data quality issues [22]. Data mining includes different techniques, starting from traditional statistical analysis and nearest-neighbour clustering to more modern techniques, such as ANN, decision trees and Bayesian networks.

Before doing the data mining process, it is necessary to execute some forms of data preprocessing for addressing problems such as noise, outliers, missing data or data from malfunctioning sensors. In some specific applications, such as those that require real-time or near real-time performance, data reduction might be needed to reduce the processing time. All data collected and transmitted to food supply chain data centres can be used by applying data mining techniques that would allow control and decision-making about every stage.

3.2.4.2 Data Visualization

Data visualization is a very important and preferable feature making easier for the user to see the patterns and relations that are essential for data analysis. Generally, people prefer graphics and visuals. As the visualization process enables us to bring various information sources together, including nonsensor sources, context can be added that informs the interpretive process. It lets us create designs that tell a story about the data [25]. Good design, particularly in the agriculture and food industry, often utilizes relative data, which connects the sensor data to peer data sets or to values that generate a fully rounded and qualified data.

As the sensor distribution increases, so does the need to visualize distributed sensor networks as WSNs. Visualization makes sensor network applications user friendly taking into consideration that the food and agriculture professionals have different levels of education, locations and workload. There are different efforts to address this. Here we can mention hiperSense, which is designed to provide scalable sensor data visualization with the ability to handle up to 6200 independent streams of data [26]. In the environmental sensing domain, an integrated system is used based on MicaZ motes such as those used to monitor soil temperature and moisture. The resulting data visualization was done through Microsoft Research SensorMap.

Another popular tool used for sensor data visualization is Google's Fusion Tables. This is a web-based application that allows users to gather, visualize and share large data tables [27].

3.2.4.3 Big Sensor Data

Sensor data is part of the huge big data generated in our planet. This is getting a lot of attention into business analytics world. Therefore, the Big data is said to be the 'new oil'. However, this analogy does not always hold true, as data is not a finite resource like oil. In fact the opposite is true, with new reserves of data being created on a daily basis [26] to IBM, mankind and its supporting infrastructure generate a massive 2.5 quintillion bytes of data daily.

4. SENSORS TECHNOLOGY APPLICATIONS IN AGRICULTURE

Sensors play an integral role in numerous modern industrial applications, including food processing and everyday monitoring of activities such as transport, air quality, medical therapeutics and many more. Sensor technology is being used for more than a century, but smart sensors, their systems and the integrated ICT applications are adapted only recently. Unimaginable progress has been made in computational capabilities, storage, energy management and a variety of form factors, connectivity options and software development environments. At the same time, advances have occurred in sensing capabilities. Recently, we see the emergence of biosensors that are now found in a variety of consumer products, such as tests for bacterial contaminations, heavy metals and other toxins.

4.1 Structured Database Solutions

As we see the information is stored in databases that allow apply queries about data elements for finding the location of products that may be contaminated. Enterprise Resource Planning (ERP) systems such as SAP can read standardized data from barcodes and RFIDs, including GTINs and GLNs [28].

Vendors and business partners can exchange data through electronic data interchange systems. Those can be accessed even through web-based portals for inputting or extracting data with other partners. The existence of different standards make it difficult to complete the tracking but the recent cloud computing and SaaS (Software as a service) solutions have reduced the cost and provide solutions to capture, record, store and share traceability data.

4.2 Radio Frequency Identification-Based Solutions

RFID offers promising capabilities for traceability in developing and developed countries and represents a more advanced alternative to older barcode systems. RFID tags use an initial signal from an RFID reader to scavenge power and store data on an event at a specific point in time. For example, let us suppose that everyday the humidity is checked in a field of rice. RFID tags do not use a power source and are less expensive than active RFID tags.

Products tagged with RFID are interfaced with the data provided by WSNs. So, the data about the humidity in the field of rice above-mentioned will be transmitted to the data control centre. On the other hand, GPS, low earth orbit satellites and motion sensors may interface with RFID tags to communicate data on location and position coordinates (latitude/longitude). Therefore, the data collected from RFID readers can be integrated as an application on a mobile device. Thus, an 'ecosystem' that includes a combination of RFIDs, WSNs, GPS, mobile devices and applications can make it possible to manage traceability across the supply chain. Product traceability recorded through such an ecosystem-based solution may range from data on logistics and postharvest practices surrounding the trees of the small-scale producers [10].

4.3 Fujitsu Cloud-Based System

Cloud-based systems are becoming very popular even in agriculture and food industry. Fujitsu provided a SaaS-based solution for agricultural production management, which is designed to support management in both agriculture and food-related industries, while also offering an innovation support service that promotes on-site ICT utilization [6].

Some applications in food industry and agriculture are described.

4.4 eData Sharing Solutions

The axTool is a knowledge and information-sharing environment, embedded in a portal published at http://www.agrixchange.eu, allowing the agrofood community to model use cases involving information sharing/exchange in the agrofood sector. A methodology for use case modelling is defined, through which agrofood specialists specify the actors involved in the use case (eg, entities in the supply chain), the activities they execute and their time sequence (eg, the workflow in the process) and the data that are exchanged in some points of the model by different entities in the use case.

AgriXchange 'Network for data Exchange in agriculture' is an EU-funded coordination and support action aiming to set up a network for developing a system for common data exchange in the agricultural sector through the following specific actions:

- establishing a platform on data exchange in agriculture in the EU;
- developing a reference framework for interoperability of data exchange.

5. CHALLENGES FOR THE FUTURE: TRANSITIONING TO THE SUSTAINABLE FOOD SYSTEMS AND THE INFORMATION AND COMMUNICATION TECHNOLOGY

It is important to note the role of the ICT in the creation of a secure and sustainable food supply chain. Maintaining this food supply chain is becoming more and more challenging due to the increased number of the nontechnological factors such as behaviours, preferences and agricultural

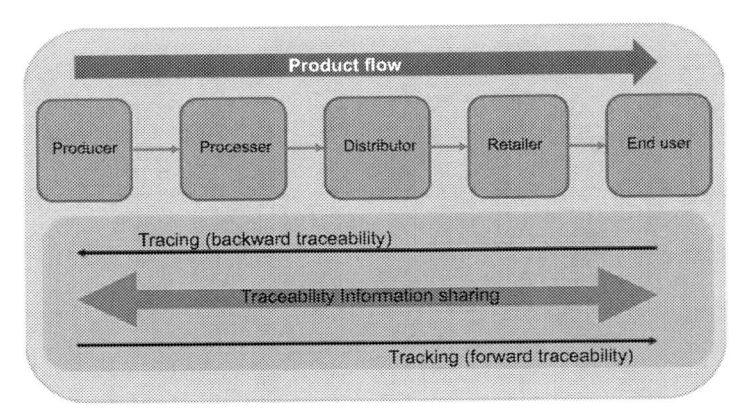

FIGURE 8 Conceptual representation of traceability information flow.

practices within different local areas. Digital technologies are very important in the creation of reliable, secure, robust and economically sustainable 'short' supply chains. They can contribute in the coordination of different segments that are part of food supply chain (Fig. 8) including regional, urban mid-range or long haul range coordination [7]. ICT is critical and can play an important role in having short supply chains, because it helps in the coordination between farmers and food producers as it offers more choices to the consumers in terms of the food accessed in their local/regional markets. ICT can contribute to the connections between the suppliers to smooth the ratio between supply and demand. In addition, ICT allows the creation of dynamic supply chains that do not need contractual agreement between supplier and customer but allow smaller, local suppliers to participate dynamically. The most important factor in the success of small supply chains is the relatively low cost of digital technologies. It is possible to create business process management systems using mobile or cloud technologies, rather than depending on large-scale industrial ICT systems that have been used in the industrialized food chain. Also, digital technologies allow the creation of digital communities and the strategic networks between individuals that participate in a short supply chain and ensure supply for customers.

Table 1 summarizes how and where digital technologies might be applied in the two main scenarios to contribute to the transition to sustainable food systems [7].

The projects for the future need to accomplish the ICT-based logistics platform in several directions:

- To improve the quality competitiveness of domestic fruits and vegetables by maximizing the customers' satisfaction and environmental protection.

TABLE 1 Role of ICT in Transitioning to Sustainable Food Systems

	Industrial Supply Chains	New, Digitally Based Supply Chains	Developed Countries
Pre-planting	Genetically Modified Organism (GMO) seeds, improved resistance to disease, improved resistance to pests, increased yields	Seeds for kitchen growing systems, hydroponics for mass market	Exchange of traditional knowledge for the upkeep of soil
In field	Sensors to detect disease, drones to analyze crops, open data to understand weather pattern, connections between communities to understand disease patterns, etc.	Sensors to detect disease, home-based information systems for growing own food, social and community connection systems	Smoothing of supply and demand between local suppliers, retailers and consumers
Production	Transportation: RFID measuring temperature of foodstuffs to reduce waste, finer grained control of supply and demand	Recipe sharing to reduce waste, details about how to store and keep foodstuffs, 3D-printing of food (fruit), in vitro production of meats	Incorporation of short supply chains to reduce waste
Transportation	Transportation: RFID measuring temperature of foodstuffs to reduce waste, finer grained control of supply and demand transport deliveries	Transportation: RFID measuring temperature of foodstuffs to reduce waste, finer grained control of supply and demand transport deliveries	Localized transport systems, 'ride sharing' for transport of produce
Marketing	Better information about the product in question for consumers	Better information about the product in question for consumers	Restructuring of supply chains to place control in consumers' hands

RFID, radio frequency identification; *ICT*, information and communication technology.

- To use the most modern hardware and software technologies according to standards to reduce delivery times; to allow continuous monitoring and real-time recording of parameters that are critical for storage of products and to ensure compliance with mandatory rules of hygiene and traceability of the productive chain.

ICT has a significant role to play in the future of food and agriculture. In the face of rising pressure from climate change, rising populations and decreasing crop yields, the world must confront the critical challenge of efficiently delivering sustainable and healthy diets to global populations.

Systems for tracking products through supply chains range from paper-based records maintained by producers, processors and suppliers to sophisticated ICT-based solutions. In addition to supporting product traceability, ICT may also support data capture, recording, storage and sharing of traceability attributes on processing, genetics, inputs, disease, pest tracking and measurement of environmental variables.

REFERENCES

[1] L. Brown, Full Planet Empty Plates: The New Geopolitics of Food Scarcity, 2012.

[2] Z. Xiaoshuan, Food Supply Chain Traceability and Sustainable Development, China Agricultural University, 2013.

[3] T.G. Karipacheri, L.D. Rios, Global Markets, Global Challenges: Improving Food Safety and Traceability while Empowering Smallholders through ICT, Module 12, Economic and Work Sector, 2014.

[4] D. Timon, S. O'Reilly, An Evaluation of Traceability Along the Irish Beef Chain, Long Terms Prospects of the Beef Industry, C. Viau, Paris, 1998.

[5] M. Muto, I. Yamano, The impact of mobile phone: panel data evidence from Uganda, World Dev. 37 (12) (1887–96).

[6] M. Sztaki, CBHU, SEVT, IBA, CCIS-cafe a Guide on ICT Solutions for Enhancing Innovation Activities in the Food Sector,, CAPINFOOD (2011).

[7] G. Berti, K. Mulligan, Industry Transformation — Horizon Scan: ICT & the Future of Food, London Imperial College, 2015.

[8] S. McCallum, Protecting the food supply: hi-tech food banks and the safety of food supply chains, DomPrep Journal, 8 (4) (2012).

[9] M. Iannetta, G. Matranga, C. Zoani, S. Canese, L. Daroda, F. Vitali, G. Zappa, Innovation in logistics and in the supply chain integrated approach, Mediterra (2014) 463–476.

[10] B. Bacheldor, Hybrid tag includes active RFID, GPS, satellite and sensors, RFID J. (2009). http://www.rfidjournal.com/article/view/4635.

[11] L. Hossain, J.D. Patrick, M.A. Rashid, Enterprise Resource Planning: Global Opportunities & Challenges, Idea Group Publishing, 2002.

[12] E.W. Hewett, Progressive challenges in horticultural supply chains: some future challenges, Proc. IV Int. Conf. Manag. Qual. Chains 1 2 (712) (2006) 39–49.

[13] J. Lua, M. Bowles, How will nanotechnology affect agricultural supply chains? International Food Agribusiness Manage. Rev. 16 (2013) 2.

[14] https://en.wikipedia.org/wiki/Sensor.

[15] A.P.F. Turner, I. Karube, G.S. Wilson, Biosensors: Fundamentals and Applications, Oxford University Press, Oxford, 1987, 770 pp.

[16] K. Campbell, Nanotechnology in the Agri-food Industry: Applications, Opportunities & Challenges, 2014.

[17] C.C. Adley, Past, present and future of sensors in food production, Foods 3 (3) (2014) 491—510.

[18] M. Rouse, Serial Peripheral Interface (SPI), Last Update: March 2011, http://whatis. techtarget.com/definition/serial-peripheral-interface-SPI.

[19] J. Eliasson, P. Lindgren, D. Jerker, A bluetooth-based sensor node for low-power ad hoc networks, J. Comput. 3 (5) (2008) 1—10.

[20] M.J. McGrath, C.N. Scanaill, Sensor Technologies, Healthcare, Wellness and Environmental Applications, Apress (2014).

[21] J. Hongbo, Prediction or not? An energy-efficient framework for clustering-based data collection in wireless sensor networks, IEEE Trans. Parallel Distrib. Syst. 22 (6) (2011) 1064—1071.

[22] P.N. Tan, "Knowledge Discovery From Sensor Data," Last Update: 1 March, 2006, http:// www.sensorsmag.com/da-control/knowledge-discovery-sensor-data-753?page_id=1.

[23] M. Swan, Sensor mania! The internet of things, wearable computing, objective metrics, and the quantified self 2.0, J. Sens. Actu. Networks 1 (2012) 217—253.

[24] Cisco, "The Internet of Things," Last Update: 2011, http://share.cisco.com/internet-of-things.html.

[25] D. McCandless, "Information Is Beautiful," Last Update: 2013, http://www. informationisbeautiful.net/tag/health/.

[26] J. Thorp, "Big Data Is Not the New Oil," Last Update: 2012, http://blogs.hbr.org/cs/2012/11/ data_humans_and_the_new_oil.html.

[27] E.S. Bradley, et al., Google Earth and Google Fusion tables in support of time-critical collaboration: mapping the deepwater horizon oil spill with the AVIRIS airborne spectrometer, Earth Science Informatics 4 (4) (2011) 169—179.

[28] http://www.ictinagriculture.org/sites/ictinagriculture.org/files/final_Module12.pdf.

Index

Printed in the United States
By Bookmasters